T0213688

Undergraduate Topics in Computer Science

'Undergraduate Topics in Computer Science' (UTiCS) delivers high-quality instructional content for undergraduates studying in all areas of computing and information science. From core foundational and theoretical material to final-year topics and applications, UTiCS books take a fresh, concise, and modern approach and are ideal for self-study or for a one- or two-semester course. The texts are all authored by established experts in their fields, reviewed by an international advisory board, and contain numerous examples and problems, many of which include fully worked solutions.

The UTiCS concept relies on high-quality, concise books in softback format, and generally a maximum of 275–300 pages. For undergraduate textbooks that are likely to be longer, more expository, Springer continues to offer the highly regarded Texts in Computer Science series, to which we refer potential authors.

More information about this series at https://link.springer.com/bookseries/7592

Liang Wang • Jianxin Zhao • Richard Mortier

OCaml Scientific Computing

Functional Programming in Data Science
and Artificial Intelligence

 Springer

Liang Wang
Computer Science and Technology
University of Cambridge
Cambridge, UK

Jianxin Zhao
Computer Science and Technology
University of Cambridge
Cambridge, UK

Richard Mortier
Computer Science and Technology
University of Cambridge
Cambridge, UK

ISSN 1863-7310 ISSN 2197-1781 (electronic)
Undergraduate Topics in Computer Science
ISBN 978-3-030-97644-6 ISBN 978-3-030-97645-3 (eBook)
https://doi.org/10.1007/978-3-030-97645-3

This Springer imprint is published by the registered company Springer Nature Switzerland AG
The registered company address is: Gewerbestrasse 11, 6330 Cham, Switzerland

To my wife Maria, our daughters Matilda and Sofia, and my parents Yulin and Fengjin.

– Liang

To my parents and sister; to their unyielding love and support. To Gcores, for being there in my darkest hour.

– Jianxin

To all those who contributed, and those on whose work we built.

– Mort

Foreword

This book is about the harmonious synthesis of functional programming and numerical computation. In recent years, Data Science has taken off in a big way, partly through the rapid increase in availability of data from many sectors of society, whether industry, government or entertainment. There are obvious benefits in operational optimisation of utilities, in a wide variety of applications in healthcare, finance, and subtle analysis of human communication.

The growth in computing power, in handheld devices, on the desk top, in the server, on the cloud combines with this data availability to allow rapid development and deployment of platforms for statistics, Machine Learning and Artificial Intelligence applications. However, many of these systems have been rapid expansions of existing software, based in traditional imperative programming languages such as Python, or C++.

In contrast, functional programming has been the preserve of more formal, or even theoretical branches of computer science until relatively recently. However, steps in the implementation of real world systems in the declarative style have revealed certain long-claimed advantages for languages like OCaml or Haskell. Not only are they more safe (e.g. against security vulnerabilities), but the also now allow for more efficient, agile, and productive development. The languages also offer the hope of system verification, which may be an important property for safety critical application areas (health informatics, as mentioned before, being one obvious area).

Real world OCaml uses include finance, operating systems and cloud applications, and now numerical processing for scientific computation and statistics. Of course, there are now also other languages targeting safer computation in this domain that are still in the procedural or imperative style, including specialised to ML/AI such Julia, or to systems, like Rust.

The starting point for this book is in numerical computation, and the exposition shows with copious examples how OCaml and the Owl library can be used to build up the basic computational units needed to build basic statistical programs. Going beyond that, the authors build up to use cases drawn from many areas of Data Science, Machine Learning, and AI. The book then delves into how to deploy at scale, using parallel, distributed, and accelerated frameworks to gain all the advantages of the model data centre/cloud computing hardware and system environments.

The book does not currently cover Bayesian Inference, nor statistical techniques such as Monte Carlo Markov Chains although it would be fairly easy to build such systems based on material here. One interesting thing to try would be to construct a Domain Specific Language for probabilistic programming based on OCaml & Owl.

This book shows how the expressiveness of OCaml, allows for succinct exposition of applications, fast and safe development of data science applications. AI and ML use in science and engineering is often exploratory, and calls for an agile approach, not just fixed choice of algorithm, with different data, repeatedly applied but rather a rapid choice of algorithm, and organisation, whether neural net architecture, and what level of distributed, parallel, or accelerated deployment. If that is the fast moving world in which you live, then this is the book for you!

Prof. Jon Crowcroft
FRS FREng
Alan Turing Institute
Cambridge, UK
December 2021

Preface

Back in the summer of 2019, we were considering the maintenance of Owl's documentation. We were glad that documentation was serving us well and growing day by day. Then it somehow occurred to us that, now that we have such comprehensive documentation to hand plus many paper drafts and blog posts, why don't we put these different pieces together into something more cohesive – a tutorial book. And that's the root of this book.

We then discussed several possible ways to organise the book. Ultimately we decided to more or less follow the outlines of traditional textbooks in scientific computing rather than how modules are organised in Owl. However, we did not wish to make it yet another data science/machine learning/numerical computing algorithms book, and so you may find this book a little bit different to others.

The book is divided into three parts, each focusing on a different area. Part I begins by introducing how basic numerical techniques are performed in OCaml, including classical mathematical topics (interpolation and quadrature), statistics, and linear algebra. It moves on from using only scalar values to multi-dimensional arrays, introducing the tensor and Ndarray, core data types in any numerical computing system. It concludes with two more classical numerical computing topics, the solution of Ordinary Differential Equations (ODEs) and Signal Processing, as well as introducing the visualisation module we use throughout this book.

Part II is dedicated to advanced optimisation techniques that are core to most current popular data science fields, such as Deep Neural Networks. We do not focus only on applications but also on the basic building blocks, starting with Algorithmic Differentiation, the most crucial building block that in turn enables Deep Neural Networks. We follow this with chapters on Optimisation and Regression, also used in building Deep Neural Networks. We then introduce Deep Neural Networks as well as topic modelling in Natural Language Processing (NLP), two advanced and currently very active fields in both industry and academia.

Part III collects a range of case studies demonstrating how you can build a complete numerical application quickly from scratch using Owl. The cases presented include computer vision and recommender systems.

The book does not assume strict ordering in reading – we hope you can simply jump to the topic that interests you most. We do assume you are familiar with the basics of programming in OCaml. Details of the Owl library are given in its documentation, but this book provides a piece-by-piece introduction. Details of how to set up an Owl programming environment are given in the Appendices.

This book is aimed at anyone with a basic knowledge of functional programming and a desire to explore the world of scientific computing, whether to glimpse the field in the round, to build applications for particular topics, or to deep-dive into how numerical systems are constructed. If you can learn something from what we have been doing in the past several years, we will be happy that this book has succeeded.

However, as it *is* a little different to many others we feel obliged to note a couple of things that this book is explicitly not trying to do.

First, it is neither about the philosophy of functional programming, nor a general introduction either to functional programming or the OCaml language specifically. Rather, it introduces how to use the numerical functionality provided in the Owl library and OCaml language to build various applications and investigate different numerical computing topics. We avoid using very advanced functional programming techniques intentionally in the book, but we do assume that readers have basic understanding of either OCaml or similar languages such as Haskell.

Second, it is not trying to deep-dive into every topic in numerical computing. Almost all the chapters in the first two parts could each be extended to an entire book in its own right – instead, we seek to give a general introduction to the world of numerical computing via functional programming, striking a balance between breadth and depth. To do this we give an overview and introduce Owl's functionality for classical topics (e.g., linear algebra, statistics, optimisation); we present interesting applications and how to use Owl to address them for selected topics (e.g., signal processing, differential equations); and we explain in detail how Owl implements several key machine learning/artificial intelligence topics (e.g., algorithmic differentiation, regression, neural networks), rather than using the library as a "black box" as many alternative presentations do.

We hope you will find this valuable, and we welcome feedback on our approach!

Liang, Jianxin, Mort
Helsinki, Beijing, Cambridge
Jan 2022

Acknowledgements

Developing a complex system then writing a book on a big topic is certainly a very challenging task. It not only requires skills, enthusiasm, persistence, but also needs strong support from families, colleagues, and communities. For years, we have received so much help from so many individuals and organisations that it is almost impossible to make an exhaustive list. Nonetheless, we would particularly like to emphasise that Owl is developed on top of an enormous amount of previous work. Without the continuous efforts of these projects and the intellectual contributions of these people over the years, it would be impossible for us to create this system and deliver this book.

Contents

Part II Advanced Data Analysis Techniques

Acronyms

AD	Algorithmic Differentiation
BERT	Bidirectional Encoder Representations from Transformers
BOW	Bag of Words
CDF	Cumulative Distribution Function
CF	Collaborative Filtering
CNN	Convolutional Neural Network
CSC	Compressed Sparse Column
CSR	Compressed Sparse Row
DFT	Discrete Fourier Transform
DNN	Deep Neural Network
FFT	Fast Fourier Transform
FST	Fast Style Transfer
GAN	Generative Adversarial Network
GEMM	General Matrix Multiply
GRU	Gated Recurrent Unit
ILP	Integer Linear Programming
IR	Information Retrieval
KKT	Karush-Kuhn-Tucker conditions
LDA	Latent Dirichlet Allocation
LSA	Latent Semantic Analysis
LSTM	Long/Short Term Memory
MAE	Mean Absolute Error
MAPE	Mean Absolute Percentage Error
MKL	Math Kernel Library
MPE	Mean Percentage Error
MSE	Mean Square Error
NLG	Natural Language Generation
NLP	Natural Language Processing
NST	Neural Style Transfer
ODE	Ordinary Differential Equations
PDE	Partial Differential Equations

PDF   Probability Density Function
PMF   Probability Mass Function
REPL   Read-Eval-Print Loop
RK   Runge-Kutta method
RMSE   Root Mean Squared Error
RNN   Recurrent Neural Network
RP-Tree   Random Projection Tree
RPN   Region Proposal Network
RSS   Residual Sum of Squares
SF   Survival Function
SVD   Singular Value Decomposition
SVM   Support Vector Machine
TF-IDF   Term Frequency–Inverse Document Frequency
VSM   Vector Space Model

Part I
Numerical Techniques

Our journey to the world of scientific computing starts here - the numerical techniques. In this part, we start getting familiar with the basic functionality of Owl library. The topics covered in this part can be categorized according to usage of the n-dimensional array. The first two chapters are about traditional mathematical fields: using mathematical functions for calculation, and statistics. They both mainly concern scalar values. The following chapter discusses about linear algebra, which is a fundamental topic in mathematics and computer science. Its core data structure is matrix, which can be seen as a 2-dimensional array. Then we further expand the dimension to introducing n-dimensional arrays. It is a key data structure that we will use throughout this book. Finally, we introduce two advanced numerical analysis topics: differential equation, and signal processing.

Chapter 1
Introduction

Abstract We begin by briefly introducing the two key topics of this book: scientific computing and functional programming. We then introduce the main tool we will use throughout this book, the Owl library. This chapter finishes with an outline of the whole book, our target audience, and how to use the book.

1.1 Scientific Computing

Scientific Computing is a rapidly evolving multidisciplinary field that uses advanced computing capabilities to understand and solve complex problems in the real world. The algorithms used in scientific computing can be generally divided into two types: numerical analysis and computer algebra (or symbolic computation). The former uses numerical approximations to solve mathematical problems, while the latter requires closed-form representations of computation, performing symbolic manipulation, to produce exact solutions.

Scientific computation is widely used in various fields in research and industry, e.g., to solve physical models to understand how systems work. Numerical computation is particularly useful when solving a system that is too complex for closed-form solutions, if they even exist. A good example is the N-body problem, from classical mechanics, where the goal is to predict the motion of N massive interacting objects. This problem arises in fields such as astrodynamics where the objects of interests are planets, satellites, etc. under the influence of gravity. Other cases include understanding the trajectories of electrons around atomic nuclei.

Consider a simplified 2-body problem with two objects interacting on a 2-dimensional plane, where one is so much more massive than the other one that it can be thought of as being static, sitting at the zero point of a Cartesian coordinate system $(0, 0)$ in the plane. In this scenario, the location of the lighter object (y_0, y_1) can be expressed through ordinary differential equations:

© The Author(s), under exclusive license to Springer Nature Switzerland AG 2022
L. Wang et al., *OCaml Scientific Computing*, Undergraduate Topics in Computer Science,
https://doi.org/10.1007/978-3-030-97645-3_1

$$y_0''(t) = \frac{y_0}{r^3} \tag{1.1}$$

$$y_1''(t) = \frac{y_1}{r^3}, \tag{1.2}$$

where $r = \sqrt{y_0^2 + y_1^2}$. The N-body problem is one of the earliest studied problems in mechanics where it was realised that numerical methods might be more powerful in solving it than classical analytical methods. We will come back to this problem in Chap. 6.

Another field, quite different in application, where scientific computing plays an important role is finance. For example, quantitative trading relies heavily on mathematical and statistical models to identify and execute opportunities in the market. A particular example is the *Black-Scholes Formula*, [1] which is perhaps one of the most famous financial formula returning a value for the appropriate price for options:

$$C = N(d_1)S_t - N(d_2)Ke^{-rt} \tag{1.3}$$

where

$$d_1 = \frac{ln\frac{S_t}{K} + (r + \frac{\sigma^2}{2})t}{\sigma\sqrt{t}} \tag{1.4}$$

$$d_2 = d_1 - \sigma\sqrt{t} \tag{1.5}$$

It states that the call option price C relies on these factors: the current stock price S_t, the strike price K, the risk-free interest rate r, the time to maturity t, and the volatility of the option σ. Here $N()$ is the cumulative distribution function of the normal distribution. This allows study of, for example, the relationship between the implied volatility of a call option and the option price.

1.2 Functional Programming

Most existing numerical or scientific computing packages follow an imperative programming paradigm, relying on statements that change a program's state. Imperative programs are modularised by building from one or more procedures or functions. This modular style is widely adopted. From the 1980s the idea of object oriented programming developed rapidly, which extends the modular programming style to include the idea of an "object" containing data alongside procedures.

Imperative programming is widely adopted in numerical computing for a good reason: most computers' hardware implementation follows an imperative design. Indeed, FORTRAN, the first cross-platform programming language and an imperative language, is still heavily used for numerical and scientific computations in various fields after first being developed at the 1950s. There is a good chance that, even if you are using modern and popular numerical libraries such as SciPy, Julia, or Matlab, they still rely on FORTRAN in the core part somewhere.

In contrast *Functional Programming* was born to perform high-level tasks. When John McCarthy designed LISP, [2] the first functional programming language, he did so to address problems in artificial intelligence. In LISP you can see the clear distinction between functional and imperative programming: while the latter uses a sequence of statements to change the state of the program, the former builds a program constructing a tree of expressions by using and composing functions.

The fundamental difference in style between these two programming paradigms lies the particularities of their underlying model of computation. Imperative programming follows the Turing Machine as noted by S. Barry Cooper and J. Van Leeuwen in their book *Alan Turing: His Work and Impact*: – "computability via Turing machines gave rise to imperative programming". On the other hand, functional programming evolved from the *lambda calculus*, [3] a formal system of computation built around application of functions. The lambda calculus was invented by Alonzo Church in the 1930s, as a formal mathematical logic system rather than a programming language. Eventually, Turing (a student of Church) proved that the lambda calculus is Turing complete. [4] We can thus say that the lambda calculus is the basis of all functional programming languages.

In contrast to imperative programming, functional programming features immutable data, first-class functions, and optimisation on tail-recursion. By using techniques such as higher order functions, currying, map-reduce, functional programming can naturally provide thread parallelism, and lazy evaluation. But for numerical computation we also require good performance – without it, no-one would be interested in using functional programming for scientific computing. We hope that by presenting Owl, built on the functional programming language OCaml, we can show that it is possible to achieve satisfactory performance and thus practical to use functional programming in scientific computation.

1.3 OCaml Numerical Library

This book is based on OCaml, a general-purpose, multi-paradigm programming language that is a dialect of ML with object-oriented features. OCaml has paid attention to its performance, using a strong static type system to prevents runtime type mismatches, and thus remove runtime type and safety checks which often negatively impact performance in dynamically typed languages. Besides performance, runtime efficiency and powerful type system, OCaml also provides many other useful features such as tail recursion, pattern matching, parametric modules, and incremental generational automatic garbage collection. The book *Real World OCaml* [5] provides a thorough introduction to this language; this book assumes you have a basic knowledge of OCaml but not more advanced topics such as monads. [6] The most advanced feature we used in this book is functor but it does not require you to understand category theory.

We will use the Owl Numerical Computing library. Owl is a software system for scientific and engineering computing developed in the Department of Computer Science & Technology, University of Cambridge. Designed and developed in OCaml, Owl inherits many useful features as a result allowing you to write succinct type-safe numerical applications in a concise functional language without sacrificing performance. OCaml speeds up the development life-cycle, reducing the cost from prototype to production use. As an example, consider expressing the formula Eq. 1.3 in OCaml code with Owl library relatively readably as:

```
open Owl_maths

let cdf x = Owl_stats.guassian_cdf ~mu:0. ~sigma:1. x

let black_scholes r s k t sigma =
  let d1 =
    (log (s/.k) +. (r +. 0.5 *. sigma *. sigma) *. t)
    *. (sqrt t) /. sigma in
  let d2 = d1 -. sigma *. (sqrt t) in
  (cdf d1) *. s -. (cdf d2) *. k *. (exp (-.r *. t))
```

A recurring theme of this book is that while these advanced applications are sophisticated, they can be built on basic numerical functionality in a well-designed scientific library. For example, you can create a deep neural network in a few lines of code using Owl:

```
open Owl
open Neural.S
open Neural.S.Graph
open Neural.S.Algodiff

let make_network input_shape =
  input input_shape
  |> lambda (fun x -> Maths.(x / F 256.))
  |> conv2d [|5;5;1;32|] [|1;1|] ~act_typ:Activation.Relu
  |> max_pool2d [|2;2|] [|2;2|]
  |> dropout 0.1
  |> fully_connected 1024 ~act_typ:Activation.Relu
  |> linear 10 ~act_typ:Activation.(Softmax 1)
  |> get_network
```

This actually invokes many basic operations such as add, div, convolution, dot which we'll cover later. The point now is that it is an example of how to dissect a complex scientific computing application into basic building blocks.

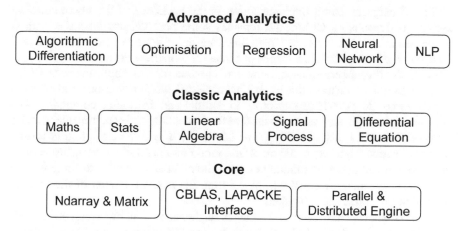

Fig. 1.1: The system architecture of the Owl library.

1.4 System Architecture

Owl is a rather complex library considering the numerous functions implemented (over 6500 functions by the end of 2020), we have strived for a modular design to make sure that the system is flexible enough to be extended in future. In the following, we will present its architecture briefly. Fig. 1.1 gives a bird's-eye view of Owl's numerical system architecture. From a perspective of numerical computing, the Owl modules can be categorised into three parts.

- The core modules are the foundation of computation. It mainly consists of the most basic data structures in Owl, namely the n-dimensional array, and the matrix. Besides, it also includes foreign function interface to other libraries (e.g., CBLAS and LAPACKE).
- Based on the core modules, Owl provides modules that provides various classic analytics, including basic mathematical and statistical functions, linear algebra, signal processing, ordinary differential equation, etc. We will talk about them in Part I of this book.
- Furthermore, based on native support for algorithmic differentiation, Owl provide a series of advanced analytics, such as optimisation, regression, neural network, natural language processing, etc. Part II of this book includes these topics.

Owl consists of both OCaml and C based implementation. The OCaml implementation, called *Base*, is the basis of all other libraries in Owl. It defines core data structures, exceptions, and part of numerical functions. Because it contains pure OCaml code so the applications built atop of it can be safely compile into native code, bytecode, JavaScript, even into unikernels. A majority of Owl's advanced functions are implemented in pure OCaml.

The C language based implementation is the backbone of the Owl numerical system. It depends on OCaml but replaces some pure OCaml functions with C implementations (e.g. vectorised math functions in the `Ndarray` module). Mixing C code into the library limits the choice of backends (e.g. browsers and MirageOS) but gives us significant performance improvement when running applications on CPU.

Though not the focus of this book, Owl can work with OCaml-based distributed engine `Actor` to extends its capability to parallel and distributed computing. The addition of the Actor subsystem makes Owl fundamentally different from mainstream numerical libraries such as SciPy [7] and Julia [8]. The core idea is to transform a user application from sequential execution mode into parallel mode (using various computation engines) with minimal efforts. The method we used is to compose two subsystems together with functors to generate the parallel version of the module defined in the numerical subsystem.

Besides what we have seen so far, Owl is further supported by several other modules. For example, **Zoo** is designed for packaging and sharing code snippets among users. This module targets small scripts and light numerical functions which may not be suitable for publishing on the heavier OPAM system. The code is distributed via gists on Github, and Zoo is able to resolve the dependency and automatically pull in and cache the code. In Part III, we will see some examples that utilise Zoo.

There are other utility modules such as plotting. Plotting is an indispensable function in modern numerical libraries. We build a `Plot` module on top of PLplot [9] which is a powerful cross-platform plotting library. However PLPlot only provides very low-level functions to interact with its multiple plotting engines, even making a simple plot involves very a lengthy and tedious control sequence. Using these low-level functions directly requires developers to understand the mechanisms in depth, which not only significantly reduces productivity but also is prone to errors. Inspired by Matlab, we implement the `Plot` module to provide developers a set of high-level APIs. The core plotting engine is very lightweight and only contains about 200 lines of code. Its core design is to cache all the plotting operations as a sequence of function closures and execute them all when we output the figure. The figures in this book are mostly generated with Owl's plotting module.

1.5 How to Run Code

"Always yield to the Hands-On Imperative."

As a book on functional programming and scientific computing, this book contains a lot of code samples for the readers to run and explore by themselves. We encourage the readers to execute the given code first, and make some changes to better learn how it works.

Most code snippets require using OCaml programming environment with the Owl library installed. Setting up the Owl environment is quite straightforward given the advancement of virtualization technology. As prerequisite, the readers need to install a series of external libraries such as OpenBLAS, plplot, eigen, etc. To make the installation even easier, we have prepared a Docker container that wraps all the necessary ingredients. All the code snippets in the book are also included. Thus the readers can just fire up a container and execute the code in whatever way they want, e.g. through the utop, an enhanced toplevel, or REPL (Read-Eval-Print Loop), for OCaml. The REPL system of Owl automatically loads all the modules, and installs multiple pretty printers for various data types. The readers can conveniently explore the world using this handy tool.

We have created a GitHub repository for this purpose. Please check our webpage[1] for detailed installation instructions.

1.6 Summary

This chapter briefly introduced the two key topics of this book, scientific computing and functional programming, before the main tool that is used in this book, the Owl library. It finished with an outline of the architecture or the Owl library, and how to run the code in this book.

References

1. Black-scholes model [Internet]. Wikipedia. Wikimedia Foundation; 2021 [cited 2022 Jan 4]. Available from: https://en.wikipedia.org/wiki/Black-Scholes_model.
2. McCarthy J. History of LISP. InHistory of programming languages 1978 Jun 1 (pp. 173-185).
3. Barendregt HP. The lambda calculus. Amsterdam: North-Holland; 1984.
4. Turing AM. Computability and λ-definability. The Journal of Symbolic Logic. 1937 Dec;2(4):153-63.
5. Minsky, Y., Madhavapeddy, A., & Hickey, J. (2013). Real World OCaml: Functional programming for the masses. " O'Reilly Media, Inc.".
6. Monad (functional programming) [Internet]. Wikipedia. Wikimedia Foundation; 2022 [cited 2022 Jan 4]. Available from: https://en.wikipedia.org/wiki/Monad_functional_programming.
7. Virtanen P, Gommers R, Oliphant TE, Haberland M, Reddy T, Cournapeau D, Burovski E, Peterson P, Weckesser W, Bright J, Van Der Walt SJ. SciPy 1.0: fundamental algorithms for scientific computing in Python. Nature methods. 2020 Mar; 17(3):261-72.
8. Bezanson J, Edelman A, Karpinski S, Shah VB. Julia: A fresh approach to numerical computing. SIAM review. 2017; 59(1):65-98.
9. LeBrun MJ, Furnish G. The PLPLOT plotting library. Institute for Fusion Studies, University of Texas, Austin, USA. 2005.

---

[1] Code repository for this book, URL:https://ocaml.xyz/springer-book-code/

Chapter 2
Numerical Algorithms

Abstract We begin our journey into the world of scientific computing. The most basic building block of any numerical computing library is a series of mathematical calculation functions. In this chapter, we introduce implementing two key mathematical techniques when working with functions: interpolation and integration. We present some key techniques, and show how they are implemented using OCaml, and specifically, how they are supported by mathematical functions provided in the *Maths* module in Owl.

2.1 Interpolation

It is often the case that we do not know the analytical form of a function f, but only some points $f_i - f(x_i)$ along it. Not knowing the analytical form means that we cannot calculate its precise value $f(x)$ at an arbitrary point x. So we have to estimate $f(x)$ by drawing a smooth curve over the domain of function f. If x is within interval of the given data points x_i, $i = 1 \ldots N$, this task is called *interpolation*; otherwise it is the much more difficult task of *extrapolation*. In this section, we use OCaml to implement interpolation. The implementation code in this chapter mostly follow that in the *Numerical Recipes* book [1].

2.1.1 Polynomial Interpolation

Perhaps the most commonly used interpolation method is *polynomial interpolation*. In a Euclidean space it is known that any two distinct points specify exactly one curve: a straight line, or a polynomial of degree 1. Similarly, any three points specify exactly one polynomial of degree 2, and so on. Formally, given n points (x_i, y_i), $i = 1 \ldots n$, in a two dimensional plane, the interpolating polynomial of degree $n - 1$ through the n points can be expressed using a Lagrange form as shown in Eq. 2.1.

L. Wang et al., *OCaml Scientific Computing*, Undergraduate Topics in Computer Science, https://doi.org/10.1007/978-3-030-97645-3_2

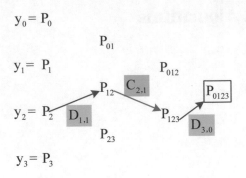

Fig. 2.1: Illustration of the Neville's Algorithm

$$P(x) = \sum_{k=1}^{n} (\prod_{j \neq k} \frac{x - x_j}{x_k - x_j}) y_k \qquad (2.1)$$

By approximating the unknown function f with such a polynomial, we can then estimate its value at any point.

While straightforward to understand, the implementation of this method turns out to be quite tricky, and also does not provide any estimate of the error in the approximation at a given point. One common implementation uses *Neville's Algorithm* [3] which builds up Eq. 2.1 recursively as illustrated in Fig. 2.1. Here $P_{i(i+1)...(i+m)}$ indicates a polynomial of degree $m - 1$ that interpolates across m points, from i to $i + m$:

$$P_{i(i+1)...(i+m)}(x) = \frac{(x - x_{i+m})P_{i(i+1)...(i+m-1)}(x) + (x_i - x)P_{(i+1)...(i+m)}(x)}{x_i - x_{i+m}} \qquad (2.2)$$

For example, suppose we are given 4 points, and we denote a polynomial of degree one that interpolates across two points 0 and 1 by P_{01}, a polynomial of degree 2 that interpolates across three points 0,1,2 by P_{012}, etc. Obviously, P_{11}, or just P_1 means a polynomial of degree 0 that goes across point 1. In other word, $P_1(x) = y_1$ for any x value. The algorithm target is to calculate $P_{0123}(x)$ for a given x value. Eq. 2.2 shows how to combine two parent polynomials into one.

In the implementation, we record the differences between two parents and one child, formally denoted by $C_{m,i}$ and $D_{m,i}$:

$$C_{m,i} = P_{i...(i+m)} - P_{i...(i+m-1)}$$
$$D_{m,i} = P_{i...(i+m)} - P_{(i+1)...(i+m)}$$

$$(2.3)$$

By combining Eq. 2.3 and Eq. 2.2, we can compute the differences at one level (defined by m, the number of points involved) using information from the previous level, as shown in Eq. 2.4.

$$C_{m+1,i} = \frac{(x_i - x)(C_{m,i+1} - D_{m,i})}{x_i - x_{i+m+1}}$$
$$D_{m+1,i} = \frac{(x_{i+m+1} - x)(C_{m,i+1} - D_{m,i})}{x_i - x_{i+m+1}}$$

$$(2.4)$$

Thus the core of this algorithm is to start from the initial value of any point, and incrementally increase it by the value of C or D calculated according to Eq. 2.4. The choice of either C or D depends on which route provides the "straightest" update toward the final target. The value of C or D that added in the last step then provides an estimate of the error in this interpolation.

The Owl_maths_interpolate module provides a polint function for interpolation and extrapolation:

```
val polint: float array -> float array
  -> float -> float * float
```

The function polint xs ys x performs polynomial interpolation based on arrays of points and values xs and ys that constrain the function, plus a given value x, and returns the estimate of the function at that point, $f(x)$, plus an error estimate dy. The implementation, following the algorithm just described, is shown below. As its name suggests, polint approximates complicated curves with the polynomial of the lowest possible degree that intersects the given points.

Neville's Algorithm: OCaml Implementation

```
let polint xs ys x =
  let n = Array.length xs in
  let c = Array.copy ys in
  let d = Array.copy ys in
  let ns = ref 0 in
  let dy = ref 0. in
  let dif = ref (abs_float (x -. xs.(0))) in

  for i = 0 to n - 1 do
    let dift = abs_float (x -. xs.(i)) in
```

```
   if dift < !dif
   then (
     ns := i;
     dif := dift)
 done;

 let y = ref ys.(!ns) in
 ns := !ns - 1;

 for m = 0 to n - 2 do
   for i = 0 to n - m - 2 do
     let ho = xs.(i) -. x in
     let hp = xs.(i + m + 1) -. x in
     let w = c.(i + 1) -. d.(i) in
     let den = ho -. hp in
     assert (den <> 0.);
     let den = w /. den in
     c.(i) <- ho *. den;
     d.(i) <- hp *. den
   done;
   if (2 * !ns) + 2 < n - m - 1
   then dy := c.(!ns + 1)
   else (
     dy := d.(!ns);
     ns := !ns - 1);
   y := !y +. !dy
 done;

 !y, !dy;;
```

To see how well this interpolation method works, consider the Gamma special function that we will introduce in Sec. 2.3. This can be seen as an interpolation solution to the integer function $y(x) = (x - 1)!$.

```
# let x = [|2; 3; 4; 5; 6|] in
  let y = Array.map (fun x -> Maths.fact (x - 1)) x;;
val y : float array = [|1.; 2.; 6.; 24.; 120.|]
```

In this example, we specify five points generated from this factorial function, and see how the approximate function obtained by interpolation compares with the Gamma function itself.

```
let f a =
  let v, _ = Owl_maths_interpolate.polint x y a in v;;
```

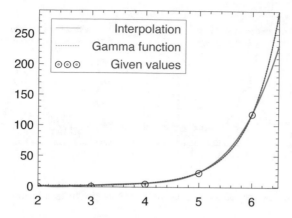

Fig. 2.2: Plot of interpolation and corresponding Gamma function.

We plot the interpolation function f and compare it to the Gamma function. As can be seen in Fig. 2.2, the interpolated line closely fits the true solution, the Gamma function, within a given domain. However, the fit when extrapolating beyond the given domain, where the x-value falls out of the given data, gets growingly less accurate.

2.1.2 Rational Interpolation

One of the problems with polynomial interpolation is that it is unstable when the specified points are equally, or near-equally, spaced. The code below shows such an example. We choose 11 points equally-spaced on the function $f(x) = (1 + x^2)^{-1}$ (represented by g), and use these points to find an interpolated polynomial (f1).

The Mat module in this code provides a data structure for expressing matrices in Owl. We will introduce this module in Chap. 5. Here we only give a glimpse of the function used in this snippet. The linspace a b n function returns a vector that contains n evenly spaced samples, calculated over the interval [a, b]. The map f m functions apply function f on matrix m. The to_array function cast a matrix to an OCaml array. Besides, the scalar_* a m functions apply basic calculation such as add, multiplication, etc. between float number a and the matrix m element-wisely.

```
let a = -5.
let b = 5.
let xs1 = Mat.linspace a b 11
let g x = 1. /. (1. +. x *. x)
let ys1 = Mat.map g xs1
```

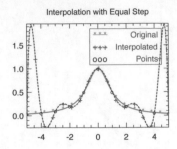 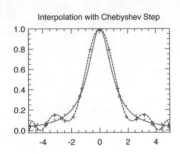

Fig. 2.3: The impact of points distances on polynomial interpolation.

```
let f1 x =
  let y, _err = Owl_maths_interpolate.polint
      (Mat.to_array xs1) (Mat.to_array ys1) x in
  y;;
```

Fig. 2.3(a) compares functions g and f1. Between each pair of points (or *nodes*). Here the interpolated polynomial changes rapidly, especially at the two ends. One solution is to use *Chebyshev nodes* [7] rather than equally spaced nodes to generate f2, as in the code shown below. Fig. 2.3(b) depicts f2, a much better approximation to the original function g than f1.

```
let xs2 = Mat.linspace 1. 10. 11
let xs2 = Mat.scalar_add
    ((a +. b) /. 2.)
    (Mat.scalar_mul
      ((b -. a) /. 2.)
      (Mat.cos (Mat.scalar_mul
        (Owl_const.pi /. 10.)
        (Mat.sub_scalar xs2 0.5)))) |> Mat.reverse
let ys2 = Mat.map g xs2

let f2 x =
  let y, _err = Owl_maths_interpolate.polint
    (Mat.to_array xs2) (Mat.to_array ys2) x in
  y;;
```

However, more often than not, in an interpolation problem, we are not free to choose sample points, and so we may end up with equally spaced nodes no matter what. In such cases *Rational Interpolation* is an appropriate alternative to the polynomial method that provides better results. A rational interpolation method expresses the interpolation method in the form of the quotient of two polynomials:

$$R(x) = \frac{\sum_{i=1}^{a} p_i \, x^i}{\sum_{j=1}^{b} q_j \, x^j} \tag{2.5}$$

Here $a + b = m$, which is the total number of given points. Similar to the Neville algorithm for polynomial interpolation, a recursive algorithm is proposed by Bulirsch and Stoer [2] to compute Eq. 2.5. Using the same definition as Eq. 2.3, this algorithm starts with one initial y value and iteratively chooses to add the C's or D's (the value difference between a child and two parents) step by step. The algorithm states the update rules as in Eq. 2.6.

$$C_{m+1,i} = \frac{D_{m,i}(C_{m,i+1} - D_{m,i})}{D_{m,i} - C_{m,i+1}}$$

$$D_{m+1,i} = \frac{C_{m,i+1}(C_{m,i+1} - D_{m,i})}{(\frac{x-x_i}{x-x_{i+m+1}})D_{m,i} - C_{m,i+1}}$$

$$\tag{2.6}$$

The OCaml implementation is shown below. It follows a similar structure to the Neville algorithm but with different updating rules.

Implementation of rational interpolation

```
let ratint xs ys x =
  let n = Array.length xs in
  let c = Array.copy ys in
  let d = Array.copy ys in
  let hh = ref (abs_float (x -. xs.(0))) in
  let y = ref 0. in
  let dy = ref 0. in
  let ns = ref 0 in
  let eps = 1e-25 in

  try
    for i = 0 to n do
      let h = abs_float (x -. xs.(i)) in
      if h = 0.
      then (
        y := ys.(i);
        dy := 0.;
        raise Owl_exception.FOUND)
      else if h < !hh
      then (
        ns := i;
        hh := h;
```

```
        c.(i) <- ys.(i);
        d.(i) <- ys.(i) +. eps)
   done;

   y := ys.(!ns);
   ns := !ns - 1;

   for m = 1 to n - 1 do
     for i = 1 to n - m do
       let w = c.(i) -. d.(i - 1) in
       let h = xs.(i + m - 1) -. x in
       let t = (xs.(i - 1) -. x) *. d.(i - 1) /. h in
       let dd = t -. c.(i) in
       if dd = 0. then failwith "Has a pole";
       let dd = w /. dd in
       d.(i - 1) <- c.(i) *. dd;
       c.(i - 1) <- t *. dd
     done
   done;

   !y, !dy
 with
 | Owl_exception.FOUND -> !y, !dy
 | e                   -> raise e;;
```

However, there are still shortcomings remained in this solution. For example, we know that any set of points in a plane define exactly one polynomial, but not all point sets have a corresponding rational function. Moreover, the poles – points where the denominator is close to 0 – tend to appear unexpectedly while performing rational interpolation.

2.2 Integration

A second key topic in numerical computing is *Integration*, perhaps one of the most basic tools in any scientist's toolbox. For example, in Physics, both calculating the displacement of objects given changing velocity and the work done to move objects using time-dependent force require computing integrals. Given a function f taking a real variable and an interval $[a, b]$ of the real line, the integral of f in this interval is:

$$\int_a^b f(x)dx$$

The result can be viewed as the sum of the signed area of the region in the Cartesian plane bounded by the curve f and the x-axis in the domain $[a, b]$. The area above the x-axis is positive (adds to the area sum) and that below the x-axis is negative (subtracts from the area sum).

It is often not possible to compute integrals analytically, so a lot of integration is performed using integral formulas. For example, $\int a\, dx = ax + C$, $\int \cos(x) dx = \sin(x) + C$, $\int e^x = e^x + C$, and many others. Some integrals are so frequently encountered in certain application areas that their solutions are provided as special functions, and we will introduce some of them in Sec. 2.3. In this section we focus solely on numerical integration methods and their implementation in Owl.

2.2.1 Basic Algorithms

Owl provides several routines to perform numerical integrations in the Owl_maths_quadrature module. For example, the code below computes $\int_1^4 x^2$ using trapz, one of the numerical integration methods:

```
# Owl_maths_quadrature.trapz (fun x -> x ** 2.) 1. 4.;;
- : float = 21.0001344681758439
```

The *fundamental theorem of calculus* states that:

If f is continuous over the interval $[a, b]$ and $F(x)$ is an anti-derivative of $f(x)$ (i.e. derivative of F equal to f), then $\int_a^b f(x) = F(a) - F(b)$.

In this case, we know one such $F(x) = \frac{x^3}{3}$. [1] We can thus verify this result using this theorem:

$$\int_1^4 x^2 = (4^3 - 1^3)/3 = 21$$

So you might be thinking, what is this trapz? Using numerical methods (or *quadrature*) to do integration dates back at least to the invention of calculus. The basic idea is to approximate an integration by summing many small areas under the curve being integrated. There are many algorithms for performing numerical integration, and the *trapezoidal* rule implemented in trapz is just one of them.

This classical method divides a to b into N equally spaced abscissas: $x_1, x_2, \ldots, x_N$. The area between each pair of adjacent x_i and x_j is approximated by a trapezoid, and the area formula becomes:

$$\int_{x_i}^{x_j} f(x) dx = h\left(\frac{f(x_i)}{2} + \frac{f(x_j)}{2}\right) + O(h^3 f'')$$

---

[1] We will discuss the topic of derivatives in detail in Chap. 8.

The error term $O(h^3 f'')$ indicates that the approximation error is related to the spacing of the abscissas h and the second order derivative of the original function. The *order* of an integration method is the lowest order polynomial whose integral it cannot accurately calculate. Extending this to a sequence of many abscissas, we obtain the extended trapezoidal rule:

$$\int_{x_1}^{x_N} f(x)dx = h(\frac{f(x_1)}{2} + f(x_2) + \ldots + f(x_{N-1}) + \frac{f(x_N)}{2}) + O(\frac{(b-a)^3 f''}{N^2}). \quad (2.7)$$

The function `trapz` implements the method shown in Eq. 2.7 with the interface:

```
val trapz : ?n:int -> ?eps:float -> (float -> float)
            -> float -> float -> float
```

Function `trapz` ~n ~eps f a b computes the integral of f on interval [a,b] using the trapezoidal rule. The argument n specifies the maximum step which defaults to 20, and `eps` is the desired fractional accuracy threshold, defaulting to 1e-6. It works by iterating for several stages, improving the accuracy at each by adding more interior points. The implementation code of `trapz` is relatively simple, as shown below.

Integration Method: Trapezoidal

```
let trapz ?(n = 20) ?(eps = 1e-6) f a b =
  let s_new = ref 0. in
  let s_old = ref 0. in
  (try
    for i = 1 to n do
      s_new := trapzd f a b i;
      if i > 5 then (
        let d = abs_float (!s_new -. !s_old) in
        let e = eps *. abs_float !s_old in
        assert (not (d < e || (!s_new = 0. && !s_old = 0.)));
        s_old := !s_new)
    done
  with
  | _ -> ());
  !s_new;;
```

The core part of this algorithm is the `trapzd` function that computes the n-th stage of refinement of the extended trapezoidal rule. At the beginning when $n = 1$, it computes only a rough estimation using the beginning and end of the domain, a

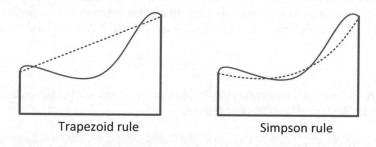

Trapezoid rule   Simpson rule

Fig. 2.4: Illustration of trapezoid and simpson integration rules.

and b: $\frac{f(a)+f(b)}{2(b-a)}$. At each subsequent step n it adds 2^{n-2} extra interior points. Its implementation is also straightforward.

```
let trapzd f a b n =
  if n = 1
  then 0.5 *. (b -. a) *. (f a +. f b)
  else (
    let m = 2. ** float_of_int (n - 1) in
    let d = (b -. a) /. m in
    let x = ref (a +. (0.5 *. d)) in
    let s = ref 0. in
    for _i = 1 to int_of_float m do
      x := !x +. d;
      s := !s +. f !x
    done;
    (0.5 *. d *. (f a +. f b)) +. (!s *. d));;
```

Besides applying `trapzd` function at each iteration, `trapz` also checks whether the integration accuracy has reached a target level, specified by `eps`. `trapz` is a quite robust integration algorithm, especially for functions that are not very smooth. That being said, setting an extremely low `eps` value can increase errors caused by representing numbers with only finite precision during the numerical computation.

Besides `trapz`, the `trapzd` method is also a core mechanism in several other integration functions such as *Simpson's Rule*. This is similar to `trapz` in its interface but different in implementation:

$$\int_{x_0}^{x_2} f(x)dx = h(\frac{f(x_0)}{3} + \frac{4f(x_1)}{3} + \frac{f(x_2)}{3}) + O(h^5 f^{(4)}). \qquad (2.8)$$

The difference between the Trapezoid method and Simpson's Rule is illustrated in Fig. 2.4. Compared to the Trapezoid method, Simpson has a higher order of error, related to the fourth derivative of function f. Rather than approximating the area between each pair of abscissas by a trapezoid, it is approximated by a trapezoid-like shape with one curved (not straight) edge. Its implementation is also similar to the previous method. Extending this to a form similar to Eq. 2.7, it can be shown that:

$$\int_{x_1}^{x_N} = \frac{4}{3} S_{2N} - \frac{S_N}{3}, \qquad (2.9)$$

where S_n is the result of executing Eq. 2.7 for n steps. Therefore, the Simpson method is often implemented iteratively as follows.

Integration Method: Simpson

```
let simpson ?(n = 20) ?(eps = 1e-6) f a b =
  let s_new = ref 0. in
  let s_old = ref 0. in
  let o_new = ref 0. in
  let o_old = ref 0. in
  (try
     for i = 1 to n do
       s_new := trapzd f a b i;
       s_old := ((4. *. !s_new) -. !o_new) /. 3.;
       if i > 5 then (
         let d = abs_float (!s_old -. !o_old) in
         let e = eps *. abs_float !o_old in
         assert (not (d < e || (!s_old = 0. && !o_old = 0.)));
         o_old := !s_old;
         o_new := !s_new)
     done
   with
   | _ -> ());
  !s_new;;
```

Another method *Romberg integration* (`romberg`) selects methods of different orders to provide both good accuracy, and normally better performance than both the `trapz` and `simpson` methods. Similar to those methods, it iteratively estimates the integral using more and more intermediate points. The key idea is that, when testing the accuracy obtained so far, it uses the interpolation method introduced in the previous section, so that the sub-area formed by each pair of adjacent abscissas is a trapezoid-like shape with one edge a higher-dimensional polynomial. In the simpson method this polynomial can only be of third order. The implementation code is shown below and again follows a similar structure to the previous two algorithm implementations.

Integration Method: Romberg

```
let romberg ?(n = 20) ?(eps = 1e-6) f a b =
  let s = Array.make (n + 1) 0. in
  let h = Array.make (n + 2) 1. in
  let rss = ref 0. in
  let k = 5 in
  (try
    for i = 0 to n - 1 do
      s.(i) <- trapzd f a b (i + 1);
      if i >= k then (
        let s' = Array.sub s (i - k) k in
        let h' = Array.sub h (i - k) k in
        let ss, dss = Owl_maths_interpolate.polint h' s' 0. in
        rss := ss;
        assert (abs_float dss > eps *. abs_float ss));
      h.(i + 1) <- 0.25 *. h.(i)
    done
  with
  | _ -> ());
  !rss;;
```

2.2.2 Gaussian Quadrature

The quadrature methods introduced so far are all based on the assumption that the abscissas are equally spaced, using $\sum_i f(x_i)h$ to appropriate the integral $\int_a^b f(x)dx$ with a small constant as the weight coefficient h. These methods work well, but it turns out that higher accuracy can be achieved by choosing different and varying weight coefficients and evaluation points. This is the basis of the *Gaussian Quadrature* method where we approximate $\int_a^b f(x)dx$ with $\sum_i f(x_i)w(x_i)$. The weight w changes with the non-equally spaced evaluation points x_i.

Let's consider an example. Suppose we need to calculate $\int_{-1}^1 \frac{1}{x^2+1}$. Basic calculus tells us that the integration result is $atan(x)|_{-1}^1$ and so we know the precise integral is:

```
# atan (1.) -. atan (-1.);;
- : float = 1.57079632
```

How well do the basic quadrature methods perform? Without loss of generality, we use 10 abscissas, and the results are shown in the second and third columns of Tbl. 2.1. Using this table, we calculate $\sum_i f(x_i)w(x_i)$, where $f(x) = \frac{1}{x^2+1}$ to obtain the result 1.56746305. Given we used only a small number of abscissas, this is a pretty decent approximation.

Table 2.1: Example of basic (left) and Gaussian (right) quadrature methods.

ID	x_i	$w(x_i)$	x_i	$w(x_i)$
0	-1	0.2	-0.97390652	0.06667134
1	-0.8	0.2	-0.86506336	0.14945134
2	-0.6	0.2	-0.86506336	0.14945134
3	-0.4	0.2	-0.43339539	0.26926671
4	-0.2	0.2	-0.14887433	0.29552422
5	0	0.2	0.14887433	0.29552422
6	0.2	0.2	0.43339539	0.26926671
7	0.4	0.2	0.67940956	0.21908636
8	0.6	0.2	0.86506336	0.14945134
9	0.8	0.2	0.97390652	0.06667134

Next we apply Gaussian Quadrature using the parameters shown in the third and fourth columns in Tbl 2.1. We will discuss these seemingly bizarre parameter choices in a moment. For now, let us calculate $\sum_i f(x_i)w(x_i)$ again. Note that these two methods use the same number of abscissas, and the only difference is the choice of weight coefficients and abscissas location. The result should be 1.57079627. Compared to the true value 1.57079632, it is accurate to the hundred-millionth part compared to the hundredth part achieved by the earlier method.

But what about those curious choice of parameters? They are generated based on something called the *Legendre Polynomials*, obtained using the gauss_legendre function, as shown in the code below. Notice that these numbers are not affected by the choice of $f(x)$, but only by the upper and lower integration limit and number of abscissas.

```
# Owl_maths_quadrature.gauss_legendre ~a:(-1.) ~b:1. 10;;
- : float array * float array =
([|-0.97; -0.86; -0.67; -0.43; -0.14; 0.14; 0.43; 0.67;
  0.86; 0.97|], [|0.06; 0.14; 0.21; 0.26; 0.29; 0.29;
  0.26; 0.21; 0.14; 0.06|])
```

The Legendre Polynomials are a system of complete and orthogonal polynomials, with the Legendre polynomial of order n defined iteratively as:

$$P_0(x) = 1$$
$$P_1(x) = x$$
$$(n + 1)P_{n+1}(x) = (2n + 1)xP_n(x) - nP_{n-1}(x)$$

$$(2.10)$$

In choosing the coefficients in Gaussian quadrature, the x_i's are the roots of the Legendre Polynomials, i.e. those x that make $P_n(x) = 0$, where n is the number of abscissas. Discussion of the theory behind the Legendre polynomials is beyond the scope of this book; see for example Chapter 4 of Numerical Methods [1] for more information.

Given these Legendre coefficients, the implementation of Gaussian quadrature is not so different from the other previous methods. In the math module in Owl, the function gaussian_fixed calculates the quadrature following the procedure shown in the previous example. Note the gauss_legendre_cache function. As the choice of Legendre coefficients is not affected by the specific function used, we can cache previously used parameters for later use, reducing the calculation time. Based on gaussian_fixed, the gaussian function iteratively tests the quadrature result using an increasing number of abscissas, stopping when adding another abscissa has a negligible affect on the result.

Implementation of Gaussian Quadrature

```
let gaussian_fixed ?(n = 10) f a b =
  let x, w =
    match n < Array.length gauss_legendre_cache with
    | true  -> gauss_legendre_cache.(n)
    | false -> gauss_legendre n
  in
  let xr = 0.5 *. (b -. a) in
  let s = ref 0. in
  for i = 0 to n - 1 do
    let c = (xr *. (x.(i) +. 1.)) +. a in
    s := !s +. (w.(i) *. f c)
  done;
  !s *. xr

let gaussian ?(n = 50) ?(eps = 1e-6) f a b =
  let s_new = ref infinity in
  let s_old = ref infinity in
  (try
    for i = 1 to n do
      s_new := gaussian_fixed ~n:i f a b;
      assert (abs_float (!s_new -. !s_old) > eps);
```

```
        s_old := !s_new
    done
  with
  | _ -> ());
  !s_new;;
```

In Sec. 2.3 we will introduce several special integral functions that solve particular integrals. One is the *sine* integral, listed in Tbl. 2.2. As an example, we can compute this integral using Gaussian quadrature to validate its precision. Setting $x = 4$, the code below shows that numerical method `gaussian` works well to approximate the true value of the integral.

```
# let f t = Maths.(div (sin t) t);;
# Owl_maths_quadrature.gaussian f 0. 4.;;
- : float = 1.75820313914469306
# Owl_maths.si 4.;;
- : float = 1.75820313894905289
```

Besides using the Legendre polynomials, the coefficients in Gaussian quadrature can be generated in many other ways including from the Jacobi polynomials, Laguerre polynomials, the Gauss-Hermite quadrature formula, and so on. Again, for more information on these methods, please refer to [1].

2.3 Special Functions

As well as basic mathematical functions from arithmetic, trigonometry, and so on, the `Maths` module contains a lot of other mathematical functions, important in various fields including mathematical analysis, physics, and so on. The implementations of these functions rely on the Cephes Mathematical Functions Library, a C language library that provides implementation of such functions of interest to scientists and engineers. In the final section of this chapter we briefly introduce some of them, listed in Tbl. 2.2, since you may find them useful from time to time.

The *Airy function*, named after the British astronomer George Airy contains two linearly independent solutions `Ai` and `Bi` to the Airy differential equation. The `Maths.airy` function accepts an float number `x` as input, and outputs four floats: `Ai`, its derivative `Ai'`, `Bi`, and its derivative `Bi'`. One application example is found in problems of quantum mechanics and non-uniform optical wave guides. It is often important to find approximate solutions to linear differential equations with spatially varying coefficients. The Airy function can be used to provide uniform semi-classical approximations in these problems [4].

The *Bessel functions* were first defined by the mathematician Daniel Bernoulli and then generalised by Friedrich Bessel. In its equation listed in Tbl. 2.2, the complex

Table 2.2: Special functions

Name	Function	Definition
Airy	`airy`	solution to $y''(x) = xy(x)$
Bessel	`j0`, `j1`, etc.	canonical solutions to $x^2 y'' + xy' + (x^2 - \alpha^2)y = 0$
Gamma	`gamma`	$\Gamma(z) = \int_0^\infty x^{z-1} e^{-x} dx$
Beta	`beta`	$B(x, y) = \int_0^1 t^{x-1}(1-t)^{y-1} dt = \frac{\Gamma(x)\Gamma(y)}{\Gamma(x+y)}$
Error	`erf`	$\frac{2}{\sqrt{\pi}} \int_0^x e^{-t^2} dt$
Elliptic Integral	`ellipe`	$f(x) = \int_c^x R(t, \sqrt{(P(t))}) dt$
Dawson Integral	`dawsn`	$D(x) = e^{-x^2} \int_0^x e^{t^2} dt$
Fresnel Integral	`fresnel`	$S(x) = \int_0^x \sin(t^2) dt$
		$C(x) = \int_0^x \cos(t^2) dt$
Sine Integral	`si`	$Si(x) = \int_0^x \frac{\sin(t)}{t} dt$

number α is called the *order* of the Bessel functions. They are important when studying wave propagation and static potentials, such as electromagnetic waves in a cylindrical waveguide. In solving cylindrical coordinate systems, Bessel functions of integer order or half integer order are often used.

Bessel functions can be divided into two kinds; both solve Bessel's differential equations, but the first is not singular at $x = 0$, unlike the second. A special case is when x is purely imaginary. In this case, the solutions are called the *modified Bessel functions*, also be categorised into both first kind and second kind. These functions are all supported in the `Maths` module, with names such as `j0` (first kind, order 0), `j1` (first kind, order 1), and `y0` (second kind, order 0), etc.

Both the Airy function and the Bessel function family are solutions to certain differential equations, a topic we will cover in Chap. 6. Fig. 2.5a plots the two solutions `Ai` and `Bi` of the Airy function; Fig. 2.5b plots the first kind Bessel function of order 0, 1, and 2. You will notice that they show similar patterns as the solutions to differential equations in Chap. 6.

The *Gamma function* is widely used in a range of areas such as fluid dynamics, geometry, and astrophysics. It is especially suitable for describing a common pattern of processes that decay exponentially in time or space. The equations in Tbl. 2.2 work for any complex number input. If that input is a positive integer n, the Gamma function simplifies to the factorial function, $\Gamma(n) = (n-1)!, n \in \mathbf{Z}^+$.

The Gamma function is arguably the most common special function. One explanation is that it has the form $f(x)e^{-h(t)}$, which conveniently describes processes that decay exponentially with given factors – very common in the physical world – and so is widely used in fields ranging from quantum physics to astrophysics and fluid dynamics [5]. In Chap. 3, we will introduce the gamma distribution in statistics, which can be used to model a variety of processes.

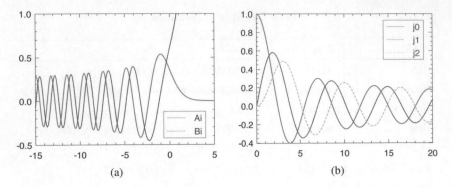

(a)   (b)

Fig. 2.5: Example of (a) the two solutions of an Airy equation, and (b) Bessel function of the first kind, with different orders.

In fact, the Gamma function is difficult to avoid even in the definition of other special functions For example, the *Beta function* is defined using the Gamma function. Proposed by Gabriele Veneziano in a paper that wound up marking the beginning of string theory [6], it is also used in the analysis of the preferential attachment process, a type of stochastic urn process that describes how the resource can be distributed among a group of individuals based on the existing resource each one has acquired. The code below illustrates the relationship between the Gamma and Beta functions, and the symmetry of the Beta function, i.e. $B(x, y) = B(y, x)$.

```
# let x = Maths.beta 3. 4.;;
val x : float = 0.0166666666666666664
# let y = Maths.((gamma 3.) *. (gamma 4.) /. (gamma (7.)));;
val y : float = 0.0166666666666666664
# Maths.beta 3. 4. == Maths.beta 4. 3.;;
- : bool = True
```

The *Error function* occurs often in probability and statistics. It is a sigmoid function whose name, *error*, was proposed due to its connection with the theory of errors in statistics. For a non-negative value x, erf x is the probability that a normal random variable Y with mean $\mu = 0$ and variance $\sigma^2 = 0.5$ falls in the domain [-x, x]. As it represents a concrete probability, the *complementary* error function 1 - erf(x) is often also used.

Finally, Elliptic integrals are used to find the perimeters of ellipses. In the elliptic integral function, R is a rational function of its two arguments, P is a polynomial of degree 3 or 4 with no repeated roots, and c is a constant. Computing the circumference of an ellipse normally requires integration calculation, but the elliptic integral function provides a ready answer. Suppose an ellipse has semi-major axis $a = 4$ and semi-minor axis $b = 3$. Its circumference can be simply calculated as $4a\, f(1 - \frac{b^2}{a^2})$, where f is the elliptic integral function.

```
# let a = 4.;;
# let b = 3.;;
# let c = 4. *. a *. Maths.(ellipe (1. -. pow (b /. a) 2.));;
val c : float = 22.1034921607095072
```

As well as the selection we have mentioned, the Maths module provides many other special integral functions including the *Dawson integral*, *Fresnel integral*, and *Sine integral*. They provide alternatives to the basic numerical integration methods we introduced in Sec. 2.2,

2.4 Summary

In this chapter we introduce how to use the mathematical functions in Owl. Specifically, we demonstrated that in showing how two important mathematical tools can be implemented in OCaml: interpolation, and integration. We also introduced a selection of important special functions that are provided in the Maths module. Functions presented in this chapter take only a single scalar input. Later, in Chap. 5, we will introduce functions whose inputs are N-dimensional arrays, including vectors and matrices.

References

1. Press WH, William H, Teukolsky SA, Saul A, Vetterling WT, Flannery BP Numerical recipes 3rd edition: The art of scientific computing. Cambridge university press; 2007 Sep 6.
2. Stoer J, Bulirsch R. Introduction to numerical analysis. Springer Science & Business Media; 2013 Mar 9.
3. Neville EH. Iterative interpolation. St. Joseph's IS Press; 1934.
4. Ghatak AK, Gallawa RL, Goyal IC. Modified Airy function and WKB solutions to the wave equation. NASA STI/Recon Technical Report N. 1991 Nov; 92:20427.
5. Artin E. The gamma function. Courier Dover Publications; 2015 Jan 28.
6. Veneziano G. Construction of a crossing-simmetric, Regge-behaved amplitude for linearly rising trajectories. Il Nuovo Cimento A (1965-1970). 1968 Sep 1; 57(1):190-7.
7. Brutman L, Passow E. Rational interpolation to |x| at the Chebyshev nodes. Bulletin of the Australian Mathematical Society. 1997 Aug;56(1):81-6.

Chapter 3
Statistics

Abstract In this chapter we turn to another indispensable tool in data analysis: statistics. Statistical analysis enables us to gain insight from data, and we group the statistical functions in Owl into three topics: descriptive statistics, distributions, and hypothesis testing. In this chapter we introduce how these basic statistics are supported in OCaml.

3.1 Random Variables

We start traditionally, by assigning probabilities to discrete *events*. An event may be drawn from either a finite or an infinite set of possible outcomes. The set of all possible outputs forms the event's *sample space*. For example, imagine rolling a normal six-sided dice. The sample space is then $S = \{1, 2, 3, 4, 5, 6\}$, and each outcome has equal probability, $\frac{1}{6}$. As you might expect, analytically determining the probability of a particular outcome becomes increasingly complex.

Assigning probabilities to events gives rise to the concept of a *Random Variable*. A *random variable* is a function that associates values between 0 and 1 with the possible outputs of events. For example, consider a coin-tossing game where a fair coin is tossed four times producing the result "heads", "heads", "tails", "heads". Suppose we are interested in the number of "heads". We might then define a random variable X denoting the number of "heads", with $X = 3$ in this particular case.

We can categorise random variables as *discrete* or *continuous* based on whether they describe events with a finite or infinite number of possible outcomes. The outcomes of a set of coin tosses is thus a commonly used example of a discrete random variable. Of course, not all coins are fair: consider for example a coin where the probability of tossing heads is $p \neq 0.5$. We toss such a coin just three times and define a random variable X that again describes the number of heads and thus can take the values 0, 1, 2, or 3.

We can calculate the possibility of each of these outcomes. Since each toss is an independent trial, the probability of three heads, $P(X = 3)$, is then p^3. If we consider

L. Wang et al., *OCaml Scientific Computing*, Undergraduate Topics in Computer Science, https://doi.org/10.1007/978-3-030-97645-3_3

the number of ways of obtaining two heads, $P(X = 2)$, we see that there are three such ways: writing H for heads and T for tails, HHT, HTH, and THH. Each of these outcomes has a probability of $p^2(1 - p)$ and so $P(X = 2) = 3p^2(1 - p)$. Similarly $P(X = 1) = 3p(1 - p)^2$, and $P(X = 0) = (1 - p)^3$.

Unlike a discrete random variable, a continuous random variable has an infinite number of possible outcomes covering all the values in some interval. Examples include height, temperature, and distance. It is worth noting that, as a result, the probability that a continuous random variable X takes any particular value is always zero.

3.1.1 Descriptive Statistics

If a random variable describes a single event, repeatedly sampling it gives rise to a *population* that can be quantitatively characterised by numerous descriptive statistics. Perhaps the most frequently encountered are the *mean* and *variance*. The mean, or *expectation*, of a population X with n elements is defined as:

$$E(X) = \frac{1}{n} \sum_i x_i, \tag{3.1}$$

where x_i is the value obtained by the i-th sample of X or the i-th element in the population. The variance describes how individual values deviate from the mean value:

$$Var(X) = \frac{1}{n} \sum_i (x_i - E(X))^2. \tag{3.2}$$

The *standard deviation* is then the square root of the variance. Intuitively, the mean represents the central value of a population, and the variance how values in the population are spread around the central expectation. The definitions above are for discrete random variables – to expand them to continuous random variables, simply exchange summation for integration.

Generalising, we can define the *n-th moment* of a random variable X as

$$M_n(X) = \int_x (x_i - c)^n f(x_i) dx, \tag{3.3}$$

where $f(x)$ is a continuous function that returns the values of the random variable X, and c is a constant. Under this definition, the mean is the first moment, and the variance the second moment. The third moment is named *skewness* and indicates the degree of asymmetry of the probability distribution of a random variable. The fourth moment is called the *kurtosis* and it describes how long a "tail" the probability distribution has. We will discuss distributions further in the following section.

Let's now look at a simple example. We first draw one hundred random numbers uniformly distributed between 0 and 10 using the `uniform_rvs` function:

```
let data = Array.init 100 (fun _ -> Stats.uniform_rvs 0. 10.);;
```

We can then use the mean function to calculate the mean of these samples and find that, as you might expect, it is about 5. There are corresponding functions to calculate the higher order moments, and the numerical results below show that these values form a quite widely spread distribution about the central value. But the distribution is not heavily skewed and it does not have an obvious tail, based on that skew and kurtosis are relatively small).

```
# Stats.mean data;;
- : float = 5.18160409659184573
# Stats.std data;;
- : float = 2.92844832850280135
# Stats.var data;;
- : float = 8.57580961271085229
# Stats.skew data;;
- : float = -0.109699186612116223
# Stats.kurtosis data;;
- : float = 1.75165078829330856
```

One further related concept is the *central moments* of a random variable. The k-th central moment is the expected value of the k-th power of the random variable's deviation from the mean, denoted as:

$$\frac{1}{n} \sum_i (x_i - E(X))^k.$$
(3.4)

For example, variance is the second central moment:

```
# Stats.central_moment 2 data;;
- : float = 8.49005151658374224
```

Finally, *order* (or *rank*) *statistics* are also often used to understand data in non-parametric statistics and inference. The k-th order statistic of a statistical sample is equal to its k-th smallest (or equivalently, largest) value.

The Stats module contains many useful order statistics, including: min and max, the smallest and largest samples; median, the central value in a sorted list of samples; first_quartile and third_quartile, the values below which a quarter and three-quarters of the sample fall, and percentile, further refining the notion of quartile so that, for example, the first quartile is the 25th percentile.

3.2 Distribution

We not only care about the possible outcomes represented by a random variable but also (more importantly) about the likelihoods of these outcomes. Consider again the previous example where we toss a coin three times, and represent the number of

heads obtained by the random variable X with possible outcomes $X \in \{0, 1, 2, 3\}$. If
we know the probability of tossing a "head", we can obtain the probability $P(X = x)$
for any possible value x. $p(x) = P(X = x)$ is the *probability mass function* giving the
distribution of X. For continuous random variables, we have the analogous concept
of the *probability density function*, or PDF (since, as noted above, the probability
that a continuous random variable obtains any particular value is always zero).

The `Stats` module supports many frequently used distributions, both discrete
and continuous. For each distribution, there is a set of related functions that use
the distribution's name as a common prefix. For example, the module provides the
functions below for the Gaussian distribution:

- `gaussian_rvs`, random number generator.
- `gaussian_pdf`, probability density function.
- `gaussian_cdf`, cumulative distribution function.
- `gaussian_ppf`, percent point function (inverse of CDF).
- `gaussian_sf`, survival function (1 - CDF).
- `gaussian_isf`, inverse survival function (inverse of SF).
- `gaussian_logpdf`, logarithmic probability density function.
- `gaussian_logcdf`, logarithmic cumulative distribution function.
- `gaussian_logsf`, logarithmic survival function.

Of these functions, the random number generator, probability density function, and
cumulative distribution function are the key functions for a given distribution.

3.2.1 Discrete Distribution

The coin tossing example above can be described using a classical discrete distribution, the *Binomial Probability Distribution*. Formally, consider a series of n
independent trials, each having contains two possible outcomes, with probabilities
p for the outcome of interest and $1 - p$ for the other outcome respectively. Then the
probability distribution of the discrete random variable X describing the number of
outcomes of interest is:

$$P(X = k) = \binom{N}{k} p^k (1 - p)^{n-k} \tag{3.5}$$

We can simulate this process of tossing coins with the `Stats.binomial_rvs`
function. For example, suppose the probability of tossing heads is 0.3, and the coin
is tossed 10 times:

```
# let _ =
    let toss = Array.make 10 0 in
    Array.map (fun _ -> Stats.binomial_rvs 0.3 1) toss;;
- : int array = [|0; 0; 0; 0; 0; 0; 0; 0; 1; 0|]
```

The equation Eq. 3.5 is the probability density function of this binomial distribution. It can be expressed in code as:

```
# let x = [|0; 1; 2; 3|];;
val x : int array = [|0; 1; 2; 3|]

# let p = Array.map (Stats.binomial_pdf ~p:0.3 ~n:3) x;;
val p : float array =
  [|0.342999999999999916; 0.440999999999999837;
    0.188999999999999918; 0.026999999999999823|]

# Array.fold_left (+.) 0. p;;
- : float = 0.999999999999999778
```

Along with PDF, another commonly used quantity is the probability of a random variable X sitting between the interval $[a, b]$, $P(a \leq X \leq b)$. This can be written as $P(X \leq b) - P(X \leq a - 1)$. Here the term $P(X \leq t)$ is called the *Cumulative Distribution Function* (CDF) of the random variable X. The CDF of the binomial distribution is:

$$p(X \leq k) = \sum_{i=0}^{k} \binom{N}{i} p^i (1 - p)^{n-i} \tag{3.6}$$

We can calculate CDF in the three coin toss problem using the function `binomial_cdf`:

```
# let x = [|0; 1; 2; 3|];;
val x : int array = [|0; 1; 2; 3|]
# let p = Array.map (Stats.binomial_cdf ~p:0.3 ~n:3) x;;
val p : float array = [|
  0.342999999999999972; 0.784; 0.973; 1.|]
```

Another widely encountered discrete distribution is the *Poisson Distribution*, with possibility distribution function:

$$P(X = k) = e^{-\lambda} \left(\frac{\lambda^k}{k!} \right), \tag{3.7}$$

where $\lambda > 0$ parameterises this distribution. This distribution is often used to express the probability of a given number of events occurring in a random interval of time, when the events are independent of one another and occur at a constant average rate, which is a very widely encountered situation. For example, one study found that the arrival time of a subway line in the New York City subway system follows a Poisson distribution [2].

3.2.2 Continuous Distribution

One of the most widely used continuous distribution is the *Gaussian* or *Normal Distribution* with probability function

$$p(x) = \frac{1}{\sigma\sqrt{2\pi}} e^{-\frac{1}{2}\left(\frac{x-\mu}{\sigma}\right)^2} \tag{3.8}$$

The parameters μ and σ represent the mean and standard deviation respectively. The Gaussian curve is symmetric centred around the mean (μ) where the probability obtains its peak, with the width of the spread either side of the mean controlled by the standard deviation (σ). When $\mu = 0$ and $\sigma = 1$, the distribution is called the *Standard Normal Distribution*. Depending on the difference of these two parameters, $p(x)$ can take different shapes. Let's look at an example:

We first draw two datasets of 999 points, each from a different Gaussian distribution, $N(\mu, \sigma^2)$. For the first we use ($\mu = 1, \sigma = 1$) and for the second we use ($\mu = 12, \sigma = 3$).

```
let x = Array.init 999
    (fun _ -> Stats.gaussian_rvs ~mu:1. ~sigma:1.);;
let y = Array.init 999
    (fun _ -> Stats.gaussian_rvs ~mu:12. ~sigma:3.);;
```

Both datasets can be visualised using histogram plots using the code below. As in Fig. 3.1(a), the second has much wider spread, and is centred to the left of the first. We also plot corresponding the probability density functions of the two data sets in Fig. 3.1(b).

```
let x' = Mat.of_array x 1 999;;
let y' = Mat.of_array y 1 999;;

let h = Plot.create ~m:1 ~n:2 "plot_02.png" in

Plot.subplot h 0 0;
Plot.set_ylabel h "frequency";
Plot.histogram ~bin:30 ~h x';
Plot.histogram ~bin:30 ~h y';

Plot.subplot h 0 1;
Plot.set_ylabel h "PDF p(x)";
Plot.plot_fun ~h (fun x -> Stats.gaussian_pdf
    ~mu:1. ~sigma:1. x) (-2.) 6.;
Plot.plot_fun ~h (fun x -> Stats.gaussian_pdf
    ~mu:12. ~sigma:3. x) 0. 25.;

Plot.output h;;
```

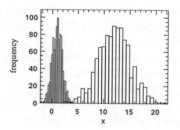 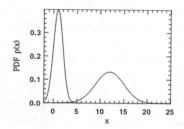

Fig. 3.1: Probability density functions of two data sets

The CDF of the Gaussian can be calculated through infinite summation, i.e. integration:

$$p(x \le k) = \frac{1}{\sqrt{2\pi}} \int_{-\infty}^{k} e^{-t^2/2} dt. \tag{3.9}$$

We can observe this function using `gaussian_cdf` with Fig. 3.2 comparing the CDF of those two sets of parameters: $\mu = 1, \sigma = 1$, and that $\mu = 12, \sigma = 3$.

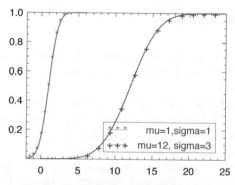

Fig. 3.2: Cumulated density functions of the two datasets

The Gaussian distribution is perhaps the most important statistical distribution, since its bell shape fits many natural phenomena ranging from exam scores in a class, the heights of a large group of people, to measurement errors in scientific evaluations, and many others.

A further, and perhaps more profound, reason for the importance of the Gaussian distribution is the *Central Limit Theorem*. Suppose we want to study a random variable X by observing its value n times, and we can compute a sample mean value m based on these n values. If this procedure is repeated k times, we can get a series of mean values $m_1, m_2, \ldots, m_k$, and they can be seen as the values for a new random variable M. The central limit theorem states that, given a sufficiently large n and under certain quite general conditions, the distribution of M approximates a Gaussian distribution regardless of the distribution of X itself. Furthermore, the mean of M's distribution is more likely to get close to the mean of X as the sample size n increases. The Central Limit Theorem thus provides the theoretical basis for validating how precise are the estimates in scientific evaluations.

Many hypothesis tests introduced in Sec. 3.5 are thereby derived from the Gaussian distribution via the central limit theorem: even if the data does not follow the Gaussian distribution, we can still use these tests for sufficiently large sample sizes.

Besides Gaussian distribution, another frequently used continuous distribution is the *Exponential Distribution*:

$$P(x) = \lambda e^{-\lambda x}(x \le 0), \tag{3.10}$$

where $\lambda > 0$ is the distribution's only parameter. The exponential distribution is often used to measure the amount of time before some specific event occurs. For example, beginning now, the time it takes before you pick up your smart phone and start browsing approximately follows an exponential distribution. It is also widely used in the field of reliability, which often needs to deal with how long a product operates normally until the next failure.

3.2.3 Special Distributions

The `Stats` module supports a wide range of distributions that are encountered sufficiently often in real world to have been named; we briefly introduce several here.

The *Gamma* distribution can be thought of as a generalised version of the exponential distribution. Where the exponential distribution measures the amount of time before some specific event occurs for the first time, the Gamma distribution measures the time before the k-th event occurs. As an example, we plot the probability density function of the Gamma distribution using `Owl_stats.gamma_pdf` using four different sets of parameters by specifying the `shape` (or k), and the `scale` (or θ) parameters, showing the result in Fig. 3.3. In fact, when $k = 1$, the PDF of the Gamma distribution is exactly the same as that of an exponential distribution.

The *Student's t-distribution* is a family of distributions whose PDFs form bell shapes that are similar to that of the Gaussian distribution. The parameter v used in the t-distribution is called its "degree of freedom", which is the sample size minus one. In fact, when v is larger than about 20, the t-distribution becomes almost identical to a Gaussian distribution. This distribution is used in many hypothesis tests (Sec. 3.5). Although the central limit theorem supports using the Gaussian distribution in the tests regardless of the distribution of given data, it does require a "large enough" sample size. When the sample size is not large though, as is often the case in the real world, the Student's t-distribution is useful.

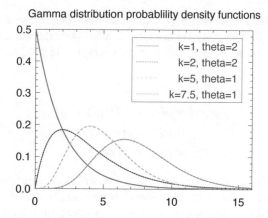

Fig. 3.3: Probability density functions of Gamma distribution

The Cauchy distribution is another family of distributions that resembles the Gaussian's bell shape, but its mean and other moments does not exist. Notwithstanding this oddity, it does have practical applications, such as modelling the ration of two Gaussian random variables, and modelling the energy distribution in the quantum mechanics, etc [4].

Besides these, the Stats module supports a wide range of other special distributions, such as the Weibull, Laplace, Gumbel, Rayleigh, etc. They can be used in various applications such as magnetic resonance imaging, communications systems engineering, speech recognition, etc.

3.3 Multiple Variables

So far we have talked about one single random variable and its distribution, but real world problems often involve multiple variables. This section investigates several aspects of this topic.

3.3.1 Joint and Conditional Probability

Joint probability is a frequently used statistical concept. For example, in a data centre, if we know the probability that the servers stop working, and the probability that the network links break, we might want to consider the probability that a data centre can function normally. The joint probability of two random variable X and Y is expressed as $p(X, Y)$, or $P(X \cap Y)$, indicating the probability of the two events happen at the same time.

There is one special case where the joint probability is intuitive to compute, where the two events are *independent*, i.e. not related with each other. In that case, the joint probability of obtaining $X = x$ and $Y = y$ is simply the product of two separate probabilities:

$$p(xy) = p(X = x \text{ AND } Y = y)$$
$$= p(X = x) * p(Y = y)$$
$$= p_X(x)p_Y(y).$$

However, intuitively many events in the real world are not totally independent of each other, giving rise to the concept of *conditional probability*. For example, consider the probability that a person wears a raincoat, and the probability that a person wears a raincoat *while it is raining*. The events "wearing a raincoat" and "it is raining" are clearly related. Formally, the probability of event X given event Y is computed as:

$$P(X|Y) = \frac{P(X \cap Y)}{P(Y)}.$$

There is no doubt that the most important application of conditional probabilities is *Bayes' Theorem*, proposed first by Thomas Bayes [1]. It can be expressed in the simple form as shown in Eq. 3.11 and provides a way to compute the conditional probability when it is not directly available.

$$P(X|Y) = \frac{P(Y|X)P(X)}{P(Y)} \tag{3.11}$$

One powerful application of this theorem is that it provides a tool to calibrate your knowledge about an event based on observed evidence. For example, a novice gambler can hardly tell if a dice is normal or loaded. If in a casino you are shown a dice and need to estimate the probability that this dice is a fake one, you might start by saying "well, I don't know, but perhaps 10% is a good starting point". Define event X to be "the dice is loaded", and you have just declared that your **prior** belief that $P(X) = 0.1$.

Suppose the dice is then thrown three times, and you obtain a six each time. You then ask yourself the same question: *given the evidence you just observed*, what is the probability that this dice is loaded? Define Y as the event "get three sixes in three throws". What you want to know is actually $P(X|Y)$. Let's calculate the different parts in Eq. 3.11.

First, let's denote with X' the event that a normal dice is used. We can easily calculate that in the normal case $P(Y|X') = 1/6^3 \approx 0.005$, and our prior implied that this case has a probability of 90% (10% chance the dice is loaded means 90% chance the dice is not loaded). We can therefore compute the $P(Y)$ as

$$P(Y) = P(Y|X)P(X) + P(Y|X')P(X').$$

Then we might assume that repeatedly throwing sixes with a loaded dice, $P(Y|X)$, is highly likely, e.g., 0.99. We can then calculate, given the observed evidence, that the dice is loaded with a probability:

$$P(X|Y) = \frac{0.99 * 0.1}{0.99 \times 0.1 + 0.005 \times 0.9} \approx 0.96. \tag{3.12}$$

This is the **posterior** obtained by applying Bayes' Theorem to our prior belief in conjunction with the observations. In this way we can repeatedly improve our knowledge of this dice as more observations are obtained. This process is widely applicable in many fields where existing theories or knowledge are tested by and must incorporate with new evidence.

3.3.2 Covariance and Correlation

We often need to understand the degree of dependency between two random variables. For example, in finance it is important for a portfolio manager to understand how one stock's return changes if another moves higher or lower, so as to optimise the performance of the portfolio. This relationship can be captured by the idea of *covariance*. The covariance of random variables X and Y is defined as expectation of the product of their deviation from mean values:

$$Cov(X,Y) = E[(X - E(X))(Y - E(Y))]$$

When the covariance is positive one variable's value tends to become larger if the other also tends to increase; when the covariance is negative the two variable's values move in opposite directions to each other. The sign of the covariance is thus a useful statistical tool to determine the relationship between the movement of two random variables. However, it s difficult to interpret its magnitude, as the magnitudes of the variables may be on quite different scales. A more useful measurement is thus the *Pearson's R* (or the Pearson correlation coefficient) which normalises the covariance by the standard deviation of the two random variables:

$$\rho(X,Y) = \frac{Cov(X,Y)}{\sigma_X \sigma_Y}.$$

The magnitude of correlation is thus bounded by $[-1, 1]$ and indicates how strong is the dependency. A correlation value 1 means the two variables are perfectly and positively correlated; -1 means they are perfectly and negatively correlated. Therefore, Pearson's R correlation measure the strength and direction of the linear relationship between two variables.

Let's look at an example. We investigate the relationship between variables X, Y, and Z where Y and Z are linearly related to X. We first build x by drawing 50 random values uniformly from an interval between 0 and 10. Both y and z are then computed by scaling x and adding different amounts of Gaussian noise.

```
let noise sigma = Stats.gaussian_rvs ~mu:0. ~sigma;;
let x = Array.init 50 (fun _ -> Stats.uniform_rvs 0. 10.);;
let y = Array.map (fun a -> 2.5 *. a +. noise 1.) x;;
let z = Array.map (fun a -> 2.5 *. a +. noise 8.) x;;
```

It is easy to see the relationship between two variables in a scatter plot as shown in Fig. 3.4. The left figure shows the functional relation between x and y whilst the right shows that between x and z. As we add more noise to z than y, the points in the second figure are diffused more widely.

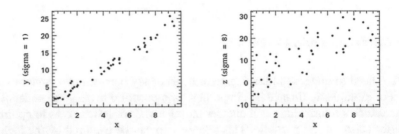

Fig. 3.4: Functional relation between x and the other two variables.

Intuitively, we can easily see there is a stronger relation between x and y from the figures. To describe this difference numerically, the Stats module provide the corrcoef function to calculates the correlations. As expected, the weaker linear relation between x and z produces a smaller correlation value than between x and y.

```
# Stats.corrcoef x y;;
- : float = 0.991145445979576656
# Stats.corrcoef x z;;
- : float = 0.692163016204755288
```

3.4 Sampling

We often wish to consider the entire set of individuals relevant to a study, which is called the *population*, to characterise the set by statistics such as mean, standard deviation etc. However, it is often difficult, if not impossible, to enumerate and measure the entire population. Consider trying to compute the mean average weight of all the grains of sand on the Earth. Clearly it is impossible to take every grain and weigh it to compute the average. Instead, we must consider taking only a *sample* that represents the population. But how should we describe such a sample?

3.4.1 Unbiased Estimator

Perhaps the easiest way, conceptually, to perform sampling is *random sampling*: simply draw members randomly from within the population of interest. But how can we be sure that our sample represents the population well? A slightly more sophisticated method *stratified random sampling*, where we first divide the population into groups, and then we randomly sample from within each group. In some cases, this is more likely to lead to a good representative sample. For example, when designing a door-to-door survey you want people from all age groups to be equally represented, and stratified random sampling might be more likely to achieve this than simply randomly sampling from among the entire population. In the latter case, perhaps some age groups would be very unlikely to respond because they were out at school or work.

After choosing a suitable sample, we next describe population using the sample. Statistics such as mean and variance are still very useful, but they pertain to the population. How can we use the statistics of the sample to describe the whole population? That depends on whether the statistic in question is an *unbiased estimator*, i.e. if its expected value is the same in both the sample and the corresponding population. For example, let's take a sample of n elements, and its mean m is:

$$m = \frac{1}{n} \sum_{i=1}^{n} x_i,$$

where x_i is an element in the sample. Denoting the population mean as μ it can be proved that: $E(m) = \mu$. That is, the sample mean is an unbiased estimator of the population mean. However, the same is not true for the variance. The sample variance is:

$$v = \frac{1}{n} \sum_{i=1}^{n} (x_i - m)^2.$$

Denoting the variance of population with σ^2, it can be shown that $E(v) = \frac{n-1}{n}\sigma^2$ — not σ^2. Thus, an unbiased estimator of population variance is not the variance of the sample v but $\frac{n}{n-1}v$.

3.4.2 Inferring Population Parameters

In the previous part, we have shown how to get the expected values of the mean and variance of the population from a sample drawn from the population. But we perhaps need to know more than just their expected values. For example, can we locate an interval in which we can be quite sure the population mean lies? This section investigates this question using the central limit theorem introduced in Sec. 3.2.2.

Recall that the central limit theorem states that if you take sufficiently large random samples from the population, the distribution of the sample means will be approximately normally distributed (almost) regardless of the population distribution. In practice, sample sizes larger than 20 can be considered "sufficiently large".

Quantitatively, suppose the population has mean μ and variance σ^2, then the mean value of all the samples X follows a normal distribution that has mean value μ, and variance $\frac{\sigma^2}{n}$. As both μ and σ are unknown, we need to estimate them. Let us *assume* that the population variance can be calculated directly from the sample variance:

$$\sigma^2 = \frac{1}{n-1} \sum_{i=1}^{n} (x_i - m)^2.$$

This assumption proves to be valid in practice when n is sufficiently large.

Now that we know X follows a normal distribution, we can use some of its nice properties. We know that 95% of the probability mass lies within the range of $[\mu - 1.96\sigma, \mu + 1.96\sigma]$. We can verify this point using the CDF function of the Normal distribution:

```
# let f = Stats.gaussian_cdf ~mu:0. ~sigma:1. in
  f 1.96 -. f (-1.96);;
- : float = 0.950004209703559
```

Therefore, for any value x of random variable X we know that:

$$P(\mu - 1.96\frac{\sigma}{\sqrt{n}} \le x \le \mu + 1.96\frac{\sigma}{\sqrt{n}}).$$

With some manipulation this becomes:

$$P(x - 1.96\frac{\sigma}{\sqrt{n}} \le \mu \le x + 1.96\frac{\sigma}{\sqrt{n}}).$$

That means given the sample mean m, the population mean μ lies within the range $[m - 1.96\frac{\sigma}{\sqrt{n}}, m + 1.96\frac{\sigma}{\sqrt{n}}]$ with 95% probability. This is the 95% *confidence interval* for μ.

Let's go back to the number *1.96*. We use this range because X is assumed to follow a normal distribution. The $\frac{x-\mu}{\sigma/\sqrt{n}}$ variable follows a standard normal distribution, often denoted z. We can check the standard normal distribution table to find the range that corresponds to 95% confidence. However, as we have explained, the normal distribution does not hold when n is small, since we are actually using $\frac{x-\mu}{\sqrt{\frac{\Sigma_i(x-m)^2}{n(n-1)}}}$

instead of the real z variable. The former is called the *standard t variable*, which follows the t-distribution with $n - 1$ degree of freedom. Therefore, if n is small, we need to look up the *t table*. For example, if $n = 17$, then the range parameter is about 2.12, which can be verified using the code below. The result should be 0.95.

```
# let f x = Stats.t_cdf x ~df:16. ~loc:0. ~scale:1. in
  f 2.12 -. f (-2.12);;
```

3.5 Hypothesis Tests

We have introduced the basic idea of *statistical inference*. While descriptive statistics concern properties of the observed data, statistical inference focuses on determining whether data has been sampled from a certain larger population. In other words, statistical inference creates propositions about a population, and hypothesis testing is an important method in inferential statistical analysis.

3.5.1 Theory

Let's think about the classic coin flipping experiment. Suppose we have a basic assumption/hypothesis that most coins are fair, yielding heads or tails with 50/50 chance. Now, if we flip a coin five times and get five heads, can we reasonably claim that this coin is not fair? With how much confidence can we make that claim? This sort of problem arises constantly in performance evaluation. If you claim that your proposed algorithm improves execution speed compared to the state-of-the-art, how can you be sure that your measurements justify your claim? These are the cases where we need hypothesis tests.

There are usually two hypotheses proposed concerning the statistical relationship between data sets:

- The *Null hypothesis* H_0 states that there is no relationship between two data sets.
- The *Alternative hypothesis* H_1 states that there is statistically significant relationship between two data sets.

The probability of an outcome assuming that a hypothesis is true is its p value, which is then compared with a threshold called the *significance level*, the probability of rejecting the null hypothesis when it is in fact true. We reject the null hypothesis when the p-value is smaller than the significance level. Commonly used significance levels are 5% and 1%.

For example, we might believe that the coin we flipped above is fair, but then our experimental result above (five heads in a row) has only a relatively low probability of about 3%. We might therefore reject the null hypothesis that the coin is fair at the 5% confidence level, with the risk of 5% that this coin is actually fair and we just got lucky (or unlucky) in flipping three heads in a row.

Note that not rejecting a hypothesis does not mean that it is accepted. For example, suppose we flip the same coin three times and get three heads. Under the hypothesis that this coin is fair, this result has a probability of 12.5% and so we cannot reject this hypothesis at a 5% confidence level. But this does not mean that this evidence leads us to believe the coin is fair! Therefore, one needs to be very careful in choosing hypothesis. We should have strong enough belief in H_0 a priori that it would need convincing evidence in data to challenge this belief.

3.5.2 Gaussian Distribution in Hypothesis Testing

One of the most common hypothesis tests is to check whether observed data come from a parameterised Gaussian distribution. It is called a "z-test". Let's see how to perform a z-test in code. We first generate two data sets, both drawn from Gaussian distributions with different parameters. The first, data_0, is drawn from $\mathcal{N}(0, 1)$, while the second, data_1 is drawn from $\mathcal{N}(3, 1)$, where $\mathcal{N}(\mu, \sigma)$ denotes a Gaussian distribution.

```
let data_0 = Array.init 10
  (fun _ -> Stats.gaussian_rvs ~mu:0. ~sigma:1.);;
let data_1 = Array.init 10
  (fun _ -> Stats.gaussian_rvs ~mu:3. ~sigma:1.);;
```

Suppose we do not know how they were generated. We might start with the hypothesis that both are drawn from the standard Normal distribution $\mathcal{N}(0, 1)$. We can now test this hypothesis using the Stats.z_test function:

```
# Stats.z_test ~mu:0. ~sigma:1. data_0;;
- : Owl_stats.hypothesis =
{Owl.Stats.reject = false; p_value = 0.289340080583773251;
 score = -1.05957041132113083}
```

The result is of a record type hypothesis containing fields indicating whether we should reject H_0 as well as the p value and the score calculated with the given data set. The z-score, or z-statistics, is a number describing how many standard deviations above or below the mean is. For example, a z-score of 1 indicates a value that is one standard deviation away from the mean. In this case, since the score is negative, it shows that the value is below the mean.

```
type hypothesis = {
  reject : bool;
  p_value : float;
  score : float;
};;
```

In the previous result we see reject = false, indicating that we should not reject the null hypothesis that the data set data_0 is drawn from $\mathcal{N}(0, 1)$. How about the second data set, data_1?

```
# Stats.z_test ~mu:0. ~sigma:1. data_1;;
- : Owl_stats.hypothesis =
{Owl.Stats.reject = true; p_value = 5.06534675819424548e-23;
 score = 9.88035435799393547}
```

As expected, in this case the p-value is very small and indicates that the null hypothesis should be rejected. That is, data_1 *is* drawn from a different distribution than the assumed $\mathcal{N}(0, 1)$.

Another commonly used test is to see if the given data follows a Student's t-distribution under the null hypothesis. It is called a "t-test". In the Stats module, the t_test function returns a decision of t-test. The parameter mu is the population mean, and alpha is the significance level.

3.5.3 Two-Sample Inferences

Another commonly needed test is to determine whether two samples come from the same population. For example, suppose we wish to measure the performance of an improvement to an algorithm to see if it really does improve things. We might collect execution time data from both the original and the improved algorithm, and then test the null hypothesis that the two samples are drawn from the same population.

Which test strategy we use depends on whether the two samples have the same size. If they do then we can simplify the test by simply taking the pairwise difference between the two sets and then testing whether the result is draw from a population with zero mean using the t-test from the previous section. In fact, the Stats module provides the paired sample t-test function specifically for this purpose. The t_test_paired ~alpha ~side x y returns a test decision for the null hypothesis that the data in x - y comes from a normal distribution with zero mean and unknown variance.

If the two samples are of different sizes, the problem gets trickier. Consider the confidence intervals of the two samples. If they do not overlap we can be fairly certain that the two samples come from different populations, allowing us to reject the null hypothesis that they come from the same population. However, if the intervals overlap we must ensure that certain other conditions hold. For example, if we can assume that the variances of both population are the same, then it can be shown that the variable:

$$\frac{\bar{x} - \bar{y}}{\sqrt{\frac{\sum_{i=1}^{a}(x_i-\bar{x})^2+\sum_{i=1}^{b}(y_i-\bar{y})^2}{a+b-2}\left(\frac{1}{a}+\frac{1}{b}\right)}} \tag{3.13}$$

is a standard t variable with $a + b - 2$ degrees of freedom, where a and b are the lengths of the samples X and Y. This is implemented as the unpaired sample t-test,

t_test_unpaired ~alpha ~side ~equal_var x y;;

which returns a test decision for the null hypothesis that the data in vectors x and y comes from independent normal distributions with equal means and equal but unknown variances. The argument equal_var indicates whether two samples have the same variance. If the two variances are not the same, we need to use *non-parametric tests* such as Welche's t-test. Advanced topics such as non-parametric test are not covered in this book.

3.5.4 More Tests

The Stats module supports many other kinds of hypothesis tests, less frequently
used than the *t*-test or *z*-test but important nonetheless:

- *Chi-Squared Variance Test*: var_test ~alpha ~side ~variance x returns a
  test decision for the null hypothesis that the data in x comes from a normal distribution with input variance, using the chi-squared variance test. The alternative
  hypothesis is that x comes from a normal distribution with a different variance.
- *Kolmogorov-Smirnov Test*: ks_test ~alpha x f returns a test decision for the
  null hypothesis that the data in vector x comes from independent random samples
  of the distribution with CDF f. The alternative hypothesis is that the data in x
  comes from a different distribution.
- *Two-sample Kolmogorov-Smirnov Test*: ks2_test ~alpha x y returns a test
  decision for the null hypothesis that the data in vectors x and y come from
  independent random samples of the same distribution.
- *Jarque-Bera Test*: jb_test ~alpha x returns a test decision for the null hypothesis that the data x comes from a normal distribution with an unknown mean and
  variance, using the Jarque-Bera test.
- *Wald-Wolfowitz Runs Test*: runs_test ~alpha ~v x returns a test decision for
  the null hypothesis that the data x comes in random order by running Wald-
  Wolfowitz runs test. The test is based on the number of runs of consecutive values
  above or below the mean of x. ~v is the reference value, the default value is the
  median of x.
- *Mann-Whitney Rank Test*: mannwhitneyu ~alpha ~side x y computes the
  Mann-Whitney rank test on samples *x* and *y*. Use the exact test if length of each
  sample less than 10 and no ties; otherwise using asymptotic normal distribution.

3.6 Summary

In this chapter we have briefly introduced several key topics in probability and statistics. Random variables and different types of distribution are two building blocks
in this chapter. Based on them, we introduced joint and conditional probabilities
when considering multiple variables, and the important Bayes' Theorem. We also
discussed the difference between covariance and correlation between two variables.
Then we moved from descriptive statistics to inferential statistics, introducing sampling, including the idea of unbiased population estimator based on a given sample,
and how to infer population parameters such as mean. Finally, we covered the basic
idea in hypothesis testing with examples.

References

1. Bayes FR. An essay towards solving a problem in the doctrine of chances. Biometrika. 1958 Dec 1;45(3-4):296-315.
2. Jagannath A, Trogdon T. Random matrices and the New York City subway system. Physical Review E. 2017 Sep 5; 96(3):030101.
3. Chandrasekaran B. Survey of network traffic models. Waschington University in St. Louis CSE. 2009 Jan;567.
4. Bhatia AS, Saggi MK, Zheng S. QPSO-CD: quantum-behaved particle swarm optimization algorithm with Cauchy distribution. Quantum Information Processing. 2020 Oct; 19(10):1-23.

Chapter 4
Linear Algebra

Abstract Linear algebra is an important area of mathematics and has many applications in numerical computing. In this chapter we first present an overview, and then focus on how to use OCaml to solve concrete problems and develop a thorough understanding of the key concepts in linear algebra. This chapter covers the classic topics in linear algebra, including Gaussian elimination, vector space, determinants, eigenvalues and eigenvectors. Before discussing these topics, we provide a brief introduction to the matrix data structure supported in OCaml.

4.1 Vectors and Matrices

Linear algebra is about solving systems of linear equations expressed in vectors and matrices, hence we begin by familiarising ourselves with these basic elements. All the matrix functions can be accessed from the `Dense.Matrix` module, with support of four different precisions: single, double, complex single, and complex double in the `Matrix.S`, `Matrix.D`, `Matrix.C` and `Matrix.Z` modules respectively. The `Mat` module is an alias of `Matrix.D` provided for convenience. Except for a handful of functions (e.g., `re` to extract the real part of complex elements) most are shared by these four sub-modules. Vectors are simply expressed as one-row matrices. Aside from dense matrix, Owl also provides an `Sparse` matrix module. We will briefly talk about it in the last section.

There are multiple functions to help create matrices with certain contents. For example, the following functions all create various 5x5 matrices:

- `Mat.empty 5 5`, a matrix with uninitialised elements
- `Mat.create 5 5 2.`, a matrix and initialise all to 2
- `Mat.zeros 5 5`, a matrix of all zeros
- `Mat.ones 5 5`, a matrix of all ones
- `Mat.eye 5`, an identity matrix
- `Mat.magic 5`, a magic square matrix

L. Wang et al., *OCaml Scientific Computing*, Undergraduate Topics in Computer Science,
https://doi.org/10.1007/978-3-030-97645-3_4

- `Mat.uniform 5 5`, a random matrix of uniform distribution
- `Mat.uniform_int 5 5`, a random integer matrix
- `Mat.sequential 5 5`, a matrix of sequential integers
- `Mat.semidef 5`, a random semi-definite matrix
- `Mat.gaussian 5 5`, a random Gaussian matrix
- `Mat.bernoulli 5 5`, a random Bernoulli matrix

In addition, to build a matrix from an OCaml array, use the `of_array` operation, e.g., `Mat.of_array [1.;2.;3.] 1 3` builds a matrix of size 1x3.

The `Mat` module supports many mathematical operations on two matrices x and y. For example, the dot product `Mat.dot x y` can also be written `Mat.(x *@ y)`, and element-wise addition `Mat.add x y` can be `Mat.(x + y)`. The element-wise approximated equality comparison `Mat.approx_equal` can be used as `Mat.(x =~ y)`. Of course, these operations are all available for the other precisions, namely in the `Matrix.S/C/Z` sub-modules.

If you perform operations between a matrix and a scalar value, you need to be careful about their order. Assuming `a` is a `float` scalar value, we can add it to each element in matrix x using `Mat.(x +$ a)` or `Mat.(a $+ x)`. Note the difference between these two operators, which have corresponding long names `Mat.scalar_add` and `Mat.add_scalar`.

There are other ready-made mathematical functions such as `Mat.abs` (absolute value) and `Mat.l2norm'`, where the ' in the name indicates that the result is a scalar value rather than a `1x1` matrix. The matrix module also contains operations such as concatenation (`Mat.concat`) and transpose (`Mat.transpose`).

Finally, another key matrix operation is to index elements of the matrix, supported via `get` and `set` to retrieve and modify matrix by specifying row and column indices, e.g.,

```
let x = Mat.uniform 5 5;;
Mat.set x 1 2 0.;;
let a = Mat.get x 0 3;;
```

The second line sets the element at (1,2) to 0, and then gets the value of the element at (0,3). For dense matrices we can also use shorthand `.%{i; j}` to access elements, rewriting the code above as:

```
x.%{1;2} <- 0.;;
let a = x.%{0;3};;
```

Matrix modifications using `set` are in-place. We can retrieve rows from a matrix x by the `rows` function. The selected rows will be used to assemble a new matrix. We can select columns similarly using `cols`.

This section provides a very brief introduction to the matrix module and the functions that will be used in the rest of this chapter. The matrix module builds on the `Ndarray` (N-dimensional array) module, enabling matrices and ndarrays in Owl to inter-operate. The `Ndarray` module and its functions, and thus the `Matrix` module, are discussed in Chap. 5.

The rest of this chapter makes use of the `Linalg` module that provides various high-level Linear Algebra APIs. As with the matrix module, it supports both real and complex number in single and double precision. Additionally, the `Linalg.Generic` module can do everything that `Matrix.S/D/C/Z` can but requires extra type information. In this chapter, we will show how to use different aspects of this module to investigate the Linear Algebra topics.

4.2 Gaussian Elimination

Solving linear equations systems is the core problem in Linear Algebra and is frequently encountered in scientific computation [1]. *Gaussian Elimination* is a classic method to do this, and it can work surprisingly well in modern numerical libraries [2]. Here is a simple example:

$$2x_1 + 2x_2 + 2x_3 = 4$$
$$2x_1 + 2x_2 + 3x_3 = 5$$
$$3x_1 + 4x_2 + 5x_3 = 7$$

$$(4.1)$$

Divide the first equation by 2:

$$x_1 + x_2 + x_3 = 2$$
$$2x_1 + 2x_2 + 3x_3 - 5$$
$$3x_1 + 4x_2 + 5x_3 = 7$$

$$(4.2)$$

Multiply the first equation by -2 and add it to the second. Then multiply the first equation by -3 and add it to the third. This gives us Eq. 4.3.

$$x_1 + x_2 + x_3 = 2$$
$$x_3 = 1$$
$$x_2 + 2x_3 = 1$$

$$(4.3)$$

$$x_1 + x_2 + x_3 = 2$$
$$x_2 + 2x_3 = 1$$
$$x_3 = 1$$

$$(4.4)$$

Finally, swapping the second and third, we get Eq. 4.4. We thus have $x_3 = 1$, and can substitute for x_3 in the second equation to obtain $x_2 = -1$. Plug both back to the first equation and we have $x_1 = 2$.

This process demonstrated the basic operations of Gaussian elimination: eliminate unknown variables until the system of linear equations is easy to solve, and then back-substitute. There are three basic operations we use: multiplication, adding one line to another, and swapping two lines. The starting equation Eq. 4.1 can be more concisely expressed with vectors:

$$x_1 \begin{bmatrix} 2 \\ 2 \\ 3 \end{bmatrix} + x_2 \begin{bmatrix} 2 \\ 2 \\ 4 \end{bmatrix} + x_3 \begin{bmatrix} 2 \\ 3 \\ 5 \end{bmatrix} = \begin{bmatrix} 4 \\ 5 \\ 7 \end{bmatrix}$$

or in the form $Ax = b$ using matrix notation:

$$\begin{bmatrix} 2 & 2 & 2 \\ 2 & 2 & 3 \\ 3 & 4 & 5 \end{bmatrix} \begin{bmatrix} x_1 \\ x_2 \\ x_3 \end{bmatrix} = \begin{bmatrix} 4 \\ 5 \\ 7 \end{bmatrix} \implies \begin{bmatrix} x_1 \\ x_2 \\ x_3 \end{bmatrix} = \begin{bmatrix} 2 \\ -1 \\ 1 \end{bmatrix}$$

Here A is a matrix, b is a column vector, and x is the unknown vector we are solving for. Matrix notation is often used to describe systems of linear equations concisely.

4.2.1 LU Factorisation

Let's look again at the Gaussian elimination example above. The final form Eq. 4.4 can be expressed with the matrix notation as:

$$\begin{bmatrix} 1 & 1 & 1 \\ 0 & 1 & 2 \\ 0 & 0 & 1 \end{bmatrix}$$

As we can see, all the elements below the diagonal of the square matrix are zero. Such a matrix is an *upper triangular matrix*, usually denoted U. Similarly, a square matrix where all the elements above the diagonal are zero is a *lower triangular matrix*, denoted L. We can use the `is_triu` and `is_tril` in the `Linalg` module to test for triangularity. The diagonal elements of U are the *pivots*. The i-th pivot is the coefficient of the i-th variable in the i-th equation at the i-th step during the elimination.

In general, a square matrix can be factorised into a dot product of lower and upper triangular matrices:

$$A = LU, \qquad\qquad (4.5)$$

and this process is called the *LU factorisation*. This embodies the process of Gaussian elimination. Consider again the initial problem of solving the linear equation $Ax = b$. One reason the LU factorisation is important is that if matrix A in $Ax = b$ is triangular, then solving it becomes straightforward, as seen in the previous example. We can use the `triangular_solve` function in `Linalg` to efficiently solve the linear equations if we already know that the matrix is triangular.

For a normal square matrix that can be factorised into LU, we can change $Ax = b$ to $LUx = b$. First we find column vector c so that $Lc = b$, then we can find x so that $Ux = c$. Both sets of triangular equations are easy to solve. The `lu` function performs LU factorisation. Use the previous example once more:

```
# let a = [|
    2.;2.;2.;
    2.;2.;3.;
    3.;4.;5.;|];;
# let a = Mat.of_array a 3 3;;

# let l, u, p = Linalg.D.lu a;;
val l : Owl_dense_matrix_d.mat =
          C0 C1 C2
R0         1  0  0
R1 0.666667  1  0
R2 0.666667  1  1
val u : Owl_dense_matrix_d.mat =
    C0        C1        C2
R0  3         4         5
R1  0 -0.666667 -0.333333
R2  0         0        -1
val p : Linalg.D.int32_mat =
    C0 C1 C2
R0  3  2  3
```

The first two returned matrices are the lower and upper triangular matrices. However, if we check the correctness of this factorisation with dot product, we find the result is somewhat wrong:

```
# let a' = Mat.dot l u;;
val a' : Mat.mat =
    C0 C1 C2
R0  3  4  5
R1  2  2  3
R2  2  2  2
```

It turns out that we need some extra row exchanges to obtain the correct answer, in case the number we want to use as the pivot might be zero. This process is called *pivoting*. It is closely related to the stability of numerical computation. Choosing improper pivots can lead to incorrect solutions [3]. Pivoting can be expressed as a permutation matrix P with the same rows as the identity matrix, each row and column having just a single 1 element. The full LU factorisation can thus be expressed as:

$$PA = LU$$

```
# let p = Mat.of_array [|
    0.;0.;1.;
    0.;1.;0.;
    1.;0.;0.;
  |] 3 3;;
```

```
# Mat.dot p a = Mat.dot l u;;
- : bool = true
```

Recall that the `lu` function also have a third output, the permutation vector p, which is an array $[|3; 2; 3|]$. How do we translate it to the previous permutation matrix? Each element p_i in the vector represents an updated identity matrix. On this identity matrix, we set (i, i) and (p_i, p_i) to zero, and then (i, p_i) and (p_i, i) to one. Multiplying these n matrices, we obtain the permutation matrix P. Here is a brief implementation of this process in OCaml:

```
let perm_vec_to_mat vec =
  let n = Array.length vec in
  let mat = ref (Mat.eye n) in
  for i = n - 1 downto 0 do
    let j = vec.(i) in
    let a = Mat.eye n in
    Mat.set a i i 0.;
    Mat.set a j j 0.;
    Mat.set a i j 1.;
    Mat.set a j i 1.;
    mat := Mat.dot a !mat
  done;
  !mat;;
```

Note that there is more than one way to perform LU factorisation. For example, for the same matrix, we can have another set of solution:

$$\begin{bmatrix} 1 & 0 & 0 \\ 0 & 0 & 1 \\ 0 & 1 & 0 \end{bmatrix} \begin{bmatrix} 2 & 2 & 2 \\ 2 & 2 & 3 \\ 3 & 4 & 5 \end{bmatrix} = \begin{bmatrix} 1 & 0 & 0 \\ 1.5 & 1 & 0 \\ 1 & 0 & 1 \end{bmatrix} \begin{bmatrix} 2 & 2 & 2 \\ 0 & 1 & 2 \\ 0 & 0 & 1 \end{bmatrix}$$

4.2.2 Inverse and Transpose

The concept of *inverse* matrix is related to the identity matrix I_n, which built via `Mat.eye n` where n is the size of the square matrix. The identity matrix is a special form of *Diagonal Matrix*, a square matrix whose non-zero elements occur only on its diagonal. We can check if a matrix is diagonal with the `is_diag` function.

```
# Mat.eye 5 |> Linalg.D.is_diag;;
```

The inverse of an $n \times n$ square matrix A is denoted as A^{-1} so that

$$AA^{-1} = I_n. \tag{4.6}$$

Not all square matrices have inverses. There are many sufficient and necessary conditions that indicate if A is invertible. One is that A has n pivots.

We use function `inv` to perform matrix inversion. It is straightforward to verify according to the definition. We use the `semidef` function to produce a matrix that is guaranteed to be invertible.

```
# let x = Mat.semidef 5 in
  let y = Linalg.D.inv x in
  Mat.(x *@ y =~ eye 5);;
- : bool = true
```

Another frequently used operation is to take the *transpose* of a matrix. Denoted as A^T, its i-th row is taken from the i-th column of the original matrix A. It has properties such as

$$(AB)^T = B^T A^T. \tag{4.7}$$

We can verify this property using the matrix function `Mat.transpose`.

```
# let flag =
    let a = Mat.uniform 4 4 in
    let b = Mat.uniform 4 4 in
    let m1 = Mat.(dot a b |> transpose) in
    let m2 = Mat.(dot (transpose b) (transpose a)) in
    Mat.(m1 =~ m2);;
```

Finally, a related special class of matrices is the *Symmetric Matrix*, those matrices equal to their own transpose. This can be verified using the `is_symmetric` function.

4.3 Vector Space

We have discussed solving the linear equations $Ax = b$ using Gaussian elimination where A is a square matrix. We now discuss further how to know if there exists one or more solutions. To answer such question, we need to be familiar with the concept of a *vector space*.

A vector space, denoted as R^n, contains all the vectors having n elements. In this space we define the addition and multiplication operations. Applying them to the vectors is called *linear combination*. A *subspace* in a vector space then is a non-empty set that any linear combination of the vectors in this subspace still stays in the same subspace.

There are four fundamental subspaces of interest when solving linear systems $Ax = b$ where A is a m by n matrix. The *column space*, a subspace of R^m, is all linear combinations of the columns of A. Similarly, the *row space* consists of all the linear combinations of the rows of A. The *nullspace* $N(A)$ contains all the vectors x such that $Ax = 0$. It is a subspace of R^n. The *left nullspace* is similarly the nullspace of A^T.

4.3.1 Rank and Basis

When discussing Gaussian elimination above, we assumed an ideal situation: matrix A is an $n \times n$ square matrix for which there exists one solution. But that is not always the case. A may not even be a square matrix. It is quite possible that these m equations are not enough to solve a n-variable linear system when $m < n$, or that there might not exist any solution when $m > n$. Besides, even when A is a square matrix, the information contained in one equation might be repeated in another if, for example, one equation is simply a multiple of another. Applying LU factorisation to such a matrix, we obtain the error, as the code below shows.

```
# let x = Mat.of_array [|
    1.; 2.; 3.;
    0.; 0.; 1.;
    0.; 0.; 2.;
  |] 3 3;;
```

```
# Linalg.D.lu x;;
Exception: Failure "LAPACKE: 2".
```

This indicates that we cannot pivot the second column, and so this matrix is singular and cannot be factorised into LU form. As can be seen in this example, we cannot expect the linear algebra functions work in all cases.

To determine the solution to a general $Ax = b$, we must understand the concept of *rank*, the number of pivots in the elimination process. For a more intuitive understanding of rank, we introduce the concept of *linear independence*. In a linear combination $\sum_{i=1}^{n} c_i v_i$ where v_i are vectors and c_i are numbers, if $\sum_{i=1}^{n} c_i v_i = 0$ only occurs when $c_i = 0$ for all i, vectors $v_1, v_2, \ldots, v_n$ are linearly independent. That is, the rank of a matrix is the number of independent rows in the matrix. We can understand rank as the number of "effective" rows in the matrix. As an example, we can check the rank of the previous matrix with `Linalg.D.rank x`. The result 2 indicates only two effective rows, therefore we are not able to factorise the matrix to find the only solution.

An application of rank produces a key concept in linear algebra: *basis*. A sequence of vectors is the basis of a vector space or subspace if the vectors are linearly independent, and all vectors in the space can be represented as a linear combination of vectors in the basis. A space can have infinitely many bases, but the number of vectors in each basis is the same and is called the *dimension* of the vector space. For example, if an $m \times n$ matrix A has rank of r, then the dimension of its null space is $n - r$, and the dimension of its column space is r. For a full-rank matrix where $r = n$, the dimension of its column space is n so all its columns can be a basis of the column space, and that the null space dimension is zero so that the only solution of $Ax = 0$ is a zero vector.

4.3.2 Orthogonality

We can relate the basis of a vector space to the Cartesian coordinate system in (for example) a three-dimensional space where every vector in the space can be represented in terms of values on the three axes, x, y and z. There are many such basis sets but the set { $x = (1, 0, 0)$, $y = (0, 1, 0)$, $z = (0, 0, 1)$ } is convenient because all members are mutually orthogonal, simplifying many problems.

Orthogonality is not limited to vectors. Two vectors a and b are orthogonal if $a^T b = 0$, and we can say two subspaces A and B are orthogonal if every vector in A is orthogonal to every vector in B. For example, the nullspace and row space of a matrix are orthogonal.

Among the bases of a subspace, when every vector is perpendicular to every other, it is called an orthogonal matrix. Moreover, if the length of each vector is normalised to one unit, it is said to be an *orthonormal basis*. We can use the `null` function to find an orthonormal basis vector x for the null space of a matrix, i.e. $Ax = 0$.

```
# let a = Mat.magic 4;;
# let x = Linalg.D.null a;;
val x : Owl_dense_matrix_d.mat =
           C0
R0 -0.223607
R1  -0.67082
R2   0.67082
R3  0.223607

# Mat.dot a x |> Mat.l2norm';;
- : float = 0
```

Having defined orthogonal and orthonormal bases, how to build them? The *QR Factorisation* is used to construct orthogonal bases in a subspace by decomposing a matrix A into the product of an orthogonal matrix Q and an upper triangular matrix R, i.e. $A = QR$. It is provided in the linear algebra module as the qr function. For example, the code below calculates the QR decomposition for an $m \times n$ matrix, returning a 3-tuple of Q, R, and the permutation vector of columns. The default value of parameter pivot is false; setting it to true allows qr to perform pivoted factorisation.

```
# let a = Mat.of_array [|
    12.; -51.;    4.;
     6.; 167.; -68.;
    -4.;  24.; -41.;
  |] 3 3;;

# let q, r, _ = Linalg.D.qr a;;

val q : Owl_dense_matrix_d.mat =
            C0           C1           C2
R0 -0.857143   0.394286    0.331429
R1 -0.428571  -0.902857   -0.0342857
R2  0.285714  -0.171429    0.942857

val r : Owl_dense_matrix_d.mat =
     C0   C1   C2
R0 -14   -21   14
R1   0  -175   70
R2   0     0  -35
```

4.3.3 Solving Ax = b

We can now discuss the general solution structure of $Ax = 0$ and $Ax = b$ where A is again an $m \times n$ matrix. There exist non-zero solution(s) to $Ax = 0$ if and only if the rank $r(A) <= n$. If $r(A) < n$, the nullspace of A is of dimension $n - r$ and the $n - r$ orthogonal basis can be found with null function. For example:

```
# let a = Mat.of_array [|
    1.;  5.;-1.;-1.;
    1.;-2.;  1.;  3.;
    3.;  8.;-1.;  1.;
    1.;-9.;  3.;  7.;
  |] 4 4;;
# Linalg.D.rank a;;
- : int = 2
```

This a rank 2 matrix, so the nullspace contains 4 - 2 = 2 vectors:

```
# Linalg.D.null a;;
- : Owl_dense_matrix_d.mat =
          C0          C1
R0 -0.851419 0.0136382
R1  0.273706  0.143885
R2 0.0762491  0.962526
R3   0.44086 -0.229465
```

These two vectors are the *fundamental system of solutions* of $Ax = 0$. All solutions of $Ax = 0$ can be expressed using the fundamental system:

$$c_1 \begin{bmatrix} -0.85 \\ 0.27 \\ 0.07 \\ 0.44 \end{bmatrix} + c_2 \begin{bmatrix} 0.013 \\ 0.14 \\ 0.95 \\ -0.23 \end{bmatrix} \tag{4.8}$$

Here c_1 and c_2 can be any constant scalars. Considering the general form $Ax = b$ where b is an $m \times 1$ vector, there exists just one solution if and only if $r(A) = r([A, b]) = n$ where $[A, b]$ is a matrix formed by concatenating A and b along the column. If $r(A) = r([A, b]) < n$, $Ax = b$ has an infinite number of solutions with the general form:

$$x_0 + c_1 x_1 + c_2 x_2 + \ldots + c_k x_k \tag{4.9}$$

where x_0 is a particular solution to $Ax = b$, and $x_1, x_2, \ldots, x_k$ constitutes the fundamental solution system of $Ax = 0$.

The function `linsolve a b` gives a particular solution, solving a linear system of equations a * x = b. By default, this uses LU factorisation with partial pivoting when a is square, otherwise it uses QR factorisation with column pivoting. The number of rows of a must be equal to the number of rows of b. If a is a upper or lower triangular matrix, the function calls the `solve_triangular` function. Here is an example.

```
# let a = Mat.of_array [|
      2.; 3.; 1.;
      1.;-2.; 4.;
      3.; 8.;-2.;
      4.;-1.; 9.;
  |] 4 3;;

# let b = Mat.of_array [|4.;-5.;13.;-6.|] 4 1;;

# let x0 = Linalg.D.linsolve a b;;
val x0 : Owl_dense_matrix_d.mat =
    C0
R0 -5
R1  4
R2  2
```

Then we use `null` to find the fundamental solution system. You can verify that matrix a is of rank 2, so that the solution system for $ax = 0$ should contain only $3 - 2 = 1$ vector.

```
# let x1 = Linalg.D.null a;;
val x1 : Owl_densematrix_d.mat =
          C0
R0 -0.816497
R1  0.408248
R2  0.408248
```

So the solutions to $Ax = b$ can be expressed as:

$$\begin{bmatrix} -1 \\ 2 \\ 0 \end{bmatrix} + c_1 \begin{bmatrix} -0.8 \\ 0.4 \\ 0.4 \end{bmatrix}$$

An important takeaway from this chapter is that to use these functions in `LinAlg` requires solid background of linear algebra knowledge. Blind application of them can easily lead to wrong or misleading answers.

4.3.4 Matrix Sensitivity

The *sensitivity* of a matrix is crucial in numerical computation, indicating, in a linear system $Ax = b$, how much x will vary if we change the A and b slightly. The *Condition Number* is then a measure of the sensitivity of a square matrix [4].

First we need to understand the *Norm* (or L2-norm) of a matrix $\|A\|$, calculated as the square root of the maximum eigenvalue of $A^T A$. (The concept of eigenvalue will be introduced in Sec. 4.5.) The norm of a matrix gives a upper limit so that for any x we can be certain that:

$$\|Ax\| \le \|A\|\|x\|. \tag{4.10}$$

Here $\|Ax\|$ and $\|x\|$ indicate the L2 norm. Effectively, $\|A\|$ bounds how much A can amplify the input x, so it can be a good indicator of sensitivity. We can calculate the norm using the `norm` function in the linear algebra module.

The most frequently used condition number is that representing the sensitivity of inverse matrix [5]. Under the definition of norm, the *condition number for inversion* of a matrix can be expressed as $\|A\|\|A^{-1}\|$, and calculated using the `cond` function. Let's look at an example:

```
# let a = Mat.of_array [|
    4.1; 2.8;
    9.7; 6.6;
  |] 2 2;;
```

```
# let c = Linalg.D.cond a;;
val c : float = 1622.99938385651058
```

Its condition number for inversion is much larger than one so a small change in A should lead to a large change of A^{-1}. Let's verify this.

```
# let a' = Linalg.D.inv a;;
val a' : Owl_dense_matrix_d.mat =
     C0   C1
R0  -66   28
R1   97  -41
```

Change the last row in the matrix slightly and re-calculating gives:

```
# let a2 = Mat.of_array [|
     4.1;    2.8;
     9.67; 6.607;
  |] 2 2;;
```

```
# let a2' = Linalg.D.inv a2;;
val a2' : Owl_dense_matrix_d.mat =
          C0       C1
R0   520.236 -220.472
R1  -761.417  322.835
```

As predicted by a large condition number for inversion, changing the matrix a only slightly causes the inverse of A to fluctuate dramatically, and so is the resulting solution vector x.

4.4 Determinants

Besides pivots, another basic quantity in linear algebra is the *determinant*. Its most important application is to decide if a square matrix A is invertible or singular. The determinant $\det(A) \neq 0$ if and only if $\|A\| = n$, i.e., matrix A is invertible.

For a square matrix A:

$$\begin{bmatrix} a_{11} & a_{12} & \cdots & a_{1n} \\ a_{21} & a_{22} & \cdots & a_{2n} \\ \vdots & \vdots & \cdots & \vdots \\ a_{n1} & a_{n2} & \cdots & a_{nn} \end{bmatrix}$$

its determinants $\det(A)$ is defined as:

$$\sum_{j_1 j_2 \cdots j_n} (-1)^{\tau(j_1 j_2 \cdots j_n)} a_{1 j_1} a_{2 j_2} \cdots a_{n j_n}$$

Here $\tau(j_1 j_2 \ldots j_n) = i_1 + i_2 + \ldots + i_{n-1}$, where i_k is the number of j_p that is smaller than j_k for $p \in [k+1, n]$. Mathematically, there are many techniques that can be used to simplify this calculation [6]. But as far as this book is concerned, it is sufficient for us to use the det function to calculate the determinant of a matrix.

We can also use it to understand the solution of $Ax = b$. If $\det(A) \neq 0$, $Ax = b$ has one and only one solution. This theorem is part of *Cramer's rule*, and these properties are widely used when finding *eigenvalues*, discussed next.

Since sometimes we care only whether the determinant is zero, we can instead use a related function logdet. It computes the logarithm of the determinant, but avoids potential overflow and underflow problems when computing the determinant of large matrices.

```
# let x = Mat.magic 5;;
# Linalg.D.det x;;
- : float = 5070000.00000000093
# Linalg.D.logdet x;;
- : float = 15.4388513755673653
```

4.5 Eigenvalues and Eigenvectors

In this section we move from studying $Ax = b$ to $Ax = \lambda x$. The pattern

$$Ax = \lambda x \tag{4.11}$$

frequently appears in scientific and engineering analyses describing the change of a dynamic system over time [7]. For an $n \times n$ square matrix, if there exists a scalar λ and a non-zero column vector x satisfying:

$$(\lambda I - A)x = 0, \tag{4.12}$$

then λ is an *eigenvalue* of A, and x is its corresponding *eigenvector*. To find the eigenvalues of a matrix A, we must find the roots of the determinant of $\lambda I - A$. $\det(\lambda I - A) = 0$ is the *characteristic equation* of A. For example, for matrix

$$A = \begin{bmatrix} 3 & 1 & 0 \\ -4 & -1 & 0 \\ 4 & -8 & 2 \end{bmatrix}$$

Its characteristic matrix $\lambda I - A$ is:

$$\begin{bmatrix} \lambda - 3 & 1 & 0 \\ -4 & \lambda + 1 & 0 \\ 4 & -8 & \lambda - 2 \end{bmatrix}$$

By the definition of determinant, $\det(\lambda I - A) = (\lambda - 1)^2 (\lambda - 2) = 0$. According to the theory of polynomials, this characteristic polynomial has just n roots in the complex space. Specifically, here we have three eigenvalues: $\lambda_1 = 1, \lambda_2 = 1, \lambda = 2$. Substituting λ_1 back into the characteristic equation we have: $(I - A)x = 0$. Therefore, we can find the fundamental solution system of $I - A$ with:

```
# let basis =
    let ia = Mat.((eye 3) - (of_array
      [|3.; 1.;0.;
        -4.;-1.;0.;
        4.;-8.;2.|] 3 3))
    in
    Linalg.D.null ia;;

val basis : Owl_dense_matrix_d.mat =
            C0
R0 -0.0496904
R1  0.0993808
R2  0.993808
```

We thus get a fundamental solution $x_0 = [-0.05, 0.1, 1]^T$. Similarly, we can calculate that the eigenvectors for the eigenvalue 2 are $[0, 0, 1]^T$.

We can use the function eig to compute the right eigenvectors v and eigenvalues w of an arbitrary square matrix x. The eigenvectors are column vectors in v, their corresponding eigenvalues have the same order in w as that in v.

```
# let eigvec, eigval =
    let a = Mat.of_array [|
        3.; 1.;0.;
        -4.;-1.;0.;
        4.;-8.;2.;
      |] 3 3 in
    Linalg.D.eig a;;

val eigvec : Owl_dense_matrix_z.mat =
          C0                C1                  C2
R0 (0, 0i)  (0.0496904, 0i)   (0.0496904, 0i)
R1 (0, 0i)  (-0.0993808, 0i)  (-0.0993808, 0i)
R2 (1, 0i)  (-0.993808, 0i)   (-0.993808, 0i)

val eigval : Owl_dense_matrix_z.mat =
          C0       C1       C2
R0 (2, 0i) (1, 0i) (1, 0i)
```

Note that the results are expressed as complex numbers. To get just the eigenvalues, we can use the eigvals function. Both functions take boolean arguments permute and scale indicating whether the input matrix should be permuted or diagonally scaled.

4.5.1 Complex Matrices

As in the previous examples, complex matrices are frequently used in calculating eigenvalues and eigenvectors, so we next re-introduce some concepts in complex spaces.

The notion of a symmetric matrix can be extended to the complex domain as the *Hermitian Matrix* denoted A^H. Rather than being equal to its transpose, it equals to its *conjugate transpose*: when transposing, each element $a + bi$ becomes its complex conjugate $a - bi$. The is_hermitian function tests if a matrix is Hermitian, as in the example below.

```
# let a = Dense.Matrix.Z.of_array [|
    {re=1.; im=0.}; {re=2.; im=(-1.)};
    {re=2.; im=1.}; {re=3.; im=0.};
  |] 2 2;;

val a : Dense.Matrix.Z.mat =
        C0        C1
R0 (1, 0i) (2, -1i)
R1 (2, 1i) (3, 0i)

# Linalg.Generic.is_hermitian a;;
- : bool = true
```

The conj function of a complex matrix performs conjugate transposition:

```
# Dense.Matrix.Z.(conj a |> transpose);;
- : Dense.Matrix.Z.mat =
         C0        C1
R0 (1, -0i) (2, -1i)
R1  (2, 1i) (3, -0i)
```

It can be shown that if a matrix A is Hermitian, for all complex vectors x, $x^H A x$ is real as are all its eigenvalues.

```
# Linalg.Z.eigvals a;;
- : Owl_dense_matrix_z.mat =
                        C0                           C1
R0 (-0.44949, 1.50231E-17i) (4.44949, 2.07021E-16i)
```

Related to the Hermitian Matrix is the *Unitary Matrix* U such that

$$U^H U = I. \tag{4.13}$$

The inverse and conjugate transpose of U are equal, and it is analogous to orthogonal vectors in the real space.

4.5.2 Similarity Transformation and Diagonalisation

For an $n \times n$ matrix A, and an invertible $n \times n$ matrix M, the matrix

$$B = M^{-1}AM \qquad (4.14)$$

is said to be *similar* to A, and M has the same eigenvalues as A. M can be thought of as a matrix that changes basis while A is a linear transformation, so $M^{-1}AM$ first changes the basis before applying the linear transformation and changing the basis back. For example, in a three dimensional space, if we can first transform from using three random vectors as the basis of linear transformation to using the standard basis $[1, 0, 0]$, $[0, 1, 0]$, $[0, 0, 1]$, we can greatly simplify solving this problem. Finding the suitable similar matrix is thus key to simplifying calculations in many scientific and engineering problems.

One type of possible simplification is to find a triangular matrix that is similar. *Schur's Lemma* declares that A can be decomposed into UTU^{-1} where U is a unitary matrix, and T is an upper triangular matrix. This decomposition can be achieved using the function schur:

```
# let a = Dense.Matrix.Z.of_array [|
    {re=1.; im=0.};    {re=1.; im=0.};
    {re=(-2.); im=0.}; {re=3.; im=0.};
  |] 2 2;;
```

```
# let t, u, eigvals = Linalg.Z.schur a;;
val t : Owl_dense_matrix_z.mat =
         C0                      C1
R0 (2, 11) (2.10381, -0.757614i)
R1 (0, 0i)              (2, -11)
val u : Owl_dense_matrix_z.mat =
                     C0                       C1
R0 (-0.408248, 0.408248i)  (0.563384, -0.590987i)
R1         (-0.816497, 0i) (-0.577185, 0.0138014i)
val eigvals : Owl_dense_matrix_z.mat =
         C0          C1
R0 (2, 1i) (2, -1i)
```

The returned result t is an upper triangular matrix, and u is an unitary matrix, verified as follows:

```
# Dense.Matrix.Z.(dot u (conj u |> transpose));;
- : Dense.Matrix.Z.mat =
                     C0                          C1
R0                   (1, 0i) (7.979E-17, 5.811E-17i)
R1 (7.979E-17, -5.811E-17i)                  (1, 0i)
```

Another very important similar transformation is *diagonalisation*. Suppose A has n linear-independent eigenvectors that are the columns of a matrix Q, then $Q^{-1}AQ$ is a diagonal matrix Λ, and the eigenvalues of A are the diagonal elements of Λ. Its inverse

$$A = Q\Lambda Q^{-1} \tag{4.15}$$

is the *Eigen Decomposition*. Analysing A's diagonal similar matrix Λ rather than A directly often greatly simplifies problems.

Note that not every matrix can be diagonalised. If none of the n eigenvalues of A are equal, then its n eigenvectors are linearly-independent and A can be diagonalised. Specifically, every real symmetric matrix can be diagonalised by an orthogonal matrix or, in a complex space, every hermitian matrix can be diagonalised by a unitary matrix.

4.6 Positive Definite Matrices

We now introduce the *Positive Definite Matrix*, unifying the three key ideas introduced so far: pivots, determinants, and eigenvalues.

4.6.1 Positive Definiteness

A matrix is *Positive Definite* if it is symmetric and

$$x^T A x > 0 \tag{4.16}$$

for all non-zero vectors x. There are several necessary and sufficient conditions that can be used to test if a symmetric matrix A is positive definite:

1. $x^T A x > 0$ for all non-zero real vectors x;
2. $\lambda_i > 0$ for all eigenvalues λ_i of A;
3. all the upper left matrices have positive determinants;
4. all the pivots without row exchange satisfy $d > 0$;
5. there exists invertible matrix B so that $A = B^T B$.

For the last, we can use the *Cholesky Decomposition* to find the matrix B. It decomposes a Hermitian positive definite matrix into the product of a lower triangular matrix and its conjugate transpose LL^H:

```
# let a = Mat.of_array [|
    4.; 12.;-16.;
   12.; 37.;-43.;
  -16.;-43.; 98.;
   |] 3 3;;
```

```
# let l = Linalg.D.chol a;;
# Mat.(dot (transpose l) l =~ a);;
- : bool = true
```

If we know that A in $Ax = b$ is hermitian and positive definite, we can instead solve $LL^H x = b$. As we have seen previously, solving a linear system expressed in triangular matrices is easy. The Cholesky decomposition is more efficient than the LU decomposition. In the LinAlg module, is_posdef function performs this test, implemented in Owl by checking if the Cholesky decomposition can be performed on the input matrix.

```
# Linalg.D.is_posdef a;;
val is_pos : bool = true
```

The definition of *semi-positive definite* is similar but requires only that $x^T A x \geq 0$ for all non-zero real vectors x. The following Eq. 4.17 exists in many engineering analyses.

$$Ax = \lambda M x \tag{4.17}$$

If A and M are positive definite, this pattern parallels $Ax = \lambda x$ where $\lambda > 0$, such as in a linear system $y' = Ax$ where

$$x = [x_1, x_2, \ldots, x_n] \tag{4.18}$$

$$y' = [\frac{dx_1}{dt}, \frac{dx_2}{dt}, \ldots, \frac{dx_n}{dt}] \tag{4.19}$$

We will see such an example in Chap. 8. In a linearised differential equation system the matrix A is called the *Jacobian matrix*. The eigenvalues decide whether the system is stable or not using its Jacobian matrix A. A theorem declares that this system is stable if and only if there exists positive and definite matrix V so that $-(VA + A^T V)$ is semi-positive definite [13].

4.6.2 Singular Value Decomposition

The singular value decomposition (SVD) is one of the most important matrix factorisations. It provides a numerically stable matrix decomposition that is both useful for many purposes and guaranteed to exist [8]. Specifically, any $m \times n$ matrix can be factorised to:

$$A = U\Sigma V^T \tag{4.20}$$

Here U is an $m \times m$ matrix whose columns are the eigenvectors of AA^T. Similarly, V is a $n \times n$ matrix, and the columns of V are eigenvectors of $A^T A$. The r (rank of A) singular values on the diagonal of the $m \times n$ diagonal matrix Σ are the square roots of the non-zero eigenvalues of both AA^T and $A^T A$, called the *singular values* of A.

(a) orignal   (b) n=10   (c) n=20   (d) n=50

Fig. 4.1: Using top-n elements in SVD to restructure image. The original image (a) can be restructured using the top-10 (b), top-20 (c), or top-50 elements (d).

SVD is closely related to eigenvector factorisation of a positive definite matrix: for a positive definite matrix, the SVD factorisation is the same as the $Q\Lambda Q^T$.

Looking more closely, Eq. 4.20 can be expressed as Eq. 4.21. Here any u_i or t_i is a column vector of length n and m respectively. As shown in Eq. 4.21, this matrix expression can be re-expressed as a summation where each $u_i v_i^T$ is an $m \times n$ matrix, as is A itself. Therefore, the SVD decomposes a matrix into a summation of r matrices, each scaled by a corresponding singular value. The larger the singular value is, the more weight is given to its corresponding matrix. Therefore, in the summation, if the singular value σ_i is quite small, the term $\sigma_i u_i v_i^T$ can be removed without affecting original matrix A too much. This is a fundamental use of SVD: it can locate the "significant" components in the matrix, and is thus often used to reduce the dimension of given data.

$$
A = [u_1, u_2, \ldots, u_m] \begin{bmatrix} \sigma_1 & 0 & \ldots & 0 & 0 & \ldots & 0 \\ 0 & \sigma_2 & \ldots & 0 & 0 & \ldots & 0 \\ \vdots & \vdots & \ldots & \vdots & \vdots & \vdots & \vdots \\ 0 & 0 & \ldots & \sigma_r & 0 & \ldots & 0 \\ \vdots & \vdots & \ldots & \vdots & \vdots & \vdots & \vdots \\ 0 & 0 & \ldots & 0 & 0 & \ldots & 0 \end{bmatrix} \begin{bmatrix} v_1^T \\ v_2^T \\ \vdots \\ v_n^T \end{bmatrix} = \sum_{i=0}^{r} \sigma_i u_i v_i^T \quad (4.21)
$$

As an example, we will now demonstrate how to use SVD to re-encode an image using less information, i.e. image compression. We will use a cat image as shown in Fig. 4.1a. First, we load the image as a matrix using the code below.

```
#use "imageUtils.ml";;
module N = Dense.Ndarray.S
module M = Dense.Matrix.S
let img_arr = load_ppm "cat_bw.ppm" |> N.get_slice [[];[];[0]];;
let shp = N.shape img_arr;;
let h, w = shp.(0), shp.(1);;
let img = N.reshape img_arr [|h; w|];;
```

The output img is a 256×256 matrix where each element is an integer between 0 and 255 indicating a pixel in this black-and-white image. For simplicity, we don't use external image libraries, but use the image-related utility functions in the file imageUtils.ml provided in the code repository of this book. It provides routines to handle the PPM image format. After obtaining image matrix img, we can apply SVD on it. The SVD is supported in the Linalg module.

```
let u, s, vt = Linalg.S.svd img;;
```

```
...
val s : Owl_dense_matrix_s.mat =
       C0       C1          C253      C254      C255
R0 32277 5315.51 ...  0.978465 0.314735 0.0553443
...
```

The output is three matrices representing U, Σ, and V^T. Both u and vt are 256×256 matrices. Note that the diagonal matrix s is represented as a vector in the output, and its elements are sorted from large to small. We can extend the vector to the full diagonal matrix using Matrix.diagm. However, using this function to extend s to a diagonal matrix is only possible when we know that the original diagonal matrix is square (as in this example), otherwise the vector contains $\min(m, n)$ diagonal elements.

Next we reconstruct the image using the first n singular values. For example, let's set $n = 10$, a relatively small value. That means we need to set every element in s to zero except for the first 10, which we do by using a *slicing* method. You may be familiar with such array manipulation techniques from systems such as NumPy. If not, don't worry – we will explain in detail in Chap. 5.

```
let l = (N.shape s).(1);;
let n = 10;;
let s' = N.copy s;;
let z = N.zeros [|1; l - n|];;
let _ = N.set_slice [[];[n; l - 1]] s' z;;
```

The new Σ diagonal values are contained in s'. We can now reconstruct the original image matrix by calculating $U\Sigma'V^T$:

```
let img' = M.(dot (dot u (diagm s')) vt);;
```

The resulting matrix img' can be saved in PPM format using our customised image save routine in imageUtils.ml.

```
let image = N.stack ~axis:2 [|img'; img'; img'|];;
let image = N.expand image 4;;
let fname = "cat_" ^ (string_of_int n) ^ ".ppm";;
let _ = save_ppm_from_arr image fname;;
```

The reconstructed figure is shown in Fig. 4.1b. Even though it has lost a lot of detail, it nonetheless provides a sketch of the original image. For the original matrix, we need to store $256 \times 256 = 65536$ values. Using SVD, each u_i, σ_i, and v_i contains 256 elements. To reconstruct the image we only need to keep the first 10 u_i, σ_i, and v_i, which means storing $10 \times (1 + 256 + 256) = 5130$ elements, just 7.8% of the original.

While saving space significantly, the reconstructed image is clearly of quite low quality. We can improve it by adding more singular values setting by setting $n = 20$ (Fig. 4.1c) and $n = 50$ (Fig. 4.1d). By the time we use 50 values, which still uses less than 20% of the original elements, the reconstructed figure is already quite close to the original.

Besides `svd`, the `Linalg` module also provides several variants such as `svdvals` returning the singular values, and `gsvd` performing a *generalised SVD* where `gsvd x y -> (u, v, q, d1, d2, r)` computes the generalised SVD of a pair of general rectangular matrices x and y. Here d1 and d2 contain the generalised singular value pairs of x and y. The shape of x is $m \times n$ and the shape of y is $p \times n$.

SVD has a wide and growing range of applications [9] [10]. The example above uses it for image compression, demonstrating its application in *dimensionality reduction*. The image's dimensionality is reduced from 256 to 10, 20, or 50 while retaining acceptable quality. Similarly, SVD can also be used to compress other high dimensional data to simplify data processing. In Chap. 12 we will see how SVD plays a crucial role in the language processing field to perform principal component analysis. Besides, the *Moore-Penrose pseudo-inverse* that works for non-invertible matrix can be implemented efficiently using SVD, and Owl provides `Linalg.pinv` function to compute this [11].

4.7 Sparse Matrices

What we have discussed so far are all based on dense matrices. But often elements are sparsely distributed in a matrix, as in the identity matrix for example. In that case the *sparse* structure might be more efficient in computing as it avoids storing a huge amount of uninformative zero elements. In Owl, sparse matrix support can be found in the `Sparse.Matrix` module. As with the dense matrix module, it supports the four types of number in the S, D, C, and Z submodules. In this final section, we briefly introduce how sparse matrices can be efficiently represented.

The most intuitive method to represent a matrix is the `(row, column, value)` triplet where, e.g., `(1,2,10)` means that the element on the first row and second column in a matrix is 10. Tuples can be sorted according to row and column values to improve access efficiency.

Another commonly used format is the *Compressed Sparse Row* (CSR) format which is similar to the triplet above but compresses row indices. For example, suppose we must represent the following matrix:

$$\begin{bmatrix} 10 & 0 & 9 & 8 \\ 0 & 7 & 0 & 0 \\ 6 & 5 & 0 & 0 \end{bmatrix}$$

The CSR format retains the same column indices and values but compresses the row indices. In Fig. 4.2, in the original triplet representation on the right-hand side, we must store 6 row indices, while in CSR on the left, that is reduced to just 3. In CSR, values in the same row share a single column index. Each element shows the index of the first value in that row in the whole value vector.

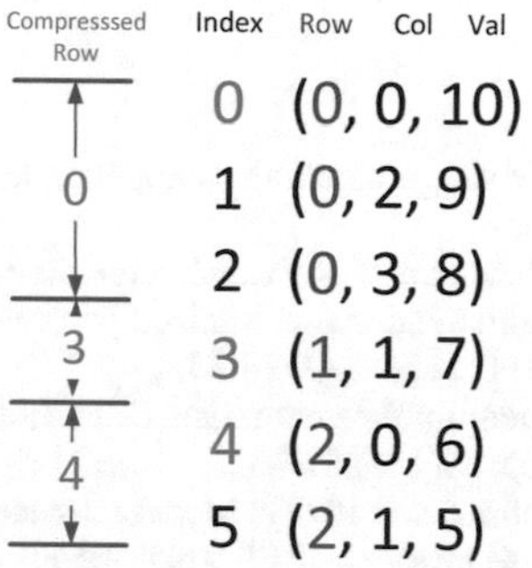

Compressed Row	Index	Row	Col	Val
	0	(0, 0, 10)		
0	1	(0, 2, 9)		
	2	(0, 3, 8)		
3	3	(1, 1, 7)		
4	4	(2, 0, 6)		
	5	(2, 1, 5)		

Fig. 4.2: Triplet and CSR sparse matrix representations

Using this representation, one can quickly locate a given element in the sparse matrix. For example, to search the (2, 1, 5) triplet, we know that we can search this element from the index 4 by checking the compressed row indices, and then look for the column number 1. The benefit of CSR is that it is more efficient at accessing row-vectors and performing row operations. Another data structure *Compressed Sparse Column* (CSC) is similar but compresses column vectors instead of rows.

The triplet is an intuitive format to express a sparse matrix. Though assignment operation is much slower in sparse matrix than in standard one, triplet formats still perform relatively well when inserting new elements. In comparison, both CSR and CSC formats are more storage-efficient allowing fast row/column access as well as arithmetic operations such as matrix-vector multiplication, but making element modification less efficient. Nevertheless, they are still widely used in current sparse matrix libraries. Eigen, an optimised C++ library for linear algebra, has implemented more versatile variants of these storage scheme for its sparse matrix representation [1].

---

[1] Sparse matrix manipulations, Eigen documentation, https://eigen.tuxfamily.org/dox/group__TutorialSparse.html.

4.8 Summary

This chapter gave an overview of several important topics of Linear Algebra. Starting from basic vector and matrix operations, we briefly introduced several key ideas from this field, including Gaussian elimination, vector spaces, determinants, Eigenvalues, Positive Definite Matrices, etc. We demonstrated these topics with examples and code using Owl. We closed with a discussion of the sparse matrix and its representation formats used.

References

1. Oliphant TE. Python for scientific computing. Computing in science & engineering. 2007 Jun 18;9(3):10-20.
2. Baboulin M, Buttari A, Dongarra J, Kurzak J, Langou J, Langou J, Luszczek P, Tomov S. Accelerating scientific computations with mixed precision algorithms. Computer Physics Communications. 2009 Dec 1; 180(12):2526-33.
3. Skeel RD. Iterative refinement implies numerical stability for Gaussian elimination. Mathematics of computation. 1980; 35(151):817-32.
4. Zames G. Feedback and optimal sensitivity: Model reference transformations, multiplicative seminorms, and approximate inverses. IEEE Transactions on automatic control. 1981 Apr; 26(2):301-20.
5. Lorber A. Error propagation and figures of merit for quantification by solving matrix equations. Analytical Chemistry. 1986 May 1; 58(6):1167-72.
6. MacDonald JK. Successive approximations by the Rayleigh-Ritz variation method. Physical Review. 1933 May 15; 43(10):830.
7. Fox RL, Kapoor MP. Rates of change of eigenvalues and eigenvectors. AIAA journal. 1968 Dec; 6(12):2426-9.
8. Stewart GW. On the early history of the singular value decomposition. SIAM review. 1993 Dec; 35(4):551-66.
9. Andrews H, Patterson C. Singular value decompositions and digital image processing. IEEE Transactions on Acoustics, Speech, and Signal Processing. 1976 Feb; 24(1):26-53.
10. Maciejewski AA, Klein CA. The singular value decomposition: Computation and applications to robotics. The International journal of robotics research. 1989 Dec; 8(6):63-79.
11. Biglieri E, Yao K. Some properties of singular value decomposition and their applications to digital signal processing. Signal Processing. 1989 Nov 1; 18(3):277-89.
12. Strang G. Linear Algebra and its Applications 1Harcourt Brace Jovanovich. San Diego, Calif. 1988; 2:98-9.
13. Cross GW. Three types of matrix stability. Linear algebra and its applications. 1978 Jun 1; 20(3):253-63.

Chapter 5
N-Dimensional Arrays

Abstract The N-dimensional array is the fundamental building block in numerical computing libraries e.g., NumPy and SciPy. It is the core dense data structure and many advanced numerical functions are built on top of it, including linear algebra, optimisation, and algorithmic differentiation. In fact, the rest of this book is built upon it. In this chapter, we introduce the `Ndarray` module and its core functions. We then pay particular attention to its two key and frequently used functionalities: slicing and broadcasting. Finally, we briefly explain the concept of a *tensor*, similar to but different from the N-dimensional array.

5.1 Ndarray

N-dimensional array (a.k.a ndarray) is the most fundamental data type in a numerical library for building scientific computing applications. Scalar-based computations are easy to understand but limited in the problems they can solve. Ndarray is the key data structure in many mainstream numerical libraries and software, including NumPy, Julia, and TensorFlow. This chapter introduces Owl's `Ndarray`, a fundamental module in OCaml numerical computing, and the basic functionalities it provides.

5.1.1 Types

We begin by looking at the data types used in `Ndarray`, which is built directly on top of OCaml's native `Bigarray` data structure. The `Bigarray` module is used to express large, multi-dimensional numerical arrays, and designed to enable efficient sharing of large numerical arrays between OCaml code and C or FORTRAN numerical libraries. More specifically, `Ndarray` is defined based on the `Bigarray.Genarray` submodule:

```
type ('a, 'b) t = ('a, 'b, c_layout) Genarray.t
```

© The Author(s), under exclusive license to Springer Nature Switzerland AG 2022
L. Wang et al., *OCaml Scientific Computing*, Undergraduate Topics in Computer Science,
https://doi.org/10.1007/978-3-030-97645-3_5

This `Genarray.t` generalized algebraic data type specifies the number type, precision, and memory layout of an multi-dimensional array. Based on different combinations of (`'a, 'b`), `Ndarray` supports four different types of data precision: single precision float (S), double precision float (D), single precision complex number (C), and double precision complex number (Z). These are all accessed by modules `Ndarray.S`, `Ndarray.D`, etc. In what follows, unless otherwise stated, we will use the double precision `Ndarray.D` module which is aliased to the `Arr` module.

The `Ndarray` module is designed to use row-based layout (`c_layout`) in its implementation. This indicates that memory addresses are continuous at the highest dimension, compared to FORTRAN layout where continuous addresses are found in the lowest dimension. From the perspective of designing Owl, our reasons for making this decision are as follows.

- Mixing memory layouts causes bugs. In particular, indexing in FORTRAN starts from 1 while indexing in C starts from 0, and many native OCaml data structures such as `Array` and `List` all index from 0. So sticking with C-layout avoids many clashes when using Owl.
- Supporting both layouts adds considerable complexity when implementing the underlying functions. Code which performs well on one layout may not do well on the other, and so many functions would require different implementations for different layouts. This would add much complexity and significantly increase the code base for only marginal benefits.
- `Ndarray` has rather different design principles comparing to `Bigarray`. The later only serves as a basic tool for operating on a chunk of memory living outside the OCaml heap, and facilitates exchanging data between different libraries (including FORTRAN ones). The numerical library we use here focuses on providing high-level numerical functions allowing programmers to write concise analytical code. The simple design and small code base outweighs the benefits of supporting both layouts.

Note that in Owl, a matrix is just a special case of an N-dimensional array, and you can add a matrix and a 2-dimensional ndarray without any problem. In fact many functions in the `Matrix` module simply calls the same functions in `Ndarray`.

5.1.2 Ndarray Creation and Properties

We start by introducing several groups of key functions provided by `Ndarray`. The first group is the creation functions that generate dense data structures. The most frequently used are probably `empty`, `create`, `zeros`, and `ones`. These functions return ndarrays of the specified shape and content.

Note that the `empty` function does not allocate any memory for an ndarray until it is accessed, so the function returns very fast after calling. The other three are hopefully self-explanatory: `zeros` and `ones` fill the allocated memory with zeros and ones respectively, whereas create function fills the memory with the specified value.

If you need random numbers, you can use another three creation functions that return ndarrays in which the elements follow specified distributions: uniform, gaussian, and bernoulli. You must provide the distribution parameters in each case, e.g., μ and σ for a Gaussian distribution.

We sometimes need to generate numbers equally spaced between bounds, for example when generating intervals or plotting figures. The sequential and linspace functions serve that purpose, producing ndarrays filled with evenly spaced numbers over a specified interval, the former with integer steps, and the latter allowing non-integer steps.

If these do not satisfy your requirements, Ndarray provides more flexible means that give you more control over initialisation via the init and init_nd functions. The first only passes 1-dimensional indices to a user-defined function, while the second passes n-dimensional indices but is much slower. The example below creates an ndarray where all the elements are even numbers:

```
# let x = Arr.init [|6;8|] (fun i -> 2. *. (float_of_int i));;
val x : Arr.arr =

    C0 C1 C2 C3 C4 C5 C6 C7
R0   0  2  4  6  8 10 12 14
R1  16 18 20 22 24 26 28 30
R2  32 34 36 38 40 42 44 46
R3  48 50 52 54 56 58 60 62
R4  64 66 68 70 72 74 76 78
R5  80 82 84 86 88 90 92 94
```

After created, there are various functions that obtain the properties of the ndarry. For example, given ndarrays x and y, the following functions are commonly used:

- shape x returns the shape of x
- num_dims x returns the number of dimensions of x
- nth_dim x n returns the size of the n-th dimension of x
- numel x returns the number of elements in x
- nnz x returns the number of non-zero elements in x
- density x returns the percentage of non-zero elements in x
- size_in_bytes x returns the size of x in bytes in memory
- same_shape x y checks whether x and y has the same shape
- kind x returns the type of ndarray x

5.1.3 Map, Fold, and Scan

Three typical generic operations are worth special attention: map, fold, and scan. The map function transforms one ndarray to another by applying the provided function to every element in the original ndarray. For example, the code below:

```
let y = Arr.map (fun a -> a +. 1.) x;;
```

adds 1 to every element in the ndarray x, then returns a new ndarray y. The map function in Ndarray is pure, always generating a fresh copy of the data structure. If your transformation function needs the index of the element being operated on, the mapi function accepts a vector index of the element being accessed.

The fold function is sometimes referred to as "reduction" in other languages. It has a named parameter called axis that specifies the axis along which you want to fold the given ndarray. The axis parameter is optional. If you do not specify one, the ndarray will be folded along all the dimensions into a single element (that is, a one-dimensional ndarray of length one). The fold function folds from left, and requires an initial value for the folding. The code below demonstrates how to implement your own sum' function. It adds all the elements in one ndarray and outputs a float number by folding the function +. with an initial value of 0.

```
let sum' ?axis x = Arr.fold ?axis ( +. ) 0. x;;
```

The functions sum, sum', prod, prod', min, min', mean, and mean' all perform foldings with various functions. The prime ending indicates that the function returns a number rather than an ndarray. As with map, if indices are required, the foldi function passes in 1-dimensional indices.

Finally, scan is analogous to a combination of map and fold, accumulating values along the specified axis but not changing the shape of the input. For example, consider generating a cumulative distribution function from a probability density function, as introduced in Chap. 3. Several functions belong to this group, such as cumsum, cumprod, cummin, cummax. Again, the scani function accepts cumulative functions that need index information.

Many common operations on ndarrays can be decomposed as a series of map, fold, and scan operations. In fact, an entire programming paradigm is built on these operations: Map-Reduce [5]. The Ndarray module has included a very comprehensive set of mathematical functions and all have been vectorised. This means you can apply them directly on an ndarray and the function will be automatically applied to every element in the ndarray. For binary math operators, there are add, sub, mul, etc. For unary operators, there are sin, cos, abs, etc.

Conceptually, all of these can be implemented using just map, fold, and scan. For example, to make your own vectorised sin function, you need only execute Arr.map sin x. However, this may well be slower than the native vectorised functions, which are implemented in C code to guarantee good performance.

5.1.4 Comparison Functions

The comparison functions in Ndarray can be divided into several groups. The first group compares two ndarrays and returns a boolean value, such as equal, less, and greater. The second compares two ndarrays and returns an 0-1 ndarray of the same shape with elements set to 1 indicating where the predicate is satisfied, and 0 if not: `elt_equal`, `elt_not_equal`, etc. The third is similar to the first, but compares an ndarray with a scalar value, and returns a boolean: `equal_scalar`, `not_equal_scalar`, etc. The fourth is similar to the second but compare an ndarray with a scalar, and returns a 0-1 ndarray: `qelt_equal_scalar`, `elt_less_scalar`, etc.

We generally recommend using the appropriate operators rather than calling these functions directly as it leads to more concise code. For example, the following code shows two ways to retain all elements greater than 0.5 and set the rest to 0 in an ndarray.

```
let y = Arr.map (fun a -> if a > 0.5 then a else 0.) x;;
let z = Arr.((x >.$ 0.5) * x);;
```

Not only does the operator version leads to more concise code, it also outperforms the first solution at the price of higher memory consumption, because the loop is performed in C rather than OCaml. Please refer to the Appendix A to learn about all these operators.

5.1.5 Iteration Functions

As with native OCaml arrays, you can use `iter` and `iteri` to iterate over all the elements in an ndarray. One common use case is to check if one or more predicates are satisfied and there is a special set of iteration functions to help you in this task: `is_zero`, `is_positive`, `is_negative`, `is_nonpositive`, `is_nonnegative`, and `is_normal`, etc. All take an ndarray as input and return a boolean value.

More complex predicates can be tested using the three functions `exists`, `not_exists`, and `for_all` while passing in arbitrarily complicated functions that return a boolean. While these functions only tell us whether the predicates are met or not, the `filter` function can return the 1-dimension indices of those elements that satisfy the predicates.

Processing 1-dimensional indices is much faster than n-dimensional indices. However, if you do need n-dimensional indices, you can use the functions `ind` and `i1d` to convert between 1- and 2-dimensional indices. Both are defined in the `Owl.Utils` module. Note that you need to pass in the original ndarray because the shape information is required for calculating index conversion.

5.1.6 Manipulation Functions

In addition to those we have introduced, the `Ndarray` module contains many useful functions to manipulate ndarrays. For example, you can tile and repeat an ndarray along a specified axis. Let's first create a sequential ndarray.

```
# let x = Arr.sequential [|3;4|];;
val x : Arr.arr =
   C0 C1 C2 C3
R0  0  1  2  3
R1  4  5  6  7
R2  8  9 10 11
```

The code below tiles x once in both dimensions.

```
# let y = Arr.tile x [|2;2|];;

val y : Arr.arr =
   C0 C1 C2 C3 C4 C5 C6 C7
R0  0  1  2  3  0  1  2  3
R1  4  5  6  7  4  5  6  7
R2  8  9 10 11  8  9 10 11
R3  0  1  2  3  0  1  2  3
R4  4  5  6  7  4  5  6  7
R5  8  9 10 11  8  9 10 11
```

The repeat function replicates each element in its adjacent cell along the specified dimension, as shown below.

```
# let z = Arr.repeat x [|2;1|];;
val z : Arr.arr =
   C0 C1 C2 C3
R0  0  1  2  3
R1  0  1  2  3
R2  4  5  6  7
R3  4  5  6  7
R4  8  9 10 11
R5  8  9 10 11
```

You can also expand the dimensionality of an ndarray, or squeeze out dimensions having only one element, or pad elements around an existing ndarray, using the functions expand, squeeze, and pad.

Another two useful functions are concatenate and split. The concatenate allows concatenation of an array of ndarrays along the specified axis, given that the dimensions other than that on which they are concatenated are equal. There are two operators associated with concatenation for matrices: @|| concatenates horizontally (i.e. along axis 1); and @= for concatenating vertically (i.e. along axis 0). split is then the inverse operation of concatenation. You can also sort an ndarray with sort, but note that modification will happen in place.

Another feature will use frequently in this book is to convert between ndarrays and normal arrays using conversion functions of_array and to_array.

5.1.7 Serialisation

In data analysis tasks we often need to save temporary results and later reload them. This process requires the serialisation and de-serialisation functions `save` and `load`. With the help of the tool `npy-ocaml`[1], we can save and load files in the NPY binary file format from NumPy. It stores all of the shape and data type information necessary to reconstruct the array correctly, even on another machine with a different architecture. It is a widely used serialisation format in numerical related tasks. Using NPY files are the same as that of normal serialisation methods, using `save_npy` and `load_npy` functions. By using these functions, we can easily interact with the Python-world data from OCaml.

5.2 Slicing

Slicing in ndarray is the operation that extractes part of an ndarray or a matrix according to a *slice definition*, an index list that indicates which indices should be accessed and in what order for each dimension of the value being sliced. Indexing and slicing are arguably the most important ndarray functions in any numerical library. Owl's flexible design significantly simplifies the code, allowing writing concise algorithms. In this section we will introduce ndarray *basic* and *fancy slicing*, which differ in how the slice is defined.

5.2.1 Slice Definition

In *basic slicing*, each dimension in the slice definition must be defined in the format of [start:stop:step]. `Ndarray` provides two functions `get_slice` and `set_slice` to retrieve and assign slice values respectively. Both accept an `int list list` as the slice definition, where every `list` element in the `int list list` is assumed to be a range. For example, [[]; [2]; [-1;3]] is equivalent to the full slice definition [R []; R [2]; R [-1;3]] in fancy slicing notation shown below.

```
type index =
  | I of int
  | L of int list
  | R of int list
```

The *fancy slicing* is defined by an `index list` that uses three type constructors. I simply specifies a numeric index, so [I 2; I 5] returns the element at position (2, 5) in a matrix. L specifies a list of indices, so [I 2; L [5;3]] returns

---

[1] NPY-OCaml package, https://github.com/LaurentMazare/npy-ocaml

a 1 x 2 matrix with the elements at (2, 5) and (2, 3) in the original matrix. Finally, R specifies a range of indices. Similarty to `get_slice`, the `get_fancy` s x retrieves a slice of x defined by s, whereas `set_fancy` s x y assigns the slice of x defined by s according to values in y. Note that y must have the same shape as that defined by s. For example, these two slides of x are equivalent:

```
let x = Arr.sequential [10; 10; 10];;
Arr.get_slice [ []; [0;8]; [3;9;2] ] x;;
Arr.get_fancy [ R[]; R[0;8]; R[3;9;2] ] x;;
```

Note that both `get_slice` and `get_fancy` return a copy rather than a view as that in NumPy, while `set_slice` and `set_fancy` modify the original data in place.

The operators for indexing and slicing are built on the extended indexing operators introduced in OCaml 4.06. In the slicing system, they are used as aliases for the slicing functions we introduced above:

- `.%{ }` : get
- `.%{ }<-` : set
- `.${ }` : get_slice
- `.${ }<-` : set_slice
- `.!{ }` : get_fancy
- `.!{ }<-` : set_fancy

For example, we can use `.%{ }` for indexing as follows:

```
open Arr;;
let x = sequential [|10; 10; 10|] in
let a = x.%{2; 3; 4} in
x.%{2; 3; 4} <- 111.;;
```

Another example uses the `.!{ }` operator for fancy slicing:

```
open Arr;;
let x = sequential [|10; 10; 10|] in
let a = x.!{L [2;2;1]; R [6;-1]; I 5} in
let b = zeros (shape a) in
x.!{L [2;2;1]; R [6;-1]; I 5} <- b;;
```

5.2.2 Conventions and Examples

These definition of slicing may look complicated. The following rules can help write correct slice definitions. These conventions can be equally applied to both basic and fancy slicing. They are essentially very similar to those in NumPy. If you already know how to slice n-dimensional arrays in NumPy, you should find them easy to follow.

- **Rule #1**: The format of the range definition follows R [start; stop; step]. start specifies the starting index, stop specifies the stopping index (inclusive), and step specifies the step size. You do not have to specify all three variables in the definition; see the following rules.
- **Rule #2**: All three variables start, stop, and step can take both positive and negative values, but step is not allowed to be zero. Positive step indicates that indices will be visited in increasing order from start to stop; and vice versa.
- **Rule #3**: For start and stop variables, positive values refer to a specific index while a negative value "a" will be translated into n + a where n is the total number of indices. E.g., [-1; 0] indicates the slice from the last index to the first.
- **Rule #4**: If you pass in an empty list R [], this will be expanded into [0; n - 1; 1] which means all the indices will be visited in increasing order with step size 1.
- **Rule #5**: If you only specify one variable such as [start], get_slice assumes that you will take one specific index by automatically extending it into [start; start; 1]. As we can see, start and stop are the same, with step size 1.
- **Rule #6**: If you only specify two variables, the slice function assumes they are [start; stop] which defines the range of indices. However, how get_slice will expand this slice definition depends. As we can see below, slice will visit the indices in different orders. If start <= stop, it will be expanded to [start; stop; 1]; otherwise it will be expanded to [start; stop; -1].
- **Rule #7**: It is not necessary to specify all the definitions for all the dimensions: get_slice function will also expand it by assuming you will take all the data in higher dimensions. E.g., suppose x has the shape [2; 3; 4], if we define the slice as [[0]], get_slice will expand the definition into [[0]; []; []].

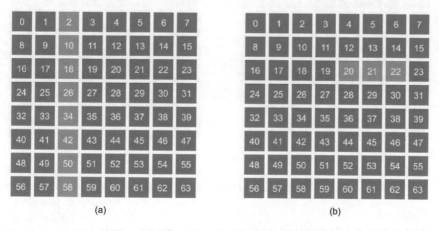

(a)   (b)

Fig. 5.1: Illustrated Examples of Slicing

That's all the rules. Let's look at some examples. All are based on an 8×8 matrix. The first is shown in Fig. 5.1(a), taking one column of the matrix using both basic and fancy slicing:

```
# Arr.get_slice [ []; [2]  ] x;;
# Arr.get_fancy [ R[]; I 2 ] x;;

- : Arr.arr =
    C0
R0  2
R1 10
R2 18
R3 26
R4 34
R5 42
R6 50
R7 58
```

The second in Fig. 5.1(b) is similar, but retrieves part of a row, again using both methods.

```
# Arr.get_slice [ [2]; [4; 6] ] x;;
# Arr.get_fancy [ I 2; R [4; 6] ] x;;

- : Arr.arr =
    C0 C1 C2
R0 20 21 22
```

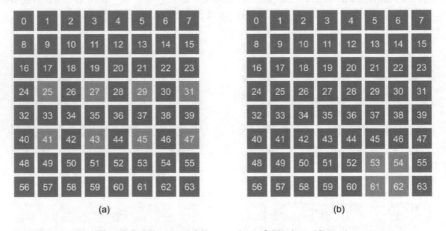

(a)   (b)

Fig. 5.2: Illustrated Examples of Slicing (Cont.)

The next example in Fig. 5.2(a) is a bit more complex. It chooses certain rows, and then chooses columns using a fixed step of 2, via fancy slicing:

```
# Arr.get_fancy [ L [3; 5]; R [1; 7; 2] ] x;;
- : Arr.arr =
   C0 C1 C2 C3
R0 25 27 29 31
R1 41 43 45 47
```

Finally, the last example in Fig. 5.2(b) obtains a sub-matrix in similar manner to examples 1 and 2, using negative integers as indices.

```
# Arr.get_fancy [ L [-2; -1]; R [-3; -2] ] x;;
- : Arr.arr =
   C0 C1
R0 53 54
R1 61 62
```

5.2.3 Advanced Usage

There are more advanced usages of slicing besides what we have just seen, which we will demonstrate in this section. Based on the same input matrix x, these examples vary the index s in Mat.get_slice s x. Mat is the matrix module built on Ndarray; all the following examples also hold for ndarrays.

- []: take all the elements
- [[2]; []]: take row 2
- [[2]]: take row 2, but only specify low dimension slice definition
- [[1;3]]: take from row 1 to 3
- [[3;1]]: take from row 3 to 1, same as the example above but in reverse order

Let's look at some more complicated examples.

- [[1;3]; [3;5]] take from row 1 to 3 and column 3 to 5, so a sub-matrix of x
- [[1;-1]; []] take from row 1 to the last row
- [[0;-1;2]] take the rows of even number indices, i.e., 0;2;4
- [[]; [1;-1;2]] take the column of odd number indices, i.e.,1;3;5...
- [[-1;0]] reverse all the rows of x
- [[-1;0]; [-1;0]] reverse all the elements of x, same as applying reverse function
- [[-2]; [0;-1;3]] take the second last row, from the first column to the last, with step size 3

The following examples show how to use slicing to achieve some even more complicated operations. We use a 5×5 sequential matrix for illustration.

```
# let x = Mat.sequential 5 5;;
val x : Mat.mat =

   C0 C1 C2 C3 C4
R0  0  1  2  3  4
R1  5  6  7  8  9
R2 10 11 12 13 14
R3 15 16 17 18 19
R4 20 21 22 23 24
```

In the first example, the `reverse` function treats a matrix as a one-dimensional vector and reverses its elements. This operation is equivalent to flipping in both vertical and horizontal directions.

```
# let reverse x =
Mat.get_slice [ [-1; 0]; [-1; 0] ] x in
  reverse x;;
- : Mat.mat =
   C0 C1 C2 C3 C4
R0 24 23 22 21 20
R1 19 18 17 16 15
R2 14 13 12 11 10
R3  9  8  7  6  5
R4  4  3  2  1  0
```

The second function rotates a matrix 90 degrees clockwise. As we can see, slicing function leads to very concise code.

```
# let rotate90 x =
 Mat.(transpose x |> get_slice [ []; [-1;0] ]) in
   rotate90 x;;
- : Mat.mat =
   C0 C1 C2 C3 C4
R0 20 15 10  5  0
R1 21 16 11  6  1
R2 22 17 12  7  2
R3 23 18 13  8  3
R4 24 19 14  9  4
```

The last function `cshift` performs right circular shift along the columns of a matrix.

```
let cshift x n =
  let c = Mat.col_num x in
  let h = Utils.Array.(range (c - n) (c - 1))
    |> Array.to_list in
  let t = Utils.Array.(range 0 (c - n -1)) |> Array.to_list in
  Mat.get_fancy [ R []; L (h @ t) ] x;;
```

Applying this to the previous x should produce the following output:

```
# cshift x 2;;
- : Mat.mat =
    C0 C1 C2 C3 C4
R0   3  4  0  1  2
R1   8  9  5  6  7
R2  13 14 10 11 12
R3  18 19 15 16 17
R4  23 24 20 21 22
```

5.3 Broadcasting

We now introduce the *broadcasting* operation. In contrast to indexing and slicing which must be explicitly invoked, broadcasting is usually implicitly invoked when certain conditions are met. This often simplifies code but can sometimes make debugging more complicated.

Many binary mathematical operators take two ndarrays as inputs, e.g., add, sub, etc. In the trivial case, the inputs should have exactly the same shape. However, we often operate on two ndarrays whose shapes do not match. How we extend the smaller one to match the bigger is called broadcasting.

Broadcasting can save unnecessary memory allocation. For example, assume we have a 1000×500 matrix x that contains 1000 samples each having 500 features. We want to add a bias vector v of shape 1×500 to each feature. As the shapes of x and v do not match, we must tile v so that it has the same shape as that of x.

```
let x = Mat.uniform 1000 500;;
let v = Mat.uniform 1 500;;
let u = Mat.tile v [|1000;1|];;
Mat.(x + u);;
```

The code above works but uses considerable memory for temporary values. High memory consumption is not desirable, particularly when running on resource-constrained devices. So we can use broadcasting to add x and v directly, without using the tile operation. The returned result is of the same shape as x.

However, the shapes of two inputs cannot be arbitrarily different in broadcasting, but are subject to some constraints. The rule is much simpler than in slicing: given two matrices/ndarrays of the same dimensionality, for each dimension, one of the following two conditions must hold: either both are equal, or one is one.

Here are some valid shapes where broadcasting can be applied between x and y.

```
x : [| 2; 1; 3 |]    y : [| 1; 1; 1 |]
x : [| 2; 1; 3 |]    y : [| 2; 1; 1 |]
x : [| 2; 1; 3 |]    y : [| 2; 3; 1 |]
...
```

Here are some invalid shapes that violate the constraint above and prevent broadcasting.

```
x : [| 2; 1; 3 |]     y : [| 1; 1; 2 |]
x : [| 2; 1; 3 |]     y : [| 3; 1; 1 |]
x : [| 2; 1; 3 |]     y : [| 3; 1; 1 |]
...
```

What if y has lower dimensionality than x? E.g., x has the shape [|2;3;4;5|] whereas y has the shape [|4;5|]. In this case, y's dimensionality is expanded to the same number as x's. Technically, two ndarrays are aligned along the highest dimension. In other words, this is done by appending 1s to the lower dimensions of y, making the new shape of y [|1;1;4;5|]. You can try expand by yourself, as shown below.

```
# let y = Arr.sequential [|4;5|];;
# let y' = Arr.expand y 4;;
# Arr.shape y';;
- : int array = [|1; 1; 4; 5|]
```

Next we show three broadcasting examples based on matrices, using both illustrations and code, so that you can better understand how the shapes are extended in broadcasting. The first example in Fig. 5.3 is a vector multiplied by scalar.

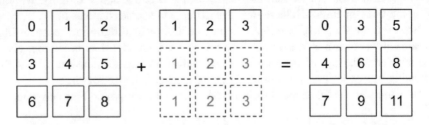

Fig. 5.3: Illustrated example of shape extension in broadcasting

```
let a = Arr.sequential [|1;3|];;
Arr.mul_scalar a 3.;;
```

The second example in Fig. 5.4 is adding a matrix and a vector.

Fig. 5.4: Illustrated example of shape extension in broadcasting (cont.)

```
let b0 = Arr.sequential [|3;3|];;
let b1 = Arr.sequential ~a:1. [|1;3|];;
Arr.add b0 b1;;
```

The third example in Fig. 5.5 is adding a column vector and a row vector.

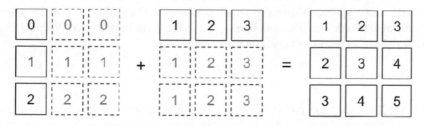

Fig. 5.5: Illustrated example of shape extension in broadcasting (cont.)

```
let c0 = Arr.sequential [|3;1|];;
let c1 = Arr.copy b1;;
Arr.add c0 c1;;
```

The broadcasting operation is transparent so it will be automatically applied if the shapes of two operators do not match and the constraints are met. Currently, the operations in Owl that support broadcasting are listed below:

- basic computation: add, sub, mul, div, pow;
- comparison operation: elt_equal, elt_not_equal, elt_less, elt_greater, elt_less_equal, elt_greater_equal;
- other: min2, max2, atan2, hypot, fmod.

5.4 Tensors

We finish this chapter by briefly introducing the concept of *tensor*. Many people use tensor and n-dimensional array interchangeably but there are differences between them. A tensor is expressed in *n*-dimensional array but must also obey certain transformation properties. For example, in a three dimension vector space we might have a vector A = [0, 1, 2]. We can find each element in this vector by a single index i, e.g. $A_1 = 1$. This vector is an object in this space, and it stays the same even if we change the standard cartesian coordinate system into other systems. If we do change the coordinate system, A must be updated accordingly. In this way we represent tensors as ndarrays, but they are not the same as ndarrays.

Consider a single vector v in a coordinate system with basis e. We can change the coordinate base to $\tilde{e}$ via a linear transformation: $\tilde{e} = Ae$ where A is a matrix. For any vector in this space using e as basis it will be transformed to $\tilde{v} = A^{-1}v$, written as:

$$\tilde{v}^i = \sum_j B^i_j v^j. \tag{5.1}$$

Here $B = A^{-1}$. We call a vector *contravector* if it changes in the opposite way to the basis. Note we use the superscript to denote the element in contravectors.

As a comparison, think about a matrix multiplication αv. The α itself forms a different vector space, the basis of which is related to the basis of v's vector space. It turns out that the direction of change of α is the same as that of e. When v uses new $\tilde{e} = Ae$, its component changes in the same way:

$$\tilde{\alpha}_j = \sum_i A^i_j \alpha_i.$$

It is called a *covector*, denoted with a subscript. We can further extend it to matrix. Consider a linear mapping L that can be represented as a matrix so that we can apply it to any vector using matrix dot product. With the change of coordinate system, it can be proved that the content of the linear map L itself is updated to:

$$\tilde{L}^i_j = \sum_{kl} B^i_k L^k_l A^l_j.$$

Again, note that we use both superscript and subscript for the linear map L, since it contains one covariant component and one contravariant component.

We can further extend this process and define the tensor with covariant and contravariant. A tensor T is an object that is invariant under a change of coordinates, and with a change of coordinates its component changes in a special way, which is shown in Eq. 5.2. Here the $ijk \ldots$ are indices of the contravariant part of the tensor and the $rst \ldots$ are that of the covariant part.

$$T^{\tilde{a}\tilde{b}\tilde{c}\ldots}_{xyz\ldots} = \sum_{ijk\ldots rst\ldots} B^a_i B^b_j B^c_k \ldots T^{ijk\ldots}_{rst\ldots} A^r_x A^s_y A^t_z \ldots \tag{5.2}$$

One of the important operations on tensors is the *tensor contraction*, applied to two tensors producing a third tensor through a summation process. Suppose we have tensor A^{abc}_{ijk} and B^{kl}_{rst}, and you can see that a pair of indices, one covariant of A and one contravariant of B, are labelled with the same letter k. This implies a summation over those two indices. As a result of summation, a new tensor C^{abcl}_{ijrst} has the indices of the tensors A and B except for the pair of indices over which the contraction operations took place. In this example, the contraction can be expressed as:

$$C^{abcl}_{ijrst} = \sum_k A^{abc}_{ijk} B^{kl}_{rst}.$$

It uses the `contract2` function in the `Ndarray` module that takes an array of `int * int` tuples to specify the pair of indices in the two input ndarrays.

```
let x = Mat.uniform 3 4;;
let y = Mat.uniform 4 5;;
```

```
let z1 = Mat.dot x y;;
let z2 = Arr.contract2 [|(1,0)|] x y;;
```

Tensor contraction is not limited to only two dimensions. For example, we have two three-dimensional array A and B, and aim to compute the matrix C so that:

$$C^i_j = \sum_{hk} A^i_{hk} B^{kh}_j \qquad (5.3)$$

The implementation is similar:

```
let x = Arr.sequential [|3;4;5|];;
let y = Arr.sequential [|4;3;2|];;
let z1 = Arr.contract2 [|(0, 1); (1, 0)|] x y;;
```

The indices mean that in contraction the 0-th dimension of x corresponds to the first dimension of y, and the first dimension of x corresponds with the 0-th dimension of y, as shown in Eq. 5.3. We can verify the result with a naive contraction summation implemented as below, and then check if the two results z1 amd z2 are the same.

```
let z2 = Arr.zeros [|5;2|];;

let _ =
  for h = 0 to 2 do
    for k = 0 to 3 do
      for i = 0 to 4 do
        for j = 0 to 1 do
          let r = (Arr.get x [|h;k;i|]) *.
            (Arr.get y [|k;h;j|]) in
          Arr.set z2 [|i;j|] ((Arr.get z2 [|i;j|]) +. r)
        done
      done
    done
  done;;
```

Tensor contraction can also express the *convolution* operations we will introduce in the signal and neural network chapters. In brief, convolution is a mathematical way of combining two signals to form a third. Each signal can be represented as a vector, matrix, or ndarray. We will see that convolution is perhaps the most important operation in current neural networks, and it can be expressed as a series of contraction computations.

As to the implementation, both the ndarray multiplication and convolution operations can be implemented using general matrix multiplication, or GEMM, a core functionality of BLAS. As we introduced in Chap. 4, GEMM is efficiently implemented in various highly optimised numerical libraries such as OpenBLAS and MKL. To achieve efficiency, the implementation of tensor contraction often requires other tensor techniques such as reshaping to transform contractions in many applications into GEMM. Note that using BLAS operations to implement tensor contraction may incur performance and storage overhead. High-performance implementation of the contraction operation has been a research topic in computation related fields [2, 3].

Contraction can also be used to implement other operations that might initially seem unrelated to matrix multiplication. For example, it can be applied on one single ndarray to perform reduction operation using the `contract1` function, as the code below demonstrates.

```
# let x = Arr.sequential [|2;2;3|];;
val x : Arr.arr =
        C0 C1 C2
R[0,0]   0   1   2
R[0,1]   3   4   5
R[1,0]   6   7   8
R[1,1]   9  10  11

# let y = Arr.contract1 [|(0,1)|] x;;
val y : Arr.arr =
  C0 C1 C2
R   9  11  13
```

Many tensor operations involve summation over particular indices. Therefore in using tensors in applications such as linear algebra and physics, the *Einstein notation* is used to simplified notations [4]. It removes the common summation notation, and any twice-repeated index in a term is summed (no index may occur three times or more in a term). For example, the matrix multiplication notation $C_{ij} = \sum_k A_{ik} B_{kj}$ can be simplified as $C = A_{ik} B_{kj}$. Eq. 5.2 can also be greatly simplified in this way.

The tensor calculus is frequently used in physics and engineering, in areas including relativity, quantum physics, and fluid mechanics. More details about the tensor calculation is beyond the scope of this book, and we refer readers to work such as Dullemond et al. [1] for a deeper understanding of this topic.

5.5 Summary

The N-dimensional array is the most fundamental data type in numerical computing, and we will use is throughout this book. This chapter explained the details of Owl's `Ndarray` module, including creation, property extraction, manipulation, serialisation, and so on. We also introduced slicing and broadcasting of ndarrays thoroughly. Mastering these topics requires practice, and this chapter provides many examples. We finished this chapter by briefly discussing the difference between tensor and ndarray.

References

1. Dullemond K, Peeters K. Introduction to tensor calculus. Kees Dullemond and Kasper Peeters. 1991:42-4.
2. Abdelfattah A, Baboulin M, Dobrev V, Dongarra J, Earl C, Falcou J, Haidar A, Karlin I, Kolev T, Masliah I, Tomov S. High-performance tensor contractions for GPUs. Procedia Computer Science. 2016 Jan 1;80:108-18.
3. Matthews DA. High-performance tensor contraction without BLAS. SIAM Journal on Scientific Computing. 2016 Jul;40.
4. Barr AH. The Einstein summation notation. An Introduction to Physically Based Modeling (Course Notes 19), pages E. 1991 Jul;1.
5. Dean J, Ghemawat S. MapReduce: simplified data processing on large clusters. Communications of the ACM. 2008 Jan 1;51(1):107-13

Chapter 6
Ordinary Differential Equations

Abstract A *differential equation* is an equation that contains a function and one or more of its derivatives. They have been studied ever since the invention of calculus, driven by applications in fields including mechanics [1], astronomy [2], geometry [3], biology [4], engineering [5], economics [6], and many more. If the function and its derivatives in a differential equation concern only one variable, we call it an *Ordinary Differential Equation* (ODE), and it can model a one-dimensional dynamical system. If there are multiple variables involved, it is a *Partial Differential Equation* (PDE). In this chapter we focus on ODE, including definition, numerical solution, and the tools provide in Owl, etc.

6.1 Defining an ODE

In general an *ordinary differential equation* can be expressed as:

$$F(x, y', y'', \ldots, y^{(n)}) = 0. \tag{6.1}$$

That is, as a function of derivatives with boundary conditions. They can be used to model dynamical systems. The initial state of the system is called its *initial values*, and these values are often known and can be represented as:

$$y|_{x=x_0} = y_0, y'|_{x=x_1} = y_1, \ldots, \tag{6.2}$$

where the y_0, y_1, etc. are known. The highest order of derivatives occurring in Eq. 6.1 is the *order* of the differential equation. A first-order differential equation can be generally expressed as:

$$\frac{dy}{dx} = f(x, y), \tag{6.3}$$

Table 6.1: Examples of solutions to certain types of ODE

Classes	Subclass
$P(y)\frac{dy}{dx} + Q(x) = 0$ $\frac{dy}{dx} + P(x)y = Q(x)$	$\int^y P(y)dy + \int^x Q(x)dx = C$ $y = e^{-\sum_{x_0}^x P(x)dx}(y_0 + \sum_{x_0}^x Q(x)e^{\sum_{x_0}^x P(x)dx}dx)$

where f is any function that contains x and y. Solving Eq. 6.1 and fitting to the given initial values as in Eq. 6.2 is called the *initial value problem*. Solving such problems is the main target of many numerical ODE solvers.

6.1.1 Exact Solutions

Many differential equations cannot be solved analytically and we must turn to numerical solvers to make progress. However, before doing so let us first look at some ODEs that do have analytical closed-form solutions.

Tbl. 6.1 shows two examples. The first row is a type of ODEs that are called the "separable equations" while the second represents ODEs called "linear first-order equations". The solutions to both forms are well-known and given in the second column. Here C is a constant determined by initial condition x_0 and y_0. $P(x)$ and $Q(x)$ are functions that contain only variable x. Note that in both types the derivative dy/dx can be expressed explicitly as a function of x and y, and therefore is called *explicit ODE* in contrast to an *implicit ODE*.

Higher order ODEs can be reduced to first order ODEs containing only y', y, and x. For example, an ODE in the form $y^{(n)} = f(x)$ can be reduced by multiple integrations on both sides. If a two-order ODE is in the form $y'' = f(x, y')$, let $y' = g(x)$, and then $y'' = p'(x)$. Put them into the original ODE, and it can be transformed as: $p' = f(x, p)$. This is a first-order ODE that can be solved by normal solutions. Suppose we must solve $y' = p = h(x, C_0)$; this explicit form of ODE can be integrated to obtain $y = \int h(x, C_0)dx + C_1$.

Please refer to classical calculus books or courses for more detail on the analytical solution of ODEs. We turn now to methods for their numerical solution.

6.1.2 Linear Systems

ODEs are often used to describe various dynamic systems. In the previous examples there is only one function, y, that changes over time. However, a real world system often contains multiple interdependent components, each is described by a function that evolves over time. Later we will talk about several ODE examples in detail, including the *two-body problem* and the *Lorenz attractor*. For now we simply observe that Eq. 6.12 and Eq. 6.13 in the sections below are quite different from the ODE forms above. For example, the Lorenz attractor system has three components that change with time: the rate of convection in the atmospheric flow, the horizontal and vertical temperature variation.

Both these systems are examples of *first-order linear systems of ODE* or just *linear systems of ODE*. Generally, if we have:

$$y(t) = \begin{bmatrix} y_1(t) \\ \vdots \\ y_n(t) \end{bmatrix}, A(t) = \begin{bmatrix} a_{11}(t) & \cdots & a_{1n}(t) \\ \vdots & \cdots & \vdots \\ a_{n1}(t) & \cdots & a_{nn}(t) \end{bmatrix}, \text{ and } g(t) = \begin{bmatrix} g_1(t) \\ \vdots \\ g_n(t) \end{bmatrix},$$

then a linear system can be expressed as:

$$y'(t) = A(t)y(t) + g(t). \tag{6.4}$$

This linear system contains n time-dependent components: $y_1(t), y_2(t), \ldots, y_n(t)$. As we will soon show, first-order linear systems are especially amenable to numerical solution. Therefore transforming a high-order single-component ODE into a linear system is sometimes necessary as we will see in the two body problem example. But let's get back to the start and examine the basics of solving ODEs numerically.

6.2 Solving ODEs Numerically

This section introduces the basic idea of solving the initial value problem numerically. Let's start with an example:

$$y' = 2xy + x, \tag{6.5}$$

where the initial value is $y(0) = 0$. This is a separable first-order ODE, and according to the Tbl. 6.1, its analytical close-form solution is shown in Eq. 6.6.

$$y = \frac{e^{x^2} - 1}{2}. \tag{6.6}$$

However, imagine we don't have the solution but want to know what is y's value when $x = 1$ (or any other value)? How can we solve it numerically?

Enter *Euler Method*, a first-order numerical procedure to solve initial value problems. The basic idea is simple: from Eq. 6.5 we know the derivative, or the "slope" at any given point on the function curve, and we know the initial value x_0 and y_0 of this function. We can then simply move from the initial point x_0 to the target x value in small steps, adjusting the direction according to the derivative at every new point. Formally, the Euler method proposes to approximate the function y using a sequence of iterative steps:

$$y_{n+1} = y_n + \Delta f(x_n, y_n), \tag{6.7}$$

where Δ is a certain step size. This method is easy to be implemented in OCaml, as shown below.

```
let x = ref 0.;;
let y = ref 0.;;
let target = 1.;;
let step = 0.001;;
let f x y = 2. *. x *. y +. x;;
while !x <= target do
  y := !y +. step *. (f !x !y);
  x := !x +. step
done;;
```

In this case, we know that the analytical solution at $x = 1$ is $0.5(e^{1^2} - 1)$:

```
# (Owl_const.e -. 1.)/. 2.;;
- : float = 0.859140914229522545
```

and the solution given by our numerical code is about 0.8591862, pretty close to the true answer.

However, although this method is easy and works well in this example, it is generally not suitable for practical applications because it is not generally very accurate (we will discuss this further shortly), nor is it stable, and nor does it provide error estimates. However, we can modify Euler's method to use a "midpoint" when making steps in an attempt to curb the error in the update process as below.

$$s_1 = f(x_n, y_n)$$
$$s_2 = f(x_n + \Delta/2, y_n + s_1\,\Delta/2)$$
$$y_{n+1} = y_n + \Delta\frac{s_1 + s_2}{2}$$

The method then becomes the *Midpoint Method*. Let's compare the performance of Euler and Midpoint in approximating the true result in Eq. 6.6:

```
let f x y = 2. *. x *. y +. x;;
let f' x = 0.5 *. (Maths.exp (x *. x) -. 1.);;

let euler step target =
  let x = ref 0. in
  let y = ref 0. in
  while !x <= target do
    y := !y +. step *. (f !x !y);
    x := !x +. step
  done;
  !y;;
```

```
let midpoint step target =
  let x = ref 0. in
  let y = ref 0. in
  while !x <= target do
    let s1 = f !x !y in
    let s2 = f (!x +. step /. 2.) (!y +. step /. 2. *. s1) in
    y := !y +. step *. (s1 +. s2) /. 2.;
    x := !x +. step
  done;
  !y;;

let _ =
  let target = 2.6 in
  let h = Plot.create "plot_rk01.png" in
  Plot.(plot_fun ~h f' 2. target);
  Plot.(plot_fun ~h (euler 0.01) 2. target);
  Plot.(plot_fun ~h (euler 0.001) 2. target);
  Plot.(plot_fun ~h (midpoint 0.01) 2. target);
  Plot.(plot_fun ~h (midpoint 0.001) 2. target);
  Plot.output h;;
```

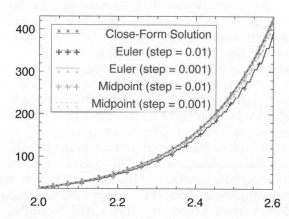

Fig. 6.1: Comparing the accuracy of Euler method and Midpoint method in approximating solution to ODE

Fig. 6.1 shows that the choice of step size indeed matters to precision. We use 0.01 and 0.001 for step size in the test, and for both cases the Midpoint method outperforms the simple Euler method.

But why stop at only one midpoint? We can continue this process and add further intermediate stages in the update sequence seeking to improve accuracy further. For example, we can do this:

$$s_1 = f(x_n, y_n)$$
$$s_2 = f(x_n + \Delta/2, y_n + s_1\Delta/2)$$
$$s_3 = f(x_n + \Delta/2, y_n + s_2\Delta/2)$$
$$s_4 = f(x_n + \Delta, y_n + s_3\Delta)$$
$$y_{n+1} = y_n + \Delta \frac{s_1 + 2s_2 + 2s_3 + s_4}{6}$$

$$(6.8)$$

In this method, each iteration computes four intermediate steps: at the initial point, at the end, and twice at the midpoints. The resulting method is often more accurate than the midpoint method.

We can continue like this but hopefully you can see the pattern now. In fact, these seemingly magical parameters are related to the terms in Taylor series expansions. In previous methods, e.g. Euler method, every time you update y_n to y_{n+1}, an error is introduced into the approximation. The *order* of a method is the exponent of the smallest power of Δ that cannot be matched. This family of methods is called the *Runge-Kutta Methods*, where the basic idea is to remove the errors order by order, using the correct set of coefficients [7]. A higher order of error indicates smaller error.

Euler's method is the most basic Runge-Kutta (RK) method, while the Midpoint method is also called the second-order Runge-Kutta Method (rk2). What Eq. 6.8 shows is a fourth-order Runge-Kutta method (rk4). It is the most frequently used RK method, and works surprisingly well in many cases. It is often a good choice especially when computing f is not expensive [8] [9].

However, as powerful as it may be, the classical rk4 is still a naive implementation. A modern ODE solver largely follows the same idea but adds more features. For example, adaptively updating the step size rather than holding it constant as in our example. Other methods also exist such as the Adams-Bashforth Method and Backward Differentiation Formula (BDF), both multi-step methods that use more information than just the derivative of the current step to compute the solution at next step, such as values from previous time steps [10]. In recent years, the Bulirsch-Stoer method is known to be both accurate and efficient computation-wise. Discussion of these advanced numerical methods and techniques are beyond the scope of this book. Please refer to Numerical Methods [11] for more information.

6.3 ODE Solvers

In practice we obviously do not want to reimplement these manual calculations every time. It's time to introduce some tools. Built on the computational functionality and ndarray data structures already introduced, the package owl-ode performs the tasks of solving the initial value problems [19].

6.3.1 Solving Linear Oscillator System

We will use a classical linear oscillator system example to demonstrate how to use the owl-ode package to solve ODE problems. More practical and interesting examples will be shown in Sec. 6.4.

Consider the following time independent linear dynamic system of two states:

$$\frac{dy}{dt} = Ay, \text{ where } A = \begin{bmatrix} 1 & -1 \\ 2 & -3 \end{bmatrix}.$$

This equation represents an oscillator system where y is the state of the system, t is time, and the initial state at $t = 0$ is $y_0 = [-1, 1]^T$. We wish to know the system state at $t = 2$. The *function* can be expressed in Owl using the matrix module.

```
let f y t =
  let a = [|[|1.; -1.|];[|2.; -3.|]|] |> Mat.of_arrays in
  Mat.(a *@ y);;
```

Next, we want to specify the *timespan* of this problem: from 0 to 2 using a step of 0.001.

```
let tspec = Owl_ode.Types.
  (T1 {t0 = 0.; duration = 2.; dt=1E-3});;
```

The last requirement for solving the problem is to have the initial values:

```
let x0 = Mat.of_array [|-1.; 1.|] 2 1;;
```

Finally we input these parameters to the rk4 solver in Owl_ode to obtain. The result shows both the steps *ts* and the system values at each step *ys*.

```
# let ts, ys = Owl_ode.Ode.odeint
  Owl_ode.Native.D.rk4 f x0 tspec ();;
val ts : Owl_dense_matrix_d.mat =
     C0   C1    C2    C3    C4      C1996 C1997 C1998 C1999 C2000
R0   0 0.001 0.002 0.003 0.004 ... 1.996 1.997 1.998 1.999     2
val ys : Owl_dense_matrix_d.mat =
     C0       C1        C2          C1999    C2000
R0  -1    -1.002 -1.00399 ... -3.56749 -3.56898
R1   1 0.995005 0.990022 ... -2.07707 -2.07798
```

6.3.2 Solver Structure

From this example, we can see that owl-ode abstracts the initial value problems as four different parts: (1) a function f to show how the system evolves in equation $y'(t) = f(y, t)$; (2) a specification of the timespan; (3) a set of system's initial values;

Table 6.2: Solvers provided by owl-ode and their types.

Classes Solvers	Type	Function	Output
Euler/Midpoint	Native	`mat -> float -> mat`	`mat * mat`
`rk4/rk23/rk45`	Native	`mat -> float -> mat`	`mat * mat`
`Cvode/Cvode_stiff`	Sundials	`arr -> float -> arr`	`arr * arr`
LSODA	ODEPACK	`mat -> float -> mat`	`mat * mat`

and most importantly, (4) a solver. Indeed, the signature of a solver clearly indicates these four parts:

```
val rk4 : (module Types.Solver
    with type state = M.arr
    and type f = M.arr -> float -> M.arr
    and type step_output = M.arr * float
    and type solve_output = M.arr * M.arr)
```

Building on this uniform abstraction, a user can choose a suitable solver and use it to solve many complex and practical ODE problems. Note that due to the difference of solvers, the requirements of different solvers varies, e.g., some require the state to be two matrices, while others process data using a more general ndarray format.

Owl-ode provides a wide range of solvers. It implements native solvers based on the basic step-by-step update idea discussed above. There are also many mature off-the-shelf tools for solving ODEs and we interface to two of them: sundials[1] and ODEPACK[2]. Both are well implemented and widely used in practical use. For example, the SciPy provides a Python wrap of the sundials, and NASA also uses its CVODE/CVODES solvers for spacecraft trajectory simulations[3].

sundials is a SUite of Nonlinear and DIfferential/ALgebraic equation Solvers. It contains six solvers, and we interface to its CVODE solver for solving initial value problems for ordinary differential equation systems. ODEPACK is a collection of FORTRAN solvers for the initial value problem for ordinary differential equation systems. We interface to its LSODA solver which is for solving the explicit form ODE.

For all these solvers, owl-ode provides an easy-to-use unified interface. Tbl. 6.2 is a table that lists all the solvers that are currently supported by owl-ode.

All solvers share similar signatures. The system evolution function takes state y and time t as input, and outputs a tuple that contains two matrices or ndarrays representing the time increment and the state of y at corresponding time step.

---

[1] SUNDIALS: SUite of Nonlinear and DIfferential/ALgebraic Equation Solvers, LLNL. URL:https://computing.llnl.gov/projects/sundials

[2] ODEPACK: Fortran ODE Solvers, LLNL. URL:https://computing.llnl.gov/casc/odepack/

[3] Uses of SUNDIALS, LLNL. URL:https://computing.llnl.gov/projects/sundials/uses-sundials

In most of the cases, ode45 is a robust choice and the first solver to try. It is robust and has medium accuracy. If the problem has a higher degree of tolerance to inaccuracy, you can also try the other native solvers which are more efficient. Alternatively, if the requirement for accuracy or stability is even higher, the Sundials/ODEPACK solvers can be used, especially when the ODEs to be solved are "stiff". We will talk about the *stiffness* of ODEs in a later section.

6.3.3 Symplectic Solver

In addition to the above, owl-ode also implements *symplectic solvers* which are a bit different to the naive solvers we have seen so far. Symplectic solvers are used to solve Hamiltonian dynamics [12] with much higher accuracy. We will illustrate this with a example.

Consider a two-body orbital system where the trajectory of the orbiting object is a circle. Such a circle can be generated by solving the linear system:

$$y_0' = y_1; y_1' = -y_0 \qquad (6.9)$$

This can also be expressed in matrix form as:

$$y' = Ay, \text{ where } A = \begin{bmatrix} 0 & 1 \\ -1 & 0 \end{bmatrix}. \qquad (6.10)$$

We can then apply the previous methods such as Euler to solve it:

```
let f y t =
  let a - [|[|0.; 1.|];[|-1.; 0.|]|] |> Mat.of_arrays in
  Mat.(a *@ y);;
let x0 = Mat.of_array [|1.; 0.|] 2 1;;
let tspec = Owl_ode.Types.(
  T1 {t0 = 0.; duration = 10.; dt=1E-2});;
let ts, ys = Owl_ode.Ode.odeint
  Owl_ode.Native.D.euler f x0 tspec ();;
```

To see the resulting trajectory, we plot the two components of ys in Fig. 6.2. An exact solution would generate a perfect circle, but here we obtain a spiral (we have used a large step size to better show the effect, but it still exists even with a small step size). Aside from using a better solver with greater precision, are there any other ways we could improve this solution?

Look again at Eq. 6.9. It exhibits an interesting pattern: it first uses the value of y_1 to update y_0', and then uses y_0 to update the value of y_1'. It thus belongs to a class of equations called *Hamiltonian systems*, often used in the classical mechanics. Such systems are represented by a set of coordinates (p, q) where the p_i's are the generalised coordinates, and the q_i's the corresponding conjugate momenta. The evolution of a Hamiltonian system is defined by the equations below:

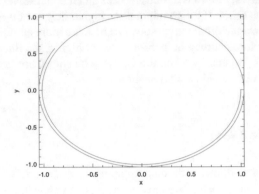

Fig. 6.2: Visualise the solution of a simple linear system

$$p' = \frac{\partial H}{\partial q}, q' = \frac{\partial H}{\partial p}.$$

Here $H(p, q)$ is a function of these two components. In the example above, p and q are just the two coordinates, and $H(x, y) = (x^2 + y^2)/2$. To solve this system, symplectic solvers are widely used in different fields, since they conserve the value of $H(p, q)$ to within the order of accuracy of the specific solver used. Most of the numerical solvers we mentioned above, including the Euler and classical RK solvers, are not symplectic solvers.

Owl_ode has implemented several symplectic solvers based on different integration algorithms: Sym_Euler, Leapfrog, PseudoLeapFrog, Ruth3, and Ruth4. Like native solvers, they have different orders of error, and they are implemented using the same basic symplectic integration algorithm, but with different parameters. If you are interested in more detail, please check the code and the paper of Candy et al. [13] on which the implementation is based. Later in this chapter we will show an example of using the symplectic solver to solve a damped harmony oscillation problem.

6.3.4 Features and Limits

One feature of owl-ode is the automatic inference of state dimensionality from initial state. For example, the native solvers take a input matrix as the state. Suppose the initial state of the system is a row vector of dimension $1 \times N$. After T time steps, the states are stacked vertically, and thus have dimensions $T \times N$. If the initial state is a column vector of shape $N \times 1$, then the stacked state after T time steps will be inferred as $N \times T$.

The temporal integration of matrices, i.e. cases where the initial state is matrix instead of vector, is also supported. If the initial state is of shape $N \times M$, then the accumulated state stacks the flattened state vertically by time steps, which is of shape $T \times (NM)$. The owl-ode provides a helper function `Native.S.to_state_array` to unpack the output state into an array of matrices.

Another feature of owl-ode is that the users can easily define a new solver by creating a module of type `Solver`. For example, to create a custom CVODE solver that has a relative tolerance of 1E-7 as opposed to the default 1E-4, we can define and use the custom_cvode as follows:

```
let custom_cvode =
  Owl_ode_sundials.cvode ~stiff:false
  ~relative_tol:1E-7 ~abs_tol:1E-4;;
let ts, xs = Owl_ode.Ode.odeint custom_cvode f x0 tspec ();;
```

Here, we use the cvode function to construct a solver `Custom_Owl_Cvode`. Similar helper functions like cvode have been also defined for native and symplectic solvers.

The owl-ode aims to run across different backends such as JavaScript. One way to do this is to use tools like js_of_ocaml to convert OCaml bytecode into JavaScript. However, this approach only supports pure OCaml implementation. The sundial and ODEPACK solvers are therefore excluded. Part of the library, the owl-ode-base contains implementations of solvers that are purely written in OCaml. These parts are therefore suitable to be used for in JavaScript and executed on web browsers.

6.4 Examples of ODE Solvers

As with many good things, mastering ODE solution requires a lot of practice. After getting to know owl-ode, in this section we will demonstrate more examples of using this tool.

6.4.1 Explicit ODE

For a starter, now we can use owl-ode to solve the motivating problem in Eq. 6.5 with some straightforward code.

```
let f y t = Mat.((2. $* y *$ t) +$ t);;
let tspec = Owl_ode.Types.(
  T1 {t0 = 0.; duration = 1.; dt=1E-3});;
let y0 = Mat.zeros 1 1;;
let solver = Owl_ode.Native.D.rk45 ~tol:1E-9 ~dtmax:10.0;;
let _, ys = Owl_ode.Ode.odeint solver f y0 tspec ();;
```

The code is quite similar to the previous example, except that we can now try another provided solver, the rk45 solver, with certain parameters specified. Don't worry about what the tol or dtmax means for now. The result is shown below. You can verify the result with Eq. 6.6, by setting x to 1 in this equation, and the numerical value of y will be close to 0.859079.

```
# Mat.transpose ys;;
- : Mat.mat =
    C0          C1          C2              C126    C127    C128
R0  0 7.62941E-06 9.91919E-05  ...  0.817586 0.83809 0.859079
```

6.4.2 Two-body Problem

In classical mechanics, the *two-body problem* is to predict the motion of two massive objects. It is assumed that the only force that is considered comes from each other, and both objects are not affected by any other object. This problem can be seen in the astrodynamics where the objects of interests are planets, satellites, etc. under the influence of only gravitation. Another case is the trajectory of an electron around an atom's atomic nucleus.

This classic problem is one of the earliest investigated mechanics problems, and was long solved from the age of Newton. It is also a typical integrable problem in classical mechanics. In this example, let's consider a simplified version of this problem. We assume that the two objects interact on a 2-dimensional plane, and one of them is so much more massive than the other one that it can be thought of as being static (think about electron and nucleus), sitting at the origin of a Cartesian coordinate system in the plane. In this system, let's consider the trajectory of the lighter object. This "one-body" problem is the basis of the two body problem. For many forces, including gravitational ones, a two-body problem can be divided into a pair of one-body problems.

Given the previous assumption and Newton's equations, it can be proved[4] that the location of the lighter object $[y_0, y_1]$ with regard to time t can be described by:

$$y_0''(t) = -\frac{y_0}{r^3}, y_1''(t) = -\frac{y_1}{r^3} \tag{6.11}$$

where $r = \sqrt{y_0^2 + y_1^2}$. This is a second-order ODE system, and so to make it solvable using our tool we must transform it into a first-order explicit ordinary differential equation system:

---

[4] Planar Two Body Problem Simulation, John Burkardt, URL: https://people.sc.fsu.edu/~jburkardt/octave_src/two_body_ode/two_body_ode.html.

$$y_0' = y_2$$
$$y_1' = y_3$$
$$y_2' = -\frac{y_0}{r^3}$$
$$y_3' = -\frac{y_1}{r^3}$$

$$(6.12)$$

Based on Eq. 6.12, we can build up our code as below:

```
let f y _t =
  let y = Mat.to_array y in
  let r = Maths.(sqrt ((sqr y.(0)) +. (sqr y.(1)))) in
  let y0' = y.(2) in
  let y1' = y.(3) in
  let y2' = -.y.(0) /. (Maths.pow r 3.) in
  let y3' = -.y.(1) /. (Maths.pow r 3.) in
  [| [|y0'; y1'; y2'; y3'|] |] |> Mat.of_arrays;;

let y0 = Mat.of_array [|-1.; 0.; 0.5; 0.5|] 1 4;;

let tspec = Owl_ode.Types.(T1 {
  t0 = 0.;
  duration = 20.;
  dt=1E-2});;

let custom_solver = Native.D.rk45 ~tol:1E-9 ~dtmax:10.0;;
```

Here y0 provides the initial status of the system: the first two numbers denote the initial location of the object, and the next two indicate the initial momentum of the object. After building the function, initial status, timespan, and solver, we can then solve the system and visualise it with the code below. The result is shown in Fig. 6.3.

```
let plot () =
  let ts, ys = Ode.odeint custom_solver f y0 tspec () in
  let h = Plot.create "two_body.png" in
  let open Plot in
  plot ~h ~spec:[ RGB (66, 133, 244); LineStyle 1 ]
    (Mat.col ys 0) (Mat.col ys 1);
  scatter ~h ~spec:[ Marker "#[0x229a]"; MarkerSize 5. ]
    (Mat.zeros 1 1) (Mat.zeros 1 1);
  text ~h ~spec:[ RGB (51,51,51)] (-.0.3) 0. "Massive Object";
  output h;;
```

An example of this simplified two-body problem is the "planet-sun" system where a planet orbits the sun. Kepler's law states that in this system the planet goes around

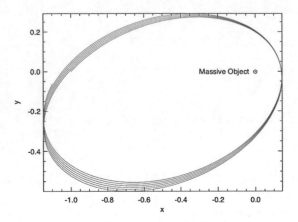

Fig. 6.3: The trajectory of lighter object orbiting the massive object in a simplified two-body problem.

the sun in an ellipse shape, with the sun at a focus of the ellipse. The orbiting trajectory in the result visually follows this theory.

6.4.3 Lorenz Attractor

Lorenz equations are one of the most thoroughly studied ODEs. This system of ODEs was proposed by Edward Lorenz in 1963 to model the flow of fluid (the air in particular) from hot area to cold area [14]. Lorenz simplified the many atmospheric factors into the simple equations below.

$$
\begin{aligned}
x'(t) &= \sigma \left(y(t) - x(t)\right) \\
y'(t) &= x(t)(\rho - z(t)) - y(t) \\
z'(t) &= x(t)y(t) - \beta \, z(t)
\end{aligned}
$$

$$(6.13)$$

Here x is proportional to the rate of convection in the atmospheric flow; y and z are proportional to the horizontal and vertical temperature variation. Parameter σ is the Prandtl number [15], and ρ is the normalised Rayleigh number. β is related to the geometry of the domain. One commonly used parameters setting is $\sigma = 10$, $\rho = 20$, and $\beta = \frac{8}{3}$.

Based on this information, we can use `owl-ode` to express the Lorenz equations with code.

```
let sigma = 10.;;
let beta = 8. /. 3.;;
let rho = 28.;;
let f y _t =
  let y = Mat.to_array y in
  let y0' = sigma *. (y.(1) -. y.(0)) in
  let y1' = y.(0) *. (rho -. y.(2)) -. y.(1) in
  let y2' = y.(0) *. y.(1) -. beta *. y.(2) in
  [| [|y0'; y1'; y2'|] |] |> Mat.of_arrays;;
```

Now, we can solve the ODE system and visualise the results. Certain details such as colour are omitted in this code snippet. In Fig. 6.4, we first show how the value of x, y and z changes with time; next we show the phase plane plots between each two of them.

```
let _ =
  let ts, ys = Ode.odeint custom_solver f y0 tspec () in
  let h = Plot.create ~m:2 ~n:2 "lorenz_01.png" in
  let open Plot in
  subplot h 0 0;
  plot ~h ts (Mat.col ys 2);
  plot ~h ts (Mat.col ys 1);
  plot ~h ts (Mat.col ys 0);
  subplot h 0 1;
  plot ~h (Mat.col ys 0) (Mat.col ys 1);
  subplot h 1 0;
  plot ~h (Mat.col ys 1) (Mat.col ys 2);
  subplot h 1 1;
  plot ~h (Mat.col ys 0) (Mat.col ys 2);
  output h;;
```

From Fig. 6.4, we can imagine that the state of the system keeps moving towards the two "voids" in a three dimensional space, jumping from one to the other. These two voids are a certain type of *attractor* in this dynamic system, towards which the system tends to evolve.

When considering the Lorenz equation, an interesting question is "what would happen if I change the initial value slightly?" For some systems, such as a pendulum, that wouldn't make much a difference, but not here. We can see that clearly with a simple experiment. Keeping function and timespan the same, let's change the initial value by just 0.1% and then solve the system again.

```
let y00 = Mat.of_array [|-1.; -1.; 1.|] 1 3;;
let y01 = Mat.of_array [|-1.001; -1.001; 1.001|] 1 3;;
let ts0, ys0 = Ode.odeint custom_solver f y00 tspec ();;
let ts1, ys1 = Ode.odeint custom_solver f y01 tspec ();;
```

To make later calculation easier, we can set the two resulting matrices to be of the same shape using slicing.

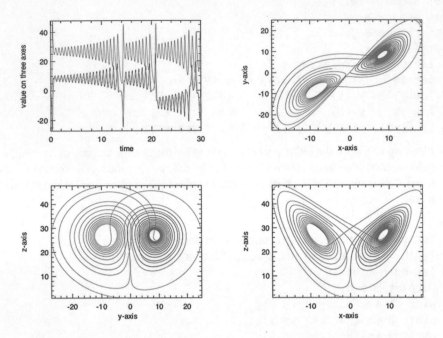

Fig. 6.4: Three components and phase plane plots of Lorenz attractor

```
let r0, c0 = Mat.shape ys0;;
let r1, c1 = Mat.shape ys1;;
let r  = if (r0 < r1) then r0 else r1;;
let ts = if (r0 < r1) then ts0 else ts1;;
let ys0 = Mat.get_slice [[0; r-1]; []] ys0;;
let ys1 = Mat.get_slice [[0; r-1]; []] ys1;;
```

Now, we can compare the Euclidean distance between the status of these two systems at certain time. Also, we show the value change of the three components after changing initial values along the time axis.

```
let _ =
  let h = Plot.create ~m:1 ~n:2 "lorenz_02.png" in
  let open Plot in
  subplot h 0 0;
  plot ~h ts (Mat.col ys1 0);
  plot ~h ts (Mat.col ys1 1);
  plot ~h ts (Mat.col ys1 2);
  subplot h 0 1;
  let diff = Mat.(
    sqr ((col ys0 0) - (col ys1 0))
```

```
    + sqr ((col ys0 1) - (col ys1 1))
    + sqr ((col ys0 2) - (col ys1 2))
    |> sqrt )
in
plot ~h ts diff;
output h;;
```

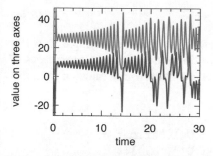 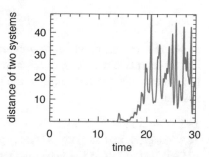

Fig. 6.5: Change the initial states on three dimension by only 0.1%, and the value of Lorenz system changes visibly.

According to Fig. 6.5, the first figure shows that, initially the systems look quite like that in Fig. 6.4, but after about 15 seconds the system state begins to change. This change is then quantified using the Euclidean distance between these two systems. Over time, the difference has dramatically increased, with no sign of convergence. Even if you extend the timespan, this conclusion still holds. Indeed, in the Lorenz system, even a tiny bit of change of the initial state can lead to a large and potentially chaotic change of future state after a while. This partly explains why weather prediction is difficult to do: you can only accurately predict the weather for a certain period of time; any farther into the future and the weather will be extremely sensitive to a tiny bit of perturbations at the beginning. As the pop culture reference has it, perhaps even the flapping of the wings of a distant butterfly several weeks earlier could be enough.

6.4.4 Damped Oscillation

Damped oscillation ODE systems appear frequently in physics and many other fields, including charge flow in electric circuits [16], sound waves [17], light waves [18], etc. One simple example is a mass attached to a vertical spring. The spring initially stretches to balance gravity; once the system reaches equilibrium, we can then study the upward displacement of the mass from its original position, x. Once the mass is in motion, at any position x, the force on the mass is proportional to displacement x, $F = -kx$ where k is a positive parameter.

This system is called a *simple harmonic oscillator*, representing an ideal case where F is the only force acting on the mass. However, in a real physical system there will also be friction, giving rise to *damped oscillation*. The frictional force is proportional to the velocity of the mass and so the total force can be expressed as $F = -kx - cv$ where c is the damping coefficient factor. As both the force and velocity can be expressed as derivatives of x over time we have, for m the mass of the object:

$$m\frac{dx}{dt^2} = -kx - c\frac{dx}{dt} \tag{6.14}$$

Recall from the previous section that the state of the system in a symplectic solver is a tuple of two matrices, representing the position and momentum coordinates of the system. We can thus express the oscillation equation with a function from this system.

```
let a = 1.0;;
let damped_noforcing a (xs, ps) _t : Owl.Mat.mat =
  Owl.Mat.((xs *$ -1.0) + (ps *$ (-1.0 *. a)));;
```

The system evolution function `damped_noforcing` takes the two matrices, position and momentum, and the float number time t as input states.[5] For simplicity, we assume the coefficients are all the same and can be set to one. Let's then solve with the symplectic solver and compare the `LeapFrog`, `ruth3`, and `Symplectic_Euler`:

```
let main () =
  let x0 = Owl.Mat.of_array [| -0.25 |] 1 1 in
  let p0 = Owl.Mat.of_array [| 0.75 |] 1 1 in
  let t0, duration = 0.0, 15.0 in
  let f = damped_noforcing a in
  let tspec = T1 { t0; duration; dt=0.1 } in
  let t, sol1, _ = Ode.odeint (module Symplectic.D.Leapfrog)
    f (x0, p0) tspec () in
  let _, sol2, _ = Ode.odeint Symplectic.D.ruth3
    f (x0, p0) tspec () in
  let _, sol3, _ =
    Ode.odeint (module Symplectic.D.Symplectic_Euler)
    f (x0, p0) tspec ()
  in
  t, sol1, sol2, sol3;;
```

Hopefully this code is not too surprising now: it first defines the initial values, the time duration, time step, and then provides this information to the solvers. Unlike native solvers, the symplectic solvers return three values instead of two. The first is the time sequence, and the next two sequences indicate how the x and p values evolve at each time point.

---

[5] Here denoted _t as it is unused in this particular evolution function.

```
let plot_sol fname t sol1 sol2 sol3 =
  let h = Plot.create fname in
  let open Plot in
  plot ~h t (Mat.col sol1 0);
  plot ~h t (Mat.col sol2 0);
  plot ~h t (Mat.col sol3 0);
  legend_on h [| "Leapfrog"; "Ruth3"; "Symplectic Euler" |];
  output h;;
```

Here we only plot the change of position *x* over time in the oscillation, and the solution provided by the three different solvers, as shown in the plotting code above. You can also try to visualise the change to the momentum *p* in a similar way. The result is shown in Fig. 6.6. You can clearly see that the displacement decreases towards equilibrium position in this damped oscillation as the energy of the system dissipates. The curves provided by three solvers are a bit different, especially at the peak of the curve, but are reasonably close most of the time.

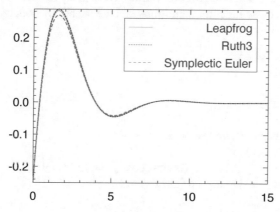

Fig. 6.6: Step response of a damped harmonic oscillator

6.5 Stiffness

Stiffness is an important concept in the numerical solution of ODE. Consider a function that has a "cliff", a point where its nearby values change rapidly. To find the solution with the methods used above may require such an extremely small stepping size that covering the whole timespan takes a very long time and a great deal of computation. Therefore, the solver of stiff equations needs to use different methods such as implicit differencing or automatic step size adjustment to be both accurate and robust [20].

The *Van der Pol equation* is a good example to show both non-stiff and stiff cases. In dynamics, the Van der Pol oscillator is a non-conservative oscillator with non-linear damping [21]. Its behaviour with time can be described with a high order ODE:

$$y'' - \mu(1 - y^2)y' + y = 0 \qquad (6.15)$$

where μ is a scalar parameter controlling the non-linearity and the strength of the damping. To make it solvable using our tool, we rewrite it as a pair of explicit first-order ODEs in a linear system:

$$y'_0 = y_1, y'_1 = \mu(1 - y_0^2)y_1 - y_0 \qquad (6.16)$$

As we will show shortly, by varying the damping parameter this group of equations can be made either non-stiff or stiff.

We provide both stiff (Owl_Cvode_Stiff) and non-stiff (Owl_Cvode) solvers by interfacing to Sundials, and the LSODA solver of ODEPACK can automatically switch between stiff and non-stiff algorithms. We will try both in the example, after starting with the basic function definitions that are shared by both cases:

```
open Owl_ode
open Owl_ode.Types
open Owl_plplot

let van_der_pol mu =
  fun y _t ->
    let y = Mat.to_array y in
    let a = y.(1) in
    let b = mu *. (1. -. Maths.sqr y.(0)) *. y.(1) -. y.(0) in
    [|[|a; b|]|] |> Mat.of_arrays;;
```

6.5.1 Solving Non-Stiff ODEs

When $\mu = 1$, the equation is non-stiff so we can use the Cvode solver from sundials:

```
let f_non_stiff = van_der_pol 1.;;
let y0 = Mat.of_array [| 0.02; 0.03 |] 1 2;;
let tspec = T1 { t0 = 0.0; dt = 0.01; duration = 30.0 };;
let ts, ys = Ode.odeint (module Owl_ode_sundials.Owl_Cvode)
  f_non_stiff y0 tspec ();;
```

To see the "non-stiffness" clearly, we can plot how the two system states change over time, and a phase plane plot of their trajectory on the plane, using the two states as x- and y-axis values. The result is shown in Fig. 6.7.

```
let () =
  let fname = "vdp_sundials_nonstiff.png" in
  let h = Plot.create ~n:2 ~m:1 fname in
  let open Plot in
  subplot h 0 0;
  plot ~h ~spec:[ LineStyle 1 ] (Mat.col ys 0) (Mat.col ys 1);
  subplot h 0 1;
  plot ~h ~spec:[ LineStyle 1 ] ts Mat.(col ys 1);
  plot ~h ~spec:[ LineStyle 3 ] ts Mat.(col ys 0);
  output h;;
```

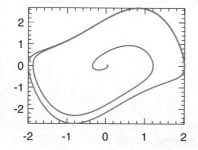 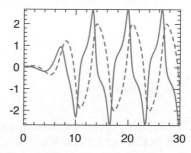

Fig. 6.7: Solving non-stiff Van der Pol equations with Sundial Cvode solver.

6.5.2 Solve Stiff ODEs

However, changing the parameter to $\mu = 1000$ makes this function *stiff*. We follow the same procedure as before, but now we use the Lsoda solver from ODEPACK and extend the timespan to 3000. From Fig. 6.8 we can see clearly the sharp cliffs that indicate the system is "stiff".

```
let f_stiff = van_der_pol 1000.;;
let y0 = Mat.of_array [| 2.; 0. |] 1 2;;
let tspec = T1 { t0 = 0.0; dt = 0.01; duration = 3000.0 };;
let () =
  let ts, ys = Ode.odeint (module Owl_ode_odepack.Lsoda)
    f_stiff y0 tspec () in
  let fname = "vdp_odepack_stiff.png" in
  let h = Plot.create fname in
  let open Plot in
  set_yrange h (-2.) 2.;
  plot ~h ~spec:[ LineStyle 1 ] ts Mat.(col ys 1);
  plot ~h ~spec:[ LineStyle 3 ] ts Mat.(col ys 0);
  output h;;
```

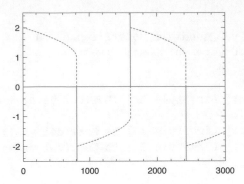

Fig. 6.8: Solving Stiff Van der Pol equations with ODEPACK LSODA solver.

6.6 Summary

This chapter discussed numerical solutions to the initial value problem for ordinary differential equations (ODEs). We presented the general definition of the ODE and the initial value problem. We explained how the numerical solutions are developed. Then we present `owl-ode`, an OCaml library that solves ODEs based on Owl. Besides implementing native solvers, it also interfaces to existing off-the-shelf open source ODE solvers from the Sundials and ODEPACK library. We demonstrated how these solvers are used to solve ODEs from several real examples, including the two-body problem, the Lorenz Attractor, and Damped Oscillation. Finally, we briefly discussed an important concept: stiffness, when solving ODE numerically, and showed how we can solve stiff and non-stiff ODEs with the example of the van der Pol equation.

References

1. Dirac PA. On the theory of quantum mechanics. Proceedings of the Royal Society of London. Series A, Containing Papers of a Mathematical and Physical Character. 1926 Oct 1;112(762):661-77.
2. Chandrasekhar S. Stochastic problems in physics and astronomy. Reviews of modern physics. 1943 Jan 1;15(1):1.
3. Mahaffy JM, Joiner KM, Zak PJ. A geometric analysis of stability regions for a linear differential equation with two delays. International Journal of Bifurcation and Chaos. 1995 Jun;5(03):779-96.
4. Agarwal P, Deniz S, Jain S, Alderremy AA, Aly S. A new analysis of a partial differential equation arising in biology and population genetics via semi analytical techniques. Physica A: Statistical Mechanics and its Applications. 2020 Mar 15;542:122769.

5. Zhang B, Zhang T, Cao Z, Zhang K. Fast new small-target detection algorithm based on a modified partial differential equation in infrared clutter. Optical engineering. 2007 Oct;46(10):106401.
6. Boucekkine R, Fabbri G, Pintus P. On the optimal control of a linear neutral differential equation arising in economics. Optimal Control Applications and Methods. 2012 Sep;33(5):511-30.
7. Hairer E, Nørsett SP, Wanner G. Solving Ordinary Differential Equations I: Nonstiff Problems [Internet]. Springer Berlin Heidelberg; 2008. Springer Series in Computational Mathematics. Available from: https://books.google.com.sg/books?id=F93u7VcSRyYC.
8. Carpenter MH, Kennedy CA, Bijl H, Viken SA, Vatsa VN. Fourth-order Runge-Kutta schemes for fluid mechanics applications. Journal of Scientific Computing. 2005 Nov; 25(1):157-94.
9. Hult J. A fourth-order Runge-Kutta in the interaction picture method for simulating supercontinuum generation in optical fibers. Journal of Lightwave Technology. 2007 Dec 1;25(12):3770-5.
10. Butcher JC. Numerical methods for ordinary differential equations. John Wiley & Sons; 2016 Aug 29.
11. Press WH, William H, Teukolsky SA, Saul A, Vetterling WT, Flannery BP. Numerical recipes 3rd edition: The art of scientific computing. Cambridge University Press; 2007 Sep 6.
12. Yoshida H. Construction of higher order symplectic integrators. Physics letters A. 1990 Nov 12;150(5-7):262-8.
13. Candy J, Rozmus W. A symplectic integration algorithm for separable Hamiltonian functions. Journal of Computational Physics. 1991 Jan 1; 92(1):230-56.
14. Lorenz System [Internet]. Wikipedia. Wikimedia Foundation; 2021 [cited 2021 Dec 23]. Available from: https://en.wikipedia.org/wiki/Lorenz_system.
15. White FM, Corfield I. Viscous fluid flow. New York: McGraw-Hill; 2006.
16. Minorsky N. Self-excited oscillations in dynamical systems possessing retarded actions.
17. Sprague MW. The single sonic muscle twitch model for the sound-production mechanism in the weakfish, Cynoscion regalis. The Journal of the Acoustical Society of America. 2000 Nov; 108(5):2430-7.
18. Haken H. Cooperative phenomena in systems far from thermal equilibrium and in nonphysical systems. Reviews of modern physics. 1975 Jan 1; 47(1):67.
19. Seri M, Kao TC. OwlDE: making ODEs first-class Owl citizens. Journal of Open Source Software. 2019 Dec 20;4(44):1812.
20. Petzold L. Automatic selection of methods for solving stiff and nonstiff systems of ordinary differential equations. SIAM journal on scientific and statistical computing. 1983 Mar;4(1):136-48.
21. Van der Pol oscillator [Internet]. Wikipedia. Wikimedia Foundation; 2021 [cited 2021Dec23]. Available from: https://en.wikipedia.org/wiki/Van_der_Pol_oscillator.

Chapter 7
Signal Processing

Abstract We rely on signals such as sound and images to convey information. Signal processing is the field concerned with analysing, generating, and transforming signals. Its applications can be found in a wide range of fields: audio processing, speech recognition, image processing, communication system, data science, etc. In this chapter we focus on Fourier Transform, the fundamental tool in signal processing and modern numerical computing. We introduce its basic idea, and then demonstrate how to perform FFT in OCaml with examples and applications. We also cover the relationship between FFT, convolution, and filters.

7.1 Discrete Fourier Transform

We begin with an example. Consider a piece of audio lasting 10 seconds. It can clearly be described in the *time domain* by plotting its sound intensity against time. On the other hand, perhaps less obviously, it can also be described in the *frequency domain*. For example, if the audio is a single A# being played for 10 seconds, it can be described with just one frequency number: 466.16 Hz. Of course, real-world sounds such as water flowing or wind whispering are not usually so pure, typically being compounded from different frequencies. How should we describe the frequencies used then?

That's where the *Fourier Transform* comes in, starting from a fundamental insight: all signals, whether sound waves or electromagnetic fields, can be described as a waveform, and *all waveforms are actually the sum of simple sinusoids of different frequencies* [23]. The Fourier Transform is the tool to break down a signal wave into different sinusoids.

More formally, we can represent a signal with the values of some quantity h as a function of time: $h(t)$, or its amplitude H as a function of frequency: $H(f)$. They are two representations of the same thing, and the Fourier Transform converts between them as shown in Fig. 7.1. These two equations show the Fourier transform, and the *inverse Fourier transform* [2], which change the signals on frequency domain back to time domain.

$$h(f) = \int H(f)e^{-2\pi i f t} df$$

$$H(f) = \int h(t)e^{2\pi i f t} dt \tag{7.1}$$

In computer-based numerical computation, signals are often represented discretely as a finite sequence of sampled data rather than as continuous values, giving rise to the *Discrete Fourier Transform* (DFT) [3]. Suppose we have represented a signal by a complex vector y that contains n elements. To get the Fourier Transform vector Y, the discrete form of Eq. 7.1 can be expressed as Eq. 7.2, where $\omega = e^{-2\pi i/n}$ and $i = \sqrt{-1}$. j and k are indices that range from 0 to $n - 1$. [1]

$$Y_k = \sum_{j=0}^{n-1} y_j \omega^{jk}$$

$$y_k = \frac{1}{n} \sum_{j=0}^{n-1} Y_k \omega^{-jk}$$

$$\tag{7.2}$$

What can we do if we know how a sound is composed? Think of a classic example where we need to remove some high pitch noise from some music. By using DFT to transform to the frequency domain we can easily find out the frequency of such noise, remove it, and convert the signal back to the time domain by using the reverse Fourier transform. Similarly, we can also remove noise from an image by applying the Fourier Transform and removing the noise frequency components, before reconstructing the image without noise by inverting the transform. We show several such examples in Sec. 7.3.

There are many many more applications of DFT than just de-noising though [4]. We all use Fourier Transform every day without knowing it: mobile phones [5], image and audio compression [6], communication networks [7], large scale numerical physics and engineering [8], etc. In many ways it is the cornerstone of computational mathematics. An important reason for its popularity is that it can be implemented very efficiently as the Fast Fourier Transform.

---

[1] The video introduction "But what is the Fourier Transform? A visual introduction" by 3Blue1Brown is a nice visual introduction to this topic.

7.2 Fast Fourier Transform

A problem with the DFT is that if you follow its definition in implementation, the computational complexity of your algorithm is $O(n^2)$: computing each of the n components of Y in Eq. 7.2 requires n multiplications and n additions. Therefore naive DFT implementation does not scale well with input size. The Fast Fourier Transform (FFT) algorithm, first formulated by Gauss in 1805 and then published by James Cooley and John Tukey in 1965 reduces its complexity to $O(n \log n)$ [9]. In essence, the FFT algorithm observes that any DFT can be represented by the sum of two sub-DFTs: one consisting of the elements of the signal in the even index positions, and the other consisting of the elements of the signal in odd positions:

$$
\begin{aligned}
Y_k &= \sum_{even\, j} \omega^{jk} y_j + \sum_{odd\, j} \omega^{jk} y_j \\
&= \sum_{j=0}^{n/2-1} \omega^{2jk} y_{2j} + \omega^k \sum_{j=0}^{n/2-1} \omega^{2jk} y_{2j+1}.
\end{aligned}
\tag{7.3}
$$

The key to enabling this step is the fact that $\omega_{2n}^2 = \omega_n$. According to Eq. 7.3, one DFT can be reduced into two DFTs each only half the original length, and then the second half is multiplied by a factor ω^k and added to the first half. The half signal can be further halved, and so on and so on. Therefore the computation can thus be recursively reduced to a logarithmic level. At the root of this recursion is the fact that a DFT on a single number returns just the number itself. As you can see, this is a typical divide-and-conquer algorithm design.

Detailed mathematical analysis of the Fourier Transform is beyond the scope of this book, and we encourage readers to consult one of the many classic textbooks on this topic such as Phillips et al [10]. We focus here on introducing how to use the FFT implementation and its applications in OCaml code. Hopefully these materials are interesting enough to encourage you to investigate more.

The implementation of the FFT module in the Owl library interfaces to the FFTPack [11] C language implementation. The core functions in the FFT module are the `fft` function and its reverse `rfft` corresponding to the two equations in Eq.7.3. This module provides these basic FFT functions as listed in Tbl. 7.1. The input to these functions are ndarrays. As with `Ndarray`, the FFT module also provides four number types: float (S), double precision(D), single precision complex(C), and double precision complex (Z). The `axis` specifies along which axis of the input ndarray a function is performed. It is set to the highest dimension if not specified. The parameter n specifies the size of output.

Table 7.1: Functions in the FFT module.

Functions	Description
fft ~axis x	one-dimensional discrete Fourier Transform
ifft ~axis x	one-dimensional inverse discrete Fourier Transform
rfft ~axis otyp x	one-dimensional discrete Fourier Transform for real input
irfft ~axis ~n otyp x	one-dimensional inverse discrete Fourier Transform for real input

7.2.1 Example: 1-D Discrete Fourier Transforms

We next show how to use these functions with some simple examples. More complex and interesting ones will follow in the next section. Let's start with the most basic fft and it inverse transform function ifft. First, we create a complex 1-D ndarray that contains 6 elements as input to fft.

```
# let a = [|1.;2.;1.;-1.;1.5;1.0|];;
# let b = Arr.of_array a [|6|]
  |> Dense.Ndarray.Generic.cast_d2z;;
val b : (Complex.t, complex64_elt) Dense.Ndarray.Generic.t =
          C0        C1        C2        C3        C4        C5
R (1, 0i) (2, 0i) (1, 0i) (-1, 0i) (1.5, 0i) (1, 0i)
```

We then apply the FFT:

```
# let c = Owl_fft.D.fft b;;
val c:(Complex.t, complex64_elt) Owl_dense_ndarray_generic.t =
R (5.5, 0i) (2.25, 0.433013i) (-2.75, 1.29904i)
  (1.5, 0i) (-2.75, -1.29904i) (2.25, -0.433013i)
```

The input and output of fft are both complex ndarrays. Note that the Fourier transform is symmetric about the y-axis as it is defined on the interval $(-\infty, \infty)$. The Fourier transform output therefore has positive (the first half) and negative frequencies (the second half). Typically, only the positive frequencies are used, so we can remove the second half output in further processing such as plotting etc.

```
# let d = Owl_fft.D.ifft c;;
val d:(Complex.t, complex64_elt) Owl_dense_ndarray_generic.t =
R (1, 1.38778E-17i) (2, 1.15186E-15i) (1, -8.65641E-17i)
  (-1, -1.52188E-15i) (1.5, 1.69831E-16i) (1, 2.72882E-16i)
```

The function ifft takes the frequency domain result c produced by fft and reconstructs the original time domain signal. Since we do not change the frequencies generated by fft, the inverse FFT should produce the same result as the original input, as in this example.

But simply manipulating arrays does not make very impressive examples. We next plot the FFT of the sum of two sine waves showing the power of FFT to separate signals of different frequencies. This example uses 600 sampling points and the sample spacing is $\frac{1}{800}$.

```
# module G = Dense.Ndarray.Generic;;
# let n = 600.;;
# let t = 1. /. 800.;;
# let x = Arr.linspace 0. (n *. t) (int_of_float n);;
val x : Arr.arr =
    C1 ... C599
R   0 0.00125209 0.00250417 0.00375626 0.00500835 ...
    0.744992 0.746244 0.747496 0.748748 0.75
```

First, we create a series of float numbers that cut the range [0, 0.75] with a step size of 0.125.

```
# let y1 = Arr.((50. *. 2. *. Owl_const.pi) $* x |> sin);;
# let y2 = Arr.(0.5 $* ((80. *. 2. *. Owl_const.pi)
    $* x |> sin));;
```

Next we create two sine signals of different frequencies: $y_1(x) = \sin(100\pi x)$, $y_2(x) = \frac{1}{2}\sin(160\pi x)$, shown in Fig. 7.1a. We then mix them together.

```
# let y = Arr.(y1 + y2) |> G.cast_d2z;;
```

The combined signal in Fig. 7.1b shows an irregular shape. Next, we apply FFT on this mixed signal:

```
# let yf = Owl_fft.D.fft y;;
val yf : (Complex.t, complex64_elt)
    Owl_dense_ndarray_generic.t =
    C1 ... C599
R (5.01874, 0i) (5.02225, 0.0182513i) (5.03281, 0.0366004i) ...
    (5.03281, -0.0366004i) (5.02225, -0.0182513i)
```

In the results, each tuple can be seen as a frequency vector in the complex space. We can plot the length of these vectors, represented by z in code below.

```
# let z = Dense.Ndarray.Z.(abs yf |> re);;
```

Finally, we plot the result using the code below. The plot is shown in Fig. 7.1c. Again, we use only half of the elements that have positive frequencies in array yf.

```
# let h = Plot.create "plot.png" in
  let xa = Arr.linspace 1. 600. 600 in
  Plot.plot ~h xa z;
  Plot.output h;;
```

The other two functions, rfft and irfft, are different to fft and ifft only in the data type: they take real numbers as input. Using them can sometimes make code cleaner. For example, in the previous example, we can simply apply rfft on a real ndarray a, and apply irfft on the result to get a real ndarray which is the same as a.

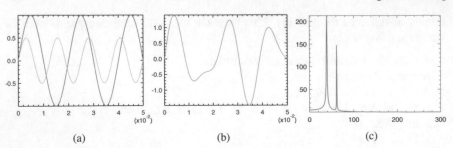

Fig. 7.1: (a) Two sine signals of different frequencies; (b) combined signal; (c)using FFT to separate two sine signals from their mixed signal.

7.3 Applications of FFTs

There are numerous applications of the FFT, but we pick three here to demonstrate the power of FFT using OCaml. The first is to find the period rules in the historical data of sunspots, and the second is about analysing the content of the dial number according to audio information. Both are inspired by those in Moler et. al. [12]. The third is about image processing. Together these applications demonstrate the wide usage of FFT in various scenarios.

7.3.1 Finding the Period of Sunspots

On the Sun's photosphere, *sunspots* appear as spots darker than the surrounding areas, and their number changes according to a certain cycle. The Sunspot Index and Long-term Solar Observations (SILS) is a world data center that preserves the sunspot data. The datasets are all available on the website of SILS. Each contains the time and a count of the sunspots measured by the "wolfer index". [13] In this example, we use the yearly data from 1700 to 2020 to determine the sunspot cycle. You might wish to try using the monthly data to the same end.

```
let data = Owl_io.read_csv ~sep:';' "sunspot_full.csv";;
let data = Array.map (fun x -> Array.map float_of_string x)
  data |> Mat.of_arrays;;
let x = Mat.get_slice [[];[0]] data;;
let y = Mat.get_slice [[];[1]] data;;
```

We can visualise the data x (date) and y (sunspot number) as in Fig. 7.2.
    The figure suggests a cycle of about 10 years, but how long is it exactly? Let's start by applying the FFT on this signal. This will show us how the observed signal is composed from signals of different cycles; then we will be able to tell which cycle is the most prominent of them. To process the data, we first remove the first element

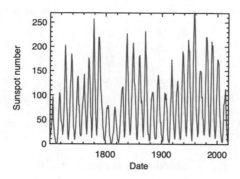

Fig. 7.2: Yearly sunspot data

of the frequency vector y' as it stores the sum of the data. The frequency is reduced
by half, since we plot only half of the coefficients.

```
let get_frequency y =
  let y' = Owl_fft.D.rfft ~axis:0 y in
  let y' = Dense.Ndarray.Z.get_slice
    [[1; (Dense.Ndarray.Z.shape y').(0) - 1];[]] y' in
  Dense.Ndarray.Z.(abs y' |> re);;
```

The frequency (cycle/year) as a unit of measurement seems a bit confusing.
To get the cyclical activity that is easier to interpret, we also plot the squared power
as a function of years/cycle. Both are plotted with code below.

```
let plot_sunspot_freq p =
  let n = (Arr.shape p).(0) in
  let f = Arr.(mul_scalar (linspace 0. 1. n) 0.5) in
  let h = Plot.create ~m:1 ~n:2 "plot_sunspot_freq.png" in
  Plot.subplot h 0 0;
  Plot.plot ~h f p;
  Plot.subplot h 0 1;
  let f' = Arr.(scalar_div 1.
    (get_slice [[1; Stdlib.(n-1)]] f)) in
  Plot.plot ~h f' p;
  Plot.output h;;
```

The result is shown in Fig. 7.3. We can see clearly that the most prominent cycle
is actually a little less than 11 years.

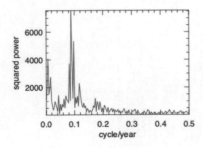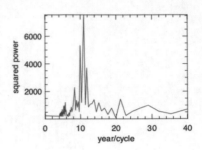

Fig. 7.3: Find sunspot cycle with FFT

7.3.2 Determine the Tone

When we dial a phone number, the sound wave can be seen as a signal. In this example, we show how to decipher which number is dialled according to the given soundwave. This example uses the data from Moler et al [12]. Let's first load and visualise them.

```
let data = Owl_io.read_csv ~sep:',' "touchtone.csv";;
let data = Array.map (fun x -> Array.map float_of_string x)
           data |> Mat.of_arrays;;
let data = Mat.div_scalar data 128.;;
```

The dataset specifies a sampling rate (fs) of 8192. We have a segment of signal that shows the touch tone of dialling a phone number. We can visualise the signal:

```
Mat.div_scalar (Mat.sequential 1 (Arr.shape data).(1)) fs;;
```

The result is shown in the left subfigure of Fig. 7.4. Apparently, according to the dense area in this signal, there are 11 digits in this phone number. The question is, which numbers?

The tune of the phone is a combination of two different frequencies. [14] All the 11 digits are composed from 7 prominent frequencies, as shown in Tbl. 7.2. This frequency keypad is specified in the Dual-tone multi-frequency signalling (DTMF) standard [15]. Therefore, we can use FFT to decompose the soundwave for each digit into its two basic frequencies, and check this table to find out the answer. Let's start by applying the FFT to the original data.

```
let yf = Owl_fft.D.rfft data;;
let y' = Dense.Ndarray.Z.(abs yf |> re);;
let n = (Arr.shape y').(1);;
let x' = Mat.linspace 0. (fs /. 2.) n;;
```

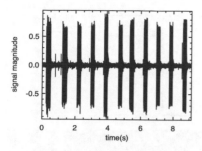
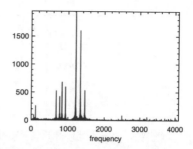

Fig. 7.4: Recording of an 11-digit number and its FFT decomposition

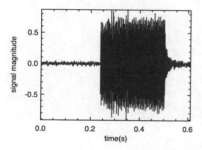
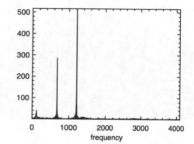

Fig. 7.5: Recording of the first digit and its FFT decomposition

We plot x' against y' similarly using the previous plotting function, and the result is shown in the right side of Fig. 7.4.

We can use the first tone as an example to find out the two frequencies from which it is composed. Let's start by getting a subset of the signal:

```
let data2 = Arr.get_slice [[];[0; 4999]] data;;
```

We then perform the same process as before producing the results shown in Fig. 7.5. It shows that the first digit is mainly composed of two frequencies, about 700 and 1200. Looking them up in Tbl. 7.2, we find that the first digit is 1. If you are interested, you can investigate the whole phone number following the same procedure.

7.3.3 Image Processing

FFT remains a powerful tool as the dimensionality of data increases. It can be used for a wide range of image processing tasks such as blurring, compression or denosing. In this section, we demonstrate an image denoising example. The basic

Table 7.2: DTMF keypad frequency combinations

	1209Hz	1336Hz	1477Hz
697Hz	1	2	3
770Hz	4	5	6
852Hz	7	8	9
941Hz	*	0	#

(a)   (b)

Fig. 7.6: Using FFT to denoise an image: (a) noisy image; (b) de-noised image.

idea is similar to the preceding case: the image signal is also composed from two dimensional signals of different frequencies, many of which play only marginal roles in forming this image. We can thus find these minor frequencies in this image via FFT that represent noise, remove them, and then recreate the image. The new image should preserve much of the original image quality while removing almost all the noise.

As input, we use a cate image as shown in Fig. 7.6a. As a tool for image processing in OCaml, we use the `imageUtils.ml` script which reads images of PPM format into an ndarray, or saves an ndarray into PPM image. The script is included in the code respository of this book.

```
#use "imageUtils.ml";;
module N = Dense.Ndarray.S
module C = Dense.Ndarray.C
```

As a first step we read the image into an ndarray using the function `load_ppm`. Since this is a grayscale image, we need only process one single channel of this array, which is a matrix having all elements in the range 0 to 255.

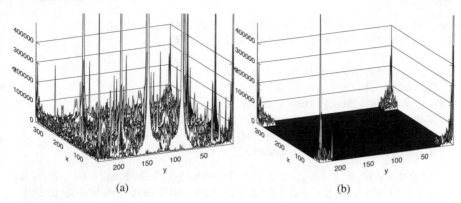

Fig. 7.7: (a) The frequency visualisation of original image, and (b) the processed data ordered by the frequency.

```
let img_arr = load_ppm "cat_noise.ppm"
  |> N.get_slice [[];[];[0]];;
let shp = N.shape img_arr;;
let h, w = shp.(0), shp.(1);;
let img = N.reshape img_arr [|h; w|]
  |> Dense.Ndarray.Generic.cast_s2c;;
```

The next step is simple: applying the `fft2` function on the image matrix. This function performs a 2-dimensional FFT on the matrix, i.e. it performs FFT on the input matrix twice, once horizontally, and once vertically. The resulting frequencies can be visualised as in Fig. 7.7a. As you can see, it contains a lot of noises.

```
let img_fft = Owl_fft.S.fft2 img;;
```

We can apply different kinds of filters on this frequency map `img_fft` to remove unwanted frequencies. We choose to use a simple one that keeps only the frequencies on the four "corners" of this matrix, setting the rest to zeros. We can achieve that with some slicing techniques.

```
let sub_length x frac =
  (float_of_int x) *. frac |> int_of_float;;

let h1 = sub_length h 0.1;;
let h2 = sub_length h 0.9;;
let w1 = sub_length w 0.1;;
let w2 = sub_length w 0.9;;
let index_0 = [ R [h1; h2]; R []];;
let index_1 = [ R [0; h1]; R [w1; w2] ];;
let index_2 = [ R [h2; h-1]; R [w1; w2] ];;
let slice_0 = C.get_fancy index_0 img_fft;;
```

```
let slice_1 = C.get_fancy index_1 img_fft;;
let slice_2 = C.get_fancy index_2 img_fft;;

let _ = C.set_fancy index_0 img_fft
  (C.shape slice_0 |> C.zeros);;
let _ = C.set_fancy index_1 img_fft
  (C.shape slice_1 |> C.zeros);;
let _ = C.set_fancy index_2 img_fft
  (C.shape slice_2 |> C.zeros);;
```

After this process, the frequency map becomes Fig. 7.7b. Finally, let's apply the reverse 2D FFT on the processed frequency map, and save the result as an image.

```
let img = Owl_fft.S.ifft2 img_fft |> C.re;;
let image = N.stack ~axis:2 [|img; img; img|];;
let image = N.expand image 4;;
let _ = save_ppm_from_arr image "cat_denoise.ppm";;
```

As shown in Fig. 7.6b, although it looks blurry, the output image removes the noise in the original image, and manages to keep most of the original information.

7.4 Filtering

In Chap. 5, we introduced the generic filter operation, allowing obtaining target data from the input according to criteria specified by functions. The concept of filtering in signal processing is similar. A filter is a device or process that removes unwanted components from a signal. Generally, any system that modifies input signals in certain ways is a filter. In this section, we introduce how FFT can be applied to perform some filtering tasks with examples.

7.4.1 Example: Smoothing

Let's start with *smoothing*, a common and quite simple filtering task. Suppose we have a segment of noisy signal such as the price of a stock. In many cases we would like to remove the extreme trends and see a long-term trend from the historical data. From Yahoo Finance [16], we acquire the stock price of Google in April 09, from 2019 to 2020, and store them in the file goog.csv (included in code respository). We can load the data into a matrix called "data" using code below:

```
let data = Owl_io.read_csv ~sep:',' "goog.csv";;
let data = Array.map (fun x ->
    Array.map float_of_string (Array.sub x 1 6))
    (Array.sub data 1 (Array.length data-1)) |> Mat.of_arrays;;
```

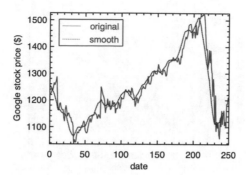

Fig. 7.8: Smoothed stock price of Google

The data contains several columns, each representing opening price, volume, high price, etc. Here we use the daily closing price as an example, which is in the fourth column.

```
let y = Mat.get_slice [[];[3]] data;;
```

To compute the moving average of this signal, we create a window with 10 elements. For this example, we only use a very simple filter which takes an average of 10 elements.

```
let filter = Mat.of_array (Array.make 10 0.1 ) 1 10;;
```

We can then slide this filter window along the input signal to smooth the data step by step.

```
let y' = Mat.mapi (fun i _ ->
    let r = Mat.get_fancy [R [i; i+9]; R []] y in
    Mat.dot filter r |> Mat.sum'
  ) (Mat.get_slice [[0; (Arr.shape y).(0) - 10]; []] y);;
```

Finally, we plot the resulting smoothed data against the original data. The results are shown in Fig. 7.8. The blue dotted line smooths the jagged original stock price line (red), and represents a general trend of the price. We suspect that the sudden drop in the beginning might be related to the COVID-19 pandemic.

7.4.2 Gaussian Filter

However, the simple filter we have used is not ideal. A common pattern in this line is that it first drops then bounces back. To get a smoother curve, we can change this

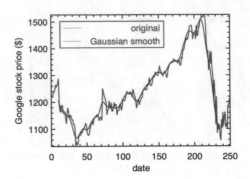

Fig. 7.9: Smoothed stock price of Google with Gaussian filtering

simple filter to another: the Gaussian filter [17]. Instead of giving equal probability to each element in the moving window, the gaussian filter assigns probability according to the gaussian distribution:

$$p(x) = e^{-\frac{x^2}{2\sigma^2}}. \tag{7.4}$$

The code below generates a simple 1-D gaussian filter. Similar to the previous simple example, the filter also needs to be normalised. For the filter window vector, its range of radius (radius) is set to truncated standard deviations, which bounds the random variable from both sides. The parameter is set to $\sigma = 3$.

```
let gaussian_kernel sigma =
  let truncate = 4. in
  let radius = truncate *. sigma +. 0.5 |> int_of_float in
  let r = float_of_int radius in
  let x = Mat.linspace (-.r) r (2 * radius + 1) in
  let f a = Maths.exp (-0.5 *. a ** 2. /. (sigma *. sigma)) in
  let x = Mat.map f x in
  Mat.(div_scalar x (sum' x));;


let filter = gaussian_kernel 3.;;
```

Computing the correlation between the filter and input data as before, we get an even smoother curve in Fig. 7.9.

Filters can generally be categorised by their usage into time domain filters and frequency domain filters. Time domain filters are used when the information is encoded in the shape of the signal's waveform, and can be used for tasks such as smoothing, waveform shaping, etc. They include filter methods such as moving average and Gaussian filters as we have shown. Frequency domain filters are used to divide a band of frequencies from signals, and their input information is in the form of sinusoids. Frequency domain filter methods include Windowed-sinc and Chebyshev etc. [18] There are many types of filters, each with different shape and application scenario, and we cannot fully cover them here. Please refer to some classical textbooks on signal processing such as Smith et al [19] for more information.

7.4.3 Signal Convolution

So far we have not used FFT functions to perform filtering. In this section we will demonstrate how they can be applied to solve our stock price problem. But first, we need to explain the idea of *convolution*. Formally it is a mathematical operation on two functions that produces a third function expressing how the shape of one is modified by the other:

$$f(t) * g(t) = \sum_{\tau=-\infty}^{\infty} f(\tau)g(t - \tau) \tag{7.5}$$

In equation Eq. 7.5, operator $*$ denotes convolution, and you can think of f as a discrete input signal with g as the filter. In fact, what we have seen in the previous example is exactly the convolution of a signal (stock price) and a filter. Note that computing $f(\tau)g(t - \tau)$ for each t requires adding all product pairs. You can see that this process is computation-heavy. It gets even more tricky when computing the convolution of two continuous signals.

The Fourier Transform can greatly reduce the complexity of computing the convolution and thus filtering. Specifically, the Convolution Theorem states that:

$$DFT(f * g) = DFT(f) \cdot DFT(g). \tag{7.6}$$

To put equation Eq. 7.6 into plain words: to obtain the DFT of two signals' convolution, we can simply get the DFT of each signal separately and then multiply them element-wise. That is, *convolution in the time domain can be expressed in multiplication in the frequency domain*. Once you have the DFT($f * g$), you can simply apply the inverse transform to get $f * g$.

Therefore, let's apply the FFT approach to the previous data and move it to the frequency domain:

```
let yf = Owl_fft.D.rfft ~axis:0 y;;
```

The resulting data yf looks like this:

```
                     C0
  R0          (312445, 0i)
  R1     (-2664.07, 17064.3i)
  R2     (-5272.52, 3899.16i)
                    ...
R125     (153.294, 68.7544i)
```

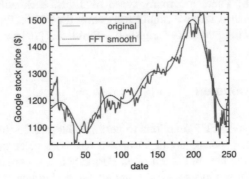

Fig. 7.10: Smoothed stock price of Google using FFT method

We keep only the five most notable frequencies, setting the rest to zero.

```
let n = (Dense.Ndarray.Z.shape yf).(0);;
let z = Dense.Ndarray.Z.zeros [|n-5; 1|];;
let _ = Dense.Ndarray.Z.set_slice [[5;n-1];[]] yf z;;
```

Then, we apply reverse FFT on this processed frequency vector and get the smoothed time domain data:

```
let y2 = Owl_fft.D.irfft ~axis:0 yf;;
```

As with the previous two methods, we can check how the smoothing approach works by plotting the filtered signal. As shown in Fig. 7.10, compared to the previous two filtering methods, FFT generates a better and smoother curve to describe the trend of the stock price.

7.4.4 FFT and Image Convolution

You might have already come across the concept of "convolution" in another context. Indeed, convolution happens to be the core idea in implementing the deep neural network (DNN). Unlike vectors, the convolution in DNN is often applied to ndarrays, but the basic process is similar to what we have described in the previous example. Starting with an ndarray input, another smaller ndarray called "kernel" slides over the input step by step. At each position, an element-wise multiplication and summation are applied, and the result is filled into corresponding position in an output ndarray. This process can be best illustrated in Fig. 7.11.

Owl has thorough support for convolution operations, including conv1d, conv2d, and conv3d. As the name suggests, they correspond to different dimensions of inputs.

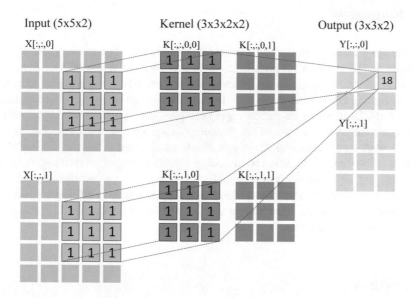

Fig. 7.11: Image convolution illustration

Other derived convolution types are also supported including dilated convolutions, transpose convolutions, and backward convolutions etc.

It's totally OK if none of this makes sense to you for now. We will explain the convolution operations and their usage in neural network in great detail in chapter 11. The point is that if you look closely you can see that image convolution is only a special high dimensional case of Eq. 7.5: a given input signal (the image), another similar but smaller filter signal (the kernel), and sliding the filter across the input signal performing element-wise multiplication and summation. We can thus use conv1d function in Owl to solve the previous simple smoothing problem:

```
let y3  = Arr.reshape y [|1;251;1|];;
let f3  = Arr.reshape filter [|10;1;1|];;
let y3' = Arr.conv1d y3 f3 [|1|];;
```

If you are interested in checking the result, the vector y3' contains the data to plot a smoothed curve. The smoothed data would be similar to that in Fig. 7.8 since the calculation is essentially the same, albeit obtained with much less code.

FFT is a popular implementation method for convolutions. There has been a lot of research on optimising and comparing its performance with other implementation methods, with practical considerations such as kernel size and implementation details of code [20, 21, 22]. We omit these technical discussions here.

7.5 Summary

This chapter revolves around a fundamental idea in signal processing: the Fourier Transform. We started with its definition, then introduced a crucial idea behind its efficient implementation, namely the Fast Fourier Transform. Owl provides support to FFT by linking to the existing FFTPack library. We showed how the FFT functions can be used in OCaml, first with some simple examples, then with three real applications: finding the frequency of sunspot events, deciphering the dial tone number according to its sound, and de-noising an image. Next, we discussed filtering in the signal process using different techniques, including simple averaging smoothing, Gaussian filtering, and FFT-based filtering. Finally we explained the relationship between the two most crucial computations in numerical applications: FFT and convolution. More about the application of convolution in neural networks will be discussed in the chapter 11 of this book.

References

1. McAulay R, Quatieri T. Speech analysis/synthesis based on a sinusoidal representation. IEEE Transactions on Acoustics, Speech, and Signal Processing. 1986 Aug; 34(4):744-54.
2. Baron Fourier JB. Théorie analytique de la chaleur. Chez Firmin Didot, père et fils; 1822.
3. Oppenheim AV. Discrete-time signal processing. Pearson Education India; 1999.
4. Bracewell RN, Bracewell RN. The Fourier transform and its applications. New York: McGraw-Hill; 1986 Feb.
5. Gupta A. Fourier transform and its application in cell phones. International Journal of Scientific and Research Publications. 2013 Jan; 3(1):1-2.
6. Strang G. Wavelet transforms versus Fourier transforms. Bulletin of the American Mathematical Society. 1993;28(2):288-305.
7. Turitsyn SK, Prilepsky JE, Le ST, Wahls S, Frumin LL, Kamalian M, Derevyanko SA. Nonlinear Fourier transform for optical data processing and transmission: advances and perspectives. Optica. 2017 Mar 20;4(3):307-22.
8. Holberg O. Computational aspects of the choice of operator and sampling interval for numerical differentiation in large-scale simulation of wave phenomena. Geophysical prospecting. 1987 Aug 1;35(6):629-55.
9. Heideman M, Johnson D, Burrus C. Gauss and the history of the fast Fourier transform. IEEE ASSP Magazine. 1984 Oct;1(4):14-21.
10. Phillips CL, Parr JM, Riskin EA, Prabhakar T. Signals, systems, and transforms. Upper Saddle River: Prentice Hall; 2003 Aug.
11. FFTPack [Internet]. FFTPACK. [cited 2021 Dec 20]. Available from: https://www.netlib.org/fftpack/.
12. Moler CB. Numerical computing with MATLAB. Society for Industrial and Applied Mathematics; 2004 Jan 1.
13. Sunspot number version 2.0: New Data and conventions [Internet]. Solar Influences Data analysis Center [cited 2021 Dec 20]. Available from: http://sidc.oma.be/silso/newdataset.
14. Borndörfer R, Eisenblätter A, Grötschel M, Martin A. Frequency assignment in cellular phone networks. Annals of Operations Research. 1998 Jan;76:73-93.
15. Dodd AZ. The essential guide to telecommunications. Prentice Hall Professional; 2002.
16. Yahoo Finance [Internet]. Yahoo! Finance [cited 2021 Dec 20]. Available from: https://finance.yahoo.com/.

17. Blinchikoff HJ, Zverev AI. Filtering in the time and frequency domains. Krieger Publishing Co., Inc.; 1986 Jun 1.
18. Widmann A, Schröger E, Maess B. Digital filter design for electrophysiological data-a practical approach. Journal of neuroscience methods. 2015 Jul 30;250:34-46.
19. Smith SW. The scientist and engineer's guide to digital signal processing. San Diego: California Technical Pub.; 1997 Jan 1.
20. Fialka O, Cadik M. FFT and convolution performance in image filtering on GPU. InTenth International Conference on Information Visualisation (IV'06) 2006 Jul 5 (pp. 609-614). IEEE.
21. Kolba D, Parks TW. A prime factor FFT algorithm using high-speed convolution. IEEE Transactions on Acoustics, Speech, and Signal Processing. 1977 Aug;25(4):281-94.
22. Vasilache N, Johnson J, Mathieu M, Chintala S, Piantino S, LeCun Y. Fast convolutional nets with fbfft: A GPU performance evaluation. arXiv preprint arXiv:1412.7580. 2014 Dec 24.
23. Oppenheim AV, Willsky AS, Nawab SH, Hernández GM. Signals & systems. Pearson Educación; 1997.

Part II
Advanced Data Analysis Techniques

After various numerical techniques, we advance our studies from linear to non-linear in this part. We also shift the topics from numerical calculations to algorithmic solutions. Specifically, this part covers a series of advanced topics which are based on a fundamental idea in modern data analysis, especially neural network: the Algorithmic Differentiation. During the introduction to this topic, we dive deep into the mechanisms of algorithmic differentiation, showing how it can be implemented step by step in the functional programming language OCaml. The next two chapters focus on the topic of optimization which is built on algorithmic differentiation, followed by the topic of regression which is then built on optimization. Based on these chapters, this part concludes with a brief overview of two state-of-the-art data analysis fields in both academics and industry: neural network, and natural language processing. For the first, we demonstrate how a functioning neural network module, based on what we have introduced so far, can be built up concisely. For the next, we focus on one specific task in neural language processing - the topic modelling.

Chapter 8
Algorithmic Differentiation

Abstract Differentiation is core to many scientific applications including maximising or minimising functions, solving systems of ODEs, and non-linear optimisation such as KKT optimality conditions. Algorithmic differentiation (AD) is a computer-friendly technique for performing differentiation that is both efficient and accurate. A recent important application of algorithmic differentiation is in machine learning and artificial intelligence. [1] Training a neural network involves two phases, namely forward and back propagation. The latter is essentially the calculation of the derivative of the whole neural network as a large function. [2] In this chapter, we will introduce this topic using an hands-on approach. Starting from the basic definition, we build up an simplified version of AD engine step by step. We then move to the AD engine in Owl to present its usage and more implementation details.

8.1 Chain Rule

We often need to study the functional relationship between variables: how much one variable changes when the value of another variable changes. For example, assume an object moves a distance of Δs in a time Δt. We can define the velocity v describing the motion of the object as the ratio of distance moved in a given time as the time considered gets smaller and smaller, eventually approaching zero:

$$v = \lim_{\Delta t \to 0} \frac{\Delta s}{\Delta t} = \frac{ds}{dt} \tag{8.1}$$

The term $\frac{ds}{dt}$ is referred to as "the **derivative** of s with respect to t".

We can extend this to higher dimensional spaces. For example, we might try to describe the temperature distribution in a solid material that has not been heated evenly. The temperature T of a point in the material can be modelled as a function of that point's cartesian coordinate (x, y, z), and we can consider how the temperature changes as within the material as shown in Eq. 8.2 [3].

© The Author(s), under exclusive license to Springer Nature Switzerland AG 2022   141
L. Wang et al., *OCaml Scientific Computing*, Undergraduate Topics in Computer Science,
https://doi.org/10.1007/978-3-030-97645-3_8

Table 8.1: A Short Table of Basic Derivatives

Functions	Derivatives
$u(x) + v(x)$	$u'(x) + v'(x)$
$Cu(x)$	$Cu'(x)$
$u(x)v(x)$	$u'(x)v(x) + u(x)v'(x)$
$\frac{u(x)}{v(x)}$	$\frac{u'(x)v(x) - u(x)v'(x)}{v^2(x)}$
$\sin(x)$	$\cos(x)$
$\cos(x)$	$-\sin(x)$
e^x	e^x
$\ln x$	$\frac{1}{x}$
$\log_a(x)$	$\frac{1}{x \ln a}$

$$\nabla T = (\frac{\partial T}{\partial x}, \frac{\partial T}{\partial y}, \frac{\partial T}{\partial z}) \tag{8.2}$$

The vector ∇T is referred to as "the *gradient* of T". Because T might vary differently with respect to the different quantities x, y, z, we refer to $\frac{\partial}{\partial x}$ as "the *partial derivative* with respect to x".

In general, the process of calculating derivatives and gradients is referred to as *differentiation*. [4] Although some differentiation involves complex manipulations and may seem daunting, the underlying procedures are often quite mechanical. One particularly important such procedure is called the *chain rule*, used to calculate the derivative of a composite function. [5] Suppose we have two functions f and g that can be composed to create a function $F(x) = f(g(x))$. The chain rule states that the derivative of F can be calculated as:

$$F'(x) = f'(g(x))g'(x) \tag{8.3}$$

This seemingly simple rule is one of the most fundamental rules in calculating derivatives. For example, let $y = x^a$, where a is a real number, and we can obtain y' using the chain rule by rewriting $y = x^a = e^{\ln x^a} = e^{a \ln x}$ and setting $u = a \ln x$ so that we have $y = e^u$. Applying the chain rule gives

$$y' = \frac{dy}{du} \cdot \frac{du}{dx} = e^u a \frac{1}{x} = ax^{a-1}$$

Tbl. 8.1 summarises how to apply the chain rule when composing functions using basic operations such as addition and multiplication, where x is a variable, u and v are functions of x, and C is a constant. These equations are like the building blocks used to differentiate more complex function. Armed with the chain rule and these basic equations, we can solve a wide range of problems.

8.2 Differentiation Methods

As functions become more complicated, manual differentiation becomes both tedious and error-prone. Fortunately, we can use computers to automate the process and there are three different ways that are widely used to do so: numerical differentiation, symbolic differentiation, and algorithmic differentiation.

8.2.1 Numerical Differentiation

Arising directly from the definition of derivative in Eq. 8.1, numerical differentiation uses a small step δ to compute an approximate value for the limit in the definition

$$f'(x) = \lim_{\delta \to 0} \frac{f(x + \delta) - f(x)}{\delta}. \tag{8.4}$$

The resulting method iterates four steps: (*i*) evaluate the given f at point x; (*ii*) choose a suitable small δ and add it to the original x; (*iii*) re-evaluate the function f at $x + \delta$; and (*iv*) calculate the derivative using Eq. 8.4. By treating the function f as a black box, then this method is straightforward to implement as long as f can be evaluated, but is unfortunately subject to errors including truncation and round-off error. [6]

Truncation error is the error caused by approximating a mathematical process. Eq. 8.4 only approximates the true value of the gradient. We can calculate the difference between Eq. 8.4 and the actual value using the Taylor expansion:

$$f(x + h) = f(x) + h f'(x) + \frac{h^2}{2} f''(\sigma_h).$$

Here h is the step size and σ_h is in the range of $[x, x + h]$. This equation can be manipulated into

$$\frac{h^2}{2} f''(\sigma_h) = f'(x) - \frac{f(x + h) - f(x)}{h}.$$

The term $\frac{h^2}{2} f''(\sigma_h)$ represents the truncation error in the approximation. For example, if $f(x) = \sin(x)$ then $f''(x) = -\sin(x)$ and if we calculate the derivative at $x = 1$ numerically using a step size of 0.01, then the truncation error will be in the range $\frac{0.01^2}{2} [\sin(1), \sin(1.01)]$.

Step size directly affects the truncation error, as can be seen by considering the effect of using a very large step size. For example, we wish to find the derivative of $f(x) = \cos(x)$ at point $x = 1$. The analytical solution gives the value $-\sin(1) = 0.84147$ but the numerical result deviates substantially from that actual value, as shown by the code below.

```
# let d =
    let _eps = 0.1 in
    let diff f x = (f (x +. _eps) -. f x) /. _eps in
    diff Maths.cos 1.;;
val d : float = -0.867061844425624506
```

Round-off error is caused by representing numbers with only finite precision during the numerical computation. Looking back at Eq. 8.4 we need to calculate $f(x + h) - f(x)$, i.e. to subtract two numbers that are almost equal which can lead to large round-off errors. For example, consider using a very small step size to avoid truncation error

```
# let d =
    let _eps = 5E-16 in
    let diff f x = (f (x +. _eps) -. f x) /. _eps in
    diff Maths.cos 1.;;
val d : float = -0.888178419700125121
```

Again, this result deviates substantially from the actual value. Indeed, if we were to use an even smaller step size, $1e - 16$, the result actually becomes 0: the round-off error is so large that the computer considers $f(x)$ and $f(x + h)$ to be equal.

Finally, due to the need to repeatedly evaluate f, performance can deteriorate significantly if f is expensive to compute. The result is that, although straightforward to implement, numerical differentiation suffers from both low accuracy and low performance.

8.2.2 Symbolic Differentiation

By manipulating the underlying mathematical expressions, symbolic differentiation obtains analytical results without numerical approximation, using rules such as those introduced in Tbl. 8.1. For example, consider the function $f(x_0, x_1, x_2) = x_0 * x_1 * x_2$. Computing ∇f symbolically, we end up with:

$$\nabla f = \left(\frac{\partial f}{\partial x_0}, \frac{\partial f}{\partial x_1}, \frac{\partial f}{\partial x_2} \right) = (x_1 * x_2, x_0 * x_2, x_1 * x_2)$$

This process significantly reduces the scope for numerical errors but the complexity of symbolic manipulation quickly grows as expressions become more complex: try using symbolic differentiation to compute the derivative of $f(x) = \prod_{i=0}^{n-1} x_i$ to test if that is true! You will quickly realise that symbolic computations may generate many very complex intermediate results, even for quite simple functions. As a result, symbolic differentiation can easily consume a huge amount of memory and CPU cycles.

In addition, in contrast to numerical differentiation, we must treat the function we are differentiating as a white box, and know exactly what is inside of it. This further increases the difficulty of using symbolic differentiation to derive arbitrary functions, even though it generates more accurate results. [7]

8.2.3 Algorithmic Differentiation

A chain-rule based technique for calculating derivatives with respect to input variables of functions defined in a computer programme, algorithmic differentiation (AD) is also known as automatic differentiation, though strictly speaking it does not fully automate differentiation and can sometimes lead to inefficient code. [8] [9]

In general though, AD efficiently generates exact results and so is highly applicable in many real world applications. It is neither numerical nor symbolic differentiation but takes the best of both being as accurate as symbolic differentiation but far more scalable. [10] According to Griewank [11], the reverse mode of AD yields any gradient vector at no more than five times the cost of evaluating the function f itself. AD has already been implemented in various popular languages, including Python [12], Julia [13], and MATLAB [14]. The rest of this chapter introduces AD. [15] [16]

8.3 How Algorithmic Differentiation Works

We start by showing how AD works using an example. We have seen how the chain rule can be applied to simple functions such as $y = x^a$, so now let us consider a more complicated function based on a sigmoid function

$$y(x_0, x_1) = \left(1 + e^{x_0 x_1 + sin(x_0)}\right)^{-1} \tag{8.5}$$

We wish to compute $\nabla y = \left(\frac{\partial y}{\partial x_0}, \frac{\partial y}{\partial x_1}\right)$. To illustrate this process we represent Eq. 8.5 as the computation graph shown in Fig. 8.1. At the right side of the figure is the final output y, and the roots of this graph are the input variables on the left side of the figure. The nodes between them are either constants or intermediate variables produced by the mathematical functions such as sine. All nodes are labelled v_i, and an edge between two nodes indicates a dependency in the computation.

Considering this graphical representation, there are two ways to apply the chain rule: forward mode and reverse mode. [1] We next consider each in turn.

---

[1] Not "forward pass" and "backward pass", which are sometimes used to describe the two phases used in reverse mode itself, or in training machine learning models.

Input variables

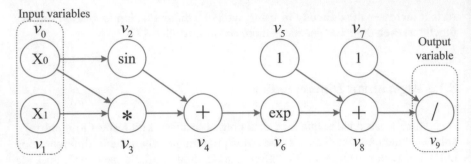

Fig. 8.1: The graph expression of a function

8.3.1 Forward Mode

We ultimately wish to calculate $\frac{\partial y}{\partial x_0}$ (and $\frac{\partial y}{\partial x_1}$, which follows similarly) but we will begin by calculating some intermediate results that will prove to be useful. Using the labels v_i to refer to the intermediate computations, we have $\frac{\partial v_0}{\partial x_0} = 1$ and $\frac{\partial v_1}{\partial x_0} = 0$ immediately. Next, consider $\frac{\partial v_3}{\partial x_0}$, which is a bit trickier and requires use of the chain rule

$$\frac{\partial v_3}{\partial x_0} = \frac{\partial(x_0 x_1)}{\partial x_0} = x_1 \frac{\partial(x_0)}{\partial x_0} + x_0 \frac{\partial(x_1)}{\partial x_0} = x_1$$

After calculating $\frac{\partial v_3}{\partial x_0}$, we proceed to compute partial derivatives of v_4, v_5, and so on to v_9 which will give us the final result $\frac{\partial y}{\partial x_0}$ we are looking for. This process starts with the input variables and ends with the output variables and so is called *forward differentiation*.[2]

We can simplify the notation by letting $\dot{v}_i = \frac{\partial v_i}{\partial x_0}$. The $\dot{v}_i$ is called the *tangent* of the function $v_i(x_0, x_1, \ldots, x_n)$ with respect to the input variable x_0, and the results of evaluating the function at each intermediate point are called *primal value*.

Let's calculate $\dot{y}$ when setting $x_0 = 1$, and $x_1 = 1$. The full forward differentiation calculation process is shown in Tbl. 8.2 where two simultaneous computation processes take place in the two computation columns: (*i*) the primal computation is the procedure defined by Eq. 8.5; and (*ii*) the tangent computation gives the derivative for each intermediate variable with regard to x_0.

The procedure in Tbl. 8.2 is illustrated in Fig. 8.2. We can validate this result with the algorithmic differentiation module `Algodiff` in Owl. Don't worry if you don't understand the code at the moment! We will cover the details later in this chapter.

```
# open Algodiff.D;;
# let f x =
    let x1 = Mat.get x 0 0 in
```

---

[2] Sometimes also known as "tangent linear" mode.

Table 8.2: The computation process of forward differentiation, shown to 2 significant figures

	Primal computation	Tangent computation
0	$v_0 = x_0 = 1$	$\dot{v}_0 = 1$
1	$v_1 = x_1 = 1$	$\dot{v}_1 = 0$
2	$v_2 = sin(v_0) = 0.84$	$\dot{v}_2 = cos(v_0) * \dot{v}_0 = 0.54 * 1 = 0.54$
3	$v_3 = v_0\, v_1 = 1$	$\dot{v}_3 = v_0\dot{v}_1 + v_1\,\dot{v}_0 = 1*0 + 1*1 = 1$
4	$v_4 = v_2 + v3 = 1.84$	$\dot{v}_4 = \dot{v}_2 + \dot{v}_3 = 1.54$
5	$v_5 = 1$	$\dot{v}_5 = 0$
6	$v_6 = \exp(v_4) = 6.30$	$\dot{v}_6 = \exp(v_4) * \dot{v}_4 = 6.30 * 1.54 = 9.70$
7	$v_7 = 1$	$\dot{v}_7 = 0$
8	$v_8 = v_5 + v_6 = 7.30$	$\dot{v}_8 = \dot{v}_5 + \dot{v}_6 = 9.70$
9	$y = v_9 = \frac{1}{v_8}$	$\dot{y} = \frac{-1}{v_8^2} * \dot{v}_8 = -0.18$

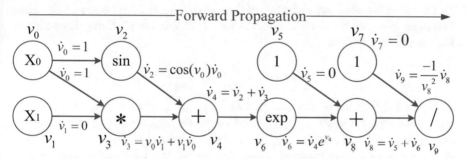

Fig. 8.2: Example of forward propagation with computational graph

```
    let x2 = Mat.get x 0 1 in
    Maths.(div (F 1.) (F 1. + exp (x1 * x2 + (sin x1))));;
val f : t -> t = <fun>
# let x = Mat.ones 1 2;;
val x : t = [Arr(1,2)]
# let _ = grad f x |> unpack_arr;;
- : A.arr =
           C0          C1
R0 -0.181974 -0.118142
```

The first value in the output ndarray -0.181974 verifies our manual calculation in the table.

8.3.2 Reverse Mode

Let us now think about this problem in the other direction, from outputs to inputs. The problem remains the same, i.e. to calculate $\frac{\partial y}{\partial x_0}$, and we will still follow the same "step by step" idea from forward mode – only that we will just calculate it backwards. For example, the graph shows that $y = v_7/v_8$ so if we only knew $\frac{\partial y}{\partial v_7}$ and $\frac{\partial y}{\partial v_8}$, we would have taken a step towards our target solution.

We first observe that $\frac{\partial y}{\partial v_9} = 1$, as y and v_9 are identical. We then compute $\frac{\partial y}{\partial v_7}$ by applying the chain rule

$$\frac{\partial y}{\partial v_7} = \frac{\partial y}{\partial v_9} * \frac{\partial v_9}{\partial v_7} = 1 * \frac{\partial v_9}{\partial v_7} = \frac{\partial (v_7/v_8)}{\partial v_7} = \frac{1}{v_8}. \tag{8.6}$$

This looks complex but we can simplify by applying a substitution

$$\bar{v}_i = \frac{\partial y}{\partial v_i}$$

for the derivative of output variable y with regard to intermediate node v_i. $\bar{v}_i$ is called "the *adjoint* of variable v_i with respect to the output variable y". Using this notation, Eq. 8.6 can be rewritten

$$\bar{v}_7 = \bar{v}_9 * \frac{\partial v_9}{\partial v_7} = 1 * \frac{1}{v_8}$$

Note the difference between tangent and adjoint. In forward mode, we know $\dot{v}_0$ and $\dot{v}_1$ and then calculate $\dot{v}_2, \dot{v}_3, \ldots$ until we get the target $\dot{v}_9$. In reverse mode, we start with $\bar{v}_9 = 1$ and calculate $\bar{v}_8, \bar{v}_7, \ldots$ until we have $\bar{v}_0 = \frac{\partial y}{\partial v_0} = \frac{\partial y}{\partial x_0}$, our target. Again, $\dot{v}_9 = \bar{v}_0$ in this example, given that we are talking about derivative with respect to x_0 when we use $\dot{v}_9$. As a result, reverse mode is also called *adjoint mode*.

With that in mind, we can now follow the complete reverse differentiation procedure. We first perform a forward pass, shown in Tbl. 8.3, to compute the required intermediate values identical to those in the Primal Computation column of Tbl. 8.2.

Note that x_0 is used in both intermediate variables v_2 and v_3 so we compute the adjoint of v_0 with respect to both v_2 (step 19) and v_3 (step 20), before accumulating them together (step 20).

As with forward mode, Fig. 8.3 depicts the reverse differentiation process in Tbl. 8.2. This result $\bar{v}_0 = -0.18$ agrees with the result when using forward mode. We can also use Owl to perform a numerical differentiation to further validate, and the code is similar to that of using algorithmic differentiation as shown above.

```
module D = Owl_numdiff_generic.Make (Dense.Ndarray.D);;
let x = Arr.ones [|2|];;
let f x =
  let x1 = Arr.get x [|0|] in
  let x2 = Arr.get x [|1|] in
  Maths.(div 1. (1. +. exp (x1 *. x2 +. (sin x1))));;
```

Table 8.3: Forward pass in the reverse mode

	Primal computation
0	$v_0 = x_0 = 1$
1	$v_1 = x_1 = 1$
2	$v_2 = sin(v_0) = 0.84$
3	$v_3 = v_0\ v_1 = 1$
4	$v_4 = v_2 + v3 = 1.84$
5	$v_5 = 1$
6	$v_6 = \exp(v_4) = 6.30$
7	$v_7 = 1$
8	$v_8 = v_5 + v_6 = 7.30$
9	$y = v_9 = \frac{1}{v_8}$

Table 8.4: The reverse pass in reverse mode

	Adjoint computation
10	$\bar{v}_9 = 1$
11	$\bar{v}_8 = \bar{v}_9 \frac{\partial(v_7/v_8)}{\partial v_8} = 1 * \frac{-v_7}{v_8^2} = \frac{-1}{7.30^2} = -0.019$
12	$\bar{v}_7 = \bar{v}_9 \frac{\partial(v_7/v_8)}{\partial v_7} = \frac{1}{v_8} = 0.137$
13	$\bar{v}_6 = \bar{v}_8 \frac{\partial v_8}{\partial v_6} = \bar{v}_8 * \frac{\partial(v_6+v_5)}{\partial v_6} = \bar{v}_8$
14	$\bar{v}_5 = \bar{v}_8 \frac{\partial v_8}{\partial v_5} = \bar{v}_8 * \frac{\partial(v_6+v_5)}{\partial v_5} = \bar{v}_8$
15	$\bar{v}_4 - \bar{v}_6 \frac{\partial v_6}{\partial v_4} - \bar{v}_8 * \frac{\partial \exp(v_4)}{\partial v_4} - \bar{v}_8 * e^{v_1}$
16	$\bar{v}_3 = \bar{v}_4 \frac{\partial v_4}{\partial v_3} = \bar{v}_4 * \frac{\partial(v_2+v_3)}{\partial v_3} = \bar{v}_4$
17	$\bar{v}_2 = \bar{v}_4 \frac{\partial v_4}{\partial v_2} = \bar{v}_4 * \frac{\partial(v_2+v_3)}{\partial v_2} = \bar{v}_4$
18	$\bar{v}_1 = \bar{v}_3 \frac{\partial v_3}{\partial v_1} = \bar{v}_3 * \frac{\partial(v_0*v_1)}{\partial v_1} = \bar{v}_4 * v_0 = \bar{v}_4$
19	$\bar{v}_{02} = \bar{v}_2 \frac{\partial v_2}{\partial v_0} = \bar{v}_2 * \frac{\partial(sin(v_0))}{\partial v_0} = \bar{v}_4 * cos(v_0)$
20	$\bar{v}_{03} = \bar{v}_3 \frac{\partial v_3}{\partial v_0} = \bar{v}_3 * \frac{\partial(v_0*v_1)}{\partial v_0} = \bar{v}_4 * v_1$
21	$\bar{v}_0 = \bar{v}_{02} + \bar{v}_{03} = \bar{v}_4(cos(v_0) + v_1) = \bar{v}_8 * e^{v_4}(0.54 + 1) = -0.019 * e^{1.84} * 1.54 = -0.18$

Then we can get the result of the differentiation at the point $(x_0, x_1) = (1, 1)$, which again agrees with both the previous results using the AD module and our table-based calculation.

```
# D.grad f x;;
- : D.arr =
          C0          C1
R -0.181973 -0.118142
```

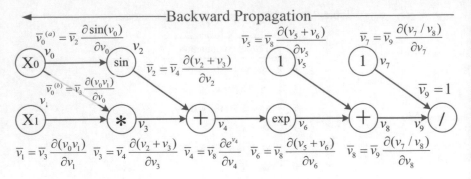

Fig. 8.3: Example of reverse accumulation with computational graph

Before we continue it is worth noting that we obtained $\frac{\partial y}{\partial x_1}$ "for free" while calculating $\frac{\partial y}{\partial x_0}$ in reverse mode. This observation will be used in the next section where we discuss how to decide whether to use forward or reverse mode in practice.

8.3.3 Forward Mode or Reverse Mode?

As both can be used to differentiate functions there is a natural question over which should be used in practice. Inevitably, the short answer is: it depends. However, a useful rule of thumb is

- if the number of input variables is far greater than that of the output variables, then use reverse mode; whereas
- if the number of output variables is far greater than that of the input variables, then use forward mode.

The rationale behind this rule of thumb is that, for each input variable, we must seed individual variables and perform one forward pass and so the number of forward passes in forward mode increases linearly as the number of inputs increases. However, in reverse mode, no matter how many inputs there are a single reverse pass gives us all the derivatives of the inputs. A real-world example where this is particularly relevant is when training deep neural networks as these typically contain very many inputs but whose output is often just a single scalar value from a loss function.

The trade-off is that reverse mode must maintain a directed computation graph in memory to propagate errors back, while forward mode does not due to the algebra of dual numbers. We will show another example of choosing between these two methods later in this chapter.

8.4 A Strawman AD Engine

Clearly you would not want to produce these tables every time you had a new function to differentiate! Far better to let the computer do it – and the "how" is the topic of this section. We will discuss how to implement both differentiation modes in OCaml, albeit using considerably simplified implementations compared to the AD module in Owl.

We will once more use the function in Eq.8.5 as an example, and limit the computations supported by our small AD engine to just sine (`sin`), exponential (`exp`), addition (`add`), division (`div`), and multiplication (`mul`).

8.4.1 Implementation of Forward Mode

Tbl. 8.2 shows that an intermediate node actually needs only know the primal and tangent values of its parents' nodes to compute its own results. Using this observation we define a data type to track these two values:

```
type df = {
  mutable p: float;
  mutable t: float;
};;

let primal df = df.p;;
let tangent df = df.t;;
```

We next define operators that take values of type `df` as input, and produce values of type `df` as output. In each case, we must define how to compute the function value `p'` and derivative value `t'` given the input data `df`. Calculation of derivatives is based on rules such as those listed in Tbl. 8.1. For example, for *sin* and *exp*:

```
let sin_ad x =
  let p = primal x in
  let t = tangent x in
  let p' = Owl_maths.sin p in
  let t' = (Owl_maths.cos p) *. t in
  {p = p'; t = t'}

let exp_ad x =
  let p = primal x in
  let t = tangent x in
  let p' = Owl_maths.exp p in
  let t' = p' *. t in
  {p = p'; t = t'}
```

Operators that accept multiple inputs require a little more thought but essentially the same procedure. First, addition:

```
let add_ad a b =
  let pa = primal a in
  let ta = tangent a in
  let pb = primal b in
  let tb = tangent b in
  let p' = pa +. pb in
  let t' = ta +. tb in
  {p=p'; t=t'}
```

Then multiplication and division:

```
let mul_ad a b =
  let pa = primal a in
  let ta = tangent a in
  let pb = primal b in
  let tb = tangent b in
  let p' = pa *. pb in
  let t' = pa *. tb +. ta *. pb in
  {p = p'; t = t'}
```

```
let div_ad a b =
  let pa = primal a in
  let ta = tangent a in
  let pb = primal b in
  let tb = tangent b in
  let p' = pa /. pb in
  let t' = (ta *. pb -. tb *. pa) /. (pb *. pb) in
  {p=p'; t=t'}
```

Having defined these functions, we next create a tiny wrapper named `diff`

```
let diff f =
  let f' x y =
    let r = f x y in
    primal r, tangent r
  in
  f'
```

...and that's all! We can now differentiate our previous example, starting with the inputs for which we know the tangent of $x1$ with regard to $x0$ is zero, as are the other constants used in the computation, while the tangent of $x0$ with regard to itself is, of course, one.

```
let x0 = {p=1.; t=1.};;
let x1 = {p=1.; t=0.};;
```

We then simply define our function f in terms of the operations given above, and apply our diff function

```
# let f x0 x1 =
  let v2 = sin_ad x0 in
  let v3 = mul_ad x0 x1 in
  let v4 = add_ad v2 v3 in
  let v5 = {p=1.; t=0.} in
  let v6 = exp_ad v4 in
  let v7 = {p=1.; t=0.} in
  let v8 = add_ad v5 v6 in
  let v9 = div_ad v7 v8 in
  v9;;
val f : df -> df -> df = <fun>
# let p, t = diff f x0 x1;;
val p : float = 0.13687741466075895
val t : float = -0.181974376561731321
```

The resulting t, $\frac{\partial y}{\partial x_0}$, is equal to that calculated earlier.

8.4.2 Implementation of Reverse Mode

Reverse mode is slightly more complex. As the previous section illustrated, forward mode needs only one pass while reverse mode requires two – a forward pass followed by a backward pass. Thus, as well as computing the primal values in the forward pass, we must also record the operations applied so we can use them in the backward pass. There are many ways this could be done, e.g., using a stack or a graph, but here we use a different approach. As ever, we begin with our data types.

```
type dr = {
  mutable p: float;
  mutable a: float ref;
  mutable adj_fun : float ref ->
    (float * dr) list -> (float * dr) list
};;

let primal dr = dr.p;;
let adjoint dr = dr.a;;
let adj_fun dr = dr.adj_fun;;
```

It is hopefully easy enough to understand p as the primal and a as the adjoint, but the adj_fun is a little more complex so we will illustrate it with an example implementing the sin operation

```
let sin_ad dr =
  let p = primal dr in
  let p' = Owl_maths.sin p in
  let adjfun' ca t =
      let r = !ca *. (Owl_maths.cos p) in
      (r, dr) :: t
  in
  {p=p'; a=ref 0.; adj_fun=adjfun'};;
```

The adj_fun can be viewed as a *placeholder* for the adjoint value that we do not
yet know during the forward pass, while t is a stack of intermediate nodes yet to
be processed in the backward pass. One way to read this definition is as saying that
once we have the adjoint value ca, we can get the new adjoint value of my parents' r
by following the derivative calculation rules and the chain rule. This result, together
with the original data dr, is pushed onto the stack t (naively implemented as an
OCaml list).

Next, we consider mul, an operation with two variables

```
let mul_ad dr1 dr2 =
  let p1 = primal dr1 in
  let p2 = primal dr2 in
  let p' = Owl_maths.mul p1 p2 in
  let adjfun' ca t =
      let r1 = !ca *. p2 in
      let r2 = !ca *. p1 in
      (r1, dr1) :: (r2, dr2) :: t
  in
  {p = p'; a = ref 0.; adj_fun = adjfun'}
```

For this operation, both parents must be added to the task stack t as multiplication
has two inputs. We also need a helper function to encapsulate input data

```
let make_reverse v =
  let a = ref 0. in
  let adj_fun _a t = t in
  {p=v; a; adj_fun}
```

With this we can now perform a forward pass on a simple example $f(x, y) =
\sin(x) * y$, evaluated at $(x, y) = (1, 2)$.

```
# let x = make_reverse 1.;;
val x : dr = {p = 1.; a = {contents = 0.}; adj_fun = <fun>}
# let y = make_reverse 2.;;
val y : dr = {p = 2.; a = {contents = 0.}; adj_fun = <fun>}
# let v = mul_ad (sin_ad x) y;;
val v : dr = {p = 1.68294196961579301;
  a = {contents = 0.}; adj_fun = <fun>}
```

At this point, after the forward pass has completed the steps described in Tbl. 8.3, we have the primal values at each intermediate node but all their adjoint values are zero as they have not yet been calculated. To also complete the steps described in Tbl. 8.4, we need two sets of information: the forward calculation values (e.g., v_8), and how to calculate the derivatives. The first is stored in the df.p field, and the second is stored in the adj_fun. We thus need a recursive function, reverse_push, that can use these sets of information to complete the reverse pass calculations.

```
let rec reverse_push = function
  | [] -> ()
  | (v, dr) :: t ->
      let aa = adjoint dr in
      let adjfun = adj_fun dr in
      aa := !aa +. v;
      let stack = adjfun aa t in
      reverse_push stack
```

Given the calculations accumulated in the forward pass, the reverse_push function replays them backwards. Starting from a list, it takes the top element dr, gets its adjoint value aa, updates it by adding its child's adjoint value v, and then obtains its adj_fun, a placeholder that awaits the adjoint value of the current node. Now that the adjoint value aa is known, we give that as input to adj_fun to calculate the new adjoint value to pass to its parents by pushing it and the parent nodes onto the calculation stack. This step is recursively executed until the stack is empty.

The other operations required by Eq. 8.5 follow similarly

```
let exp_ad dr =
  let p = primal dr in
  let p' = Owl_maths.exp p in
  let adjfun' ca t =
      let r = !ca *. (Owl_maths.exp p) in
      (r, dr) :: t
  in
  {p=p'; a=ref 0.; adj_fun=adjfun'};;

let add_ad dr1 dr2 =
  let p1 = primal dr1 in
  let p2 = primal dr2 in
  let p' = Owl_maths.add p1 p2 in
  let adjfun' ca t =
    let r1 = !ca in
    let r2 = !ca in
    (r1, dr1) :: (r2, dr2) :: t
  in
  {p = p'; a = ref 0.; adj_fun = adjfun'};;
```

```
let div_ad dr1 dr2 =
  let p1 = primal dr1 in
  let p2 = primal dr2 in
  let p' = Owl_maths.div p1 p2 in
  let adjfun' ca t =
      let r1 = !ca /. p2 in
      let r2 = !ca *. (-.p1) /. (p2 *. p2) in
      (r1, dr1) :: (r2, dr2) :: t in
  {p = p'; a = ref 0.; adj_fun = adjfun'};;
```

We can then write the reverse mode differentiation function diff that runs a forward pass followed by a backward pass

```
let diff f =
  let f' x =
    let r = f x in (* the forward pass *)
    reverse_push [(1., r)]; (* the backward pass *)
    let x0, x1 = x in
    primal x0, !(adjoint x0), primal x1, !(adjoint x1)
  in f'
```

We can then derive the function f for reverse mode differentiation

```
let f (x0, x1) =
  let v2 = sin_ad x0 in
  let v3 = mul_ad x0 x1 in
  let v4 = add_ad v2 v3 in
  let v5 = make_reverse 1. in
  let v6 = exp_ad v4 in
  let v7 = make_reverse 1. in
  let v8 = add_ad v5 v6 in
  let v9 = div_ad v7 v8 in
  v9;;
```

Finally, with all these parts to hand, we can perform the differentiation having encapsulated the input variables, in just three lines of code. Again, the resulting adjoint values are just as expected.

```
# let x1 = make_reverse 1.;;
  val x1 : dr = {p = 1.; a = {contents = 0.}; adj_fun = <fun>}
# let x0 = make_reverse 1.;;
  val x0 : dr = {p = 1.; a = {contents = 0.}; adj_fun = <fun>}
# let pri_x0, adj_x0, pri_x1, adj_x1 = diff f (x0, x1);;
val pri_x0 : float = 1.
val adj_x0 : float = -0.181974376561731321
val pri_x1 : float = 1.
val adj_x1 : float = -0.118141988016545588
```

8.4.3 A Unified Implementation

We have shown how to implement forward and reverse AD separately, from scratch, but we often need a system that supports both modes. We start by combining the two record data types df and dr into a new data type t with the accompanying operations:

```
type t =
  | DF of float * float
  | DR of float * float ref * adjoint
and adjoint = float ref -> (float * t) list -> (float * t) list

let primal = function
  | DF (p, _) -> p
  | DR (p, _, _) -> p

let tangent = function
  | DF (_, t) -> t
  | DR (_, _, _) -> failwith "error: no tangent for DR"

let adjoint = function
  | DF (_, _) ->  failwith "error: no adjoint for DF"
  | DR (_, a, _) -> a
```

We then give functions to pack the variables of both modes into the same type t so that we can write a unified differentiation function.

```
let make_forward p a - DF (p, a)

let make_reverse p =
    let a = ref 0. in
    let adj_fun _a t = t in
    DR (p, a, adj_fun)

let rec reverse_push = function
  | [] -> ()
  | (v, x) :: t ->
    (match x with
     | DR (_, a, adjfun) ->
       a := !a +. v;
       let stack = adjfun a t in
       reverse_push stack
     | _ -> failwith "error: unsupported type")
```

Using this new data type, we can then simply combine the previous forward and reverse mode implementations into a single function using pattern matching. For example, the *sin* function becomes

```
let sin_ad x =
  let ff = Owl_maths.sin in
  let df p t = (Owl_maths.cos p) *. t in
  let dr p a = !a *. (Owl_maths.cos p) in
  match x with
  | DF (p, t) ->
    let p' = ff p in
    let t' = df p t in
    DF (p', t')
  | DR (p, _, _) ->
    let p' = ff p in
    let adjfun' a t =
      let r = dr p a in
      (r, x) :: t
    in
    DR (p', ref 0., adjfun')
```

Similarly, for the multiplication operator

```
let mul_ad xa xb =
  let ff = Owl_maths.mul in
  let df pa pb ta tb = pa *. tb +. ta *. pb in
  let dr pa pb a = !a *. pb, !a *. pa in
  match xa, xb with
  | DF (pa, ta), DF (pb, tb) ->
    let p' = ff pa pb in
    let t' = df pa pb ta tb in
    DF (p', t')
  | DR (pa, _, _), DR (pb, _, _) ->
    let p' = ff pa pb in
    let adjfun' a t =
      let ra, rb = dr pa pb a in
      (ra, xa) :: (rb, xb) :: t
    in
    DR (p', ref 0., adjfun')
  | _, _ -> failwith "unsupported op"
```

While this approach certainly works, it rapidly becomes clear that we are writing a considerable amount of boilerplate code. This is generally worth avoiding, something that OCaml's powerful mechanisms for abstraction can often help to achieve. So, before moving on, let us take a moment to think about how to categorise different math functions to see if we can find some structure to exploit to allow us to stop writing so much boilerplate code.

One way to categorise maths functions is by their *arity*, namely the number of input variables. Some are unary (one input, e.g., *sin x*), some are binary functions (two inputs, e.g., *mul x y*), and so on. Each shares an identical code structure and we can use OCaml's powerful module system to abstract over all operations of a given arity to avoid writing too much boilerplate code.

We will express each operation as an OCaml module, containing three core operations: (*i*) ff, to calculate the primal value of the function; (*ii*) df, to calculate the tangent value in the forward mode; and (*iii*) dr, to calculate the adjoint value in the reverse mode. The following code uses sin and mul as examples of unary and binary operations

```
module Sin = struct
  let ff = Owl_maths.sin
  let df p t = (Owl_maths.cos p) *. t
  let dr p a = !a *. (Owl_maths.cos p)
end

module Mul = struct
  let ff = Owl_maths.mul
  let df pa pb ta tb = pa *. tb +. ta *. pb
  let dr pa pb a = !a *. pb, !a *. pa
end
```

Modules of different arity need their own functions to calculate primal, tangent, and adjoint values, which we can provide using modules as parameters

```
let unary_op (module U: Unary) = fun x ->
  match x with
  | DF (p, t) ->
    lct p' = U.ff p in
    let t' = U.df p t in
    DF (p', t')
  | DR (p, _, _) ->
    let p' = U.ff p in
    let adjfun' a t =
      let r = U.dr p a in
      (r, x) :: t
    in
    DR (p', ref 0., adjfun')


let binary_op (module B: Binary) = fun xa xb ->
  match xa, xb with
  | DF (pa, ta), DF (pb, tb) ->
    let p' = B.ff pa pb in
    let t' = B.df pa pb ta tb in
    DF (p', t')
  | DR (pa, _, _), DR (pb, _, _) ->
```

```
    let p' = B.ff pa pb in
    let adjfun' a t =
      let ra, rb = B.dr pa pb a in
      (ra, xa) :: (rb, xb) :: t
    in
    DR (p', ref 0., adjfun')
  | _, _ -> failwith "unsupported op"
```

Each such function takes a module as input and returns a function that can be used in either forward or backward mode. This procedure is performed with a concise line of code for each operation. Using our *sin* and *mul* examples from before

```
let sin_ad = unary_op (module Sin : Unary)
let mul_ad = binary_op (module Mul : Binary)
```

Finally, to complete this example, the diff function should also be implemented in a unified way. In this implementation we focus on the tangent and adjoint value of x0 only.

```
let diff f =
  let f' x =
    let x0, x1 = x in
    match x0, x1 with
    | DF (_, _), DF (_, _) ->
      f x |> tangent
    | DR (_, _, _), DR (_, _, _) ->
      let r = f x in
      reverse_push [(1., r)];
      !(adjoint x0)
    | _, _ -> failwith "error: unsupported operator"
  in
  f'
```

And we are done! Now that both the forward and reverse modes are combined, and a large amount of extra code is removed by using modules, we can turn back to our earlier example and perform diff in the new unified differentiation system.

First, the forward mode:

```
# let x0 = make_forward 1. 1.;;
val x0 : t = DF (1., 1.)
# let x1 = make_forward 1. 0.;;
val x1 : t = DF (1., 0.)
# let f_forward x =
    let x0, x1 = x in
    let v2 = sin_ad x0 in
    let v3 = mul_ad x0 x1 in
    let v4 = add_ad v2 v3 in
```

```
    let v5 = make_forward 1. 0. in
    let v6 = exp_ad v4 in
    let v7 = make_forward 1. 0. in
    let v8 = add_ad v5 v6 in
    let v9 = div_ad v7 v8 in
    v9;;
val f_forward : t * t -> t = <fun>
# diff f_forward (x0, x1)
- : float = -0.181974376561731321
```

With only a minor change in the construction of the variables, we can also provide the reverse mode

```
# let x0 = make_reverse 1.;;
val x0 : t = DR (1., {contents = 0.}, <fun>)
# let x1 = make_reverse 1.;;
val x1 : t = DR (1., {contents = 0.}, <fun>)
# let f_reverse x =
    let x0, x1 = x in
    let v2 = sin_ad x0 in
    let v3 = mul_ad x0 x1 in
    let v4 = add_ad v2 v3 in
    let v5 = make_reverse 1. in
    let v6 = exp_ad v4 in
    let v7 = make_reverse 1. in
    let v8 = add_ad v5 v6 in
    let v9 = div_ad v7 v8 in
    v9;;
val f_reverse : t * t -> t = <fun>
# diff f_reverse (x0, x1)
- : float = -0.181974376561731321
```

Again, the results are consistent with the values computed earlier. We have successfully created a small unified differentiation system that supports both forward and reverse modes.

8.5 Forward and Reverse Propagation API

We have introduced algorithmic differentiation, describing its underlying theory and a simple illustrative implementation. We now examine Owl's more realistic implementation in the `Algodiff.Generic` module, which we have already met briefly in Secs. 8.3.1 and 8.3.2 when validating the our manual forward and reverse differentiation examples.

Algodiff.Generic is a functor that accepts an Ndarray module, and by providing Dense.Ndarray.S and Dense.Ndarray.D we produce AD modules supporting float32 and float64 precision respectively.

```
module S = Owl_algodiff_generic.Make
  (Owl_algodiff_primal_ops.S)
module D = Owl_algodiff_generic.Make
  (Owl_algodiff_primal_ops.D)
```

We will use the double precision Algodiff.D module in what follows.

8.5.1 Expressing Computation

We will start by expressing our earlier example, Eq. 8.5, using the AD module.

```
module AD = Algodiff.D
```

```
let f x =
  let x0 = AD.Mat.get x 0 0 in
  let x1 = AD.Mat.get x 0 1 in
  AD.Maths.((F 1.) / (F 1. + exp (sin x0 + x0 * x1)))
```

We use the operators provided by the AD module in the function definition, including the get operation and math operations. All parameters are of the same type, AD.t allowing the AD module to handle types including scalars, matrices, or ndarrays. F is the type constructor for float numbers while Arr is for arrays, and there are corresponding packing and unpacking functions for AD.t type.

```
open AD
```

```
# let input = Arr (Dense.Matrix.D.ones 1 2);;
# let result = f input |> unpack_flt;;
val result : float = 0.13687741466075895
```

Although this number packing mechanism is slightly cumbersome, it is still straightforward to perform forward and reverse mode differentiation, as shown in the next two examples.

8.5.2 Example: Forward Mode

Forward mode is accessed using the make_forward and tangent functions. It encapsulates the input as a value of type t, and the forward pass is produced simply by running f x.

```
open AD
```

```
# let x = make_forward input (F 1.) (tag ());;
# let y = f x;;
# let y' = tangent y;;
```

All the derivatives are ready once the forward pass is finished, stored as tangent values in y and retrieved using the tangent function.

8.5.3 Example: Reverse Mode

As in our strawman example, reverse mode consists of two phases, make_reverse, initialising the reverse mode pass, and reverse_prop, propagating errors via the backward pass.

```
open AD
```

```
# let x = Mat.ones 1 2;;
# let x' = make_reverse x (tag ());;
# let y = f x';;                      (* forward pass *)
# let _ = reverse_prop (F 1.) y;;  (* backward pass *)
# let y' = adjval x';;

- : A.arr =
          C0          C1
R0 -0.181974 -0.118142
```

The make_reverse function initialises the reverse mode. It does two things: (*i*) it wraps x, a dense matrix type, into a value of type t for Algodiff to process; (*ii*) it generates a unique tag for the input so that input numbers can be nested (discussed further in Sec. 8.7.1). Calling f x' constructs the computation graph of f, capturing the result in y. Finally, reverse_prop propagates the error back to the inputs. The gradient of f is stored in the adjacent value of x', and we can retrieve the value using adjval function.

8.6 High-Level Functions

So far we have introduced the low-level functions of the AD module, which allow a great deal of flexibility in controlling the flow of computation for building more sophisticated modules utilising differentiation such as implementing a neural network. However, it is often the case that such a degree of control is not required and, for example, you do not really want to worry about whether you are using forward or reverse mode as long as the result is correct. For such times, the AD module provides high-level APIs built on forward and reverse mode but with simplified interfaces. We introduce these high-level APIs in this section.

8.6.1 Derivative and Gradient

The simplest and most commonly used function for differentiating is `diff` in the AD module. Given a function `f` that maps one scalar value to another, we can calculate its derivative at point `x` by `diff f x`. For example, given the triangular function `tanh`, we can easily calculate its derivative at position $x = 0.1$

```
open Algodiff.D
```

```
let f x = Maths.(tanh x);;
let d = diff f (F 0.1);;
```

This also supports calculating higher-order derivatives – differentials of differentials, and so on – simply by chaining multiple calls to `diff`. For example, the first, second, third and fourth order derivatives of `f` are obtained as follows

```
let f x = Maths.(tanh x);;
let f1 = diff f;;
let f2 = diff f1;;
let f3 = diff f2;;
let f4 = diff f3;;
```

We can generalise derivatives of scalars to gradients of multivariate functions. For a function that maps a vector input to a scalar, we can use the `grad` function to calculate its gradient at a given point. For example, in three dimensional space, the gradient at each point on a surface consists of three elements representing the partial derivative along the x, y, and z axes. This vector shows the direction in which the function has the largest magnitude change.

One important application of gradient is in *gradient descent*, a widely used technique for finding the minimum value of a function. [17] The underlying idea is simple: at any point on a given surface we calculate the gradient to find the direction of the maximal change at this point, and then move in this direction by a small step. [18] We repeat this process until we can no longer find a vector that would reduce the function's value – a local optimum. We will discuss this in greater details in Chap. 10 and Chap. 9.

8.6.2 Jacobian

As gradient generalises derivatives from scalars to vectors, the *Jacobian* generalises gradient from vectors to matrices: `grad` is applied to functions mapping vectors to scalars while `jacobian` on the other hand is applied to functions that map vectors to vectors. If we assume the function f takes an input vector of length n, and produces an output vector of length m, then the Jacobian is defined as:

$$\mathbf{J}(y) = \begin{bmatrix} \frac{\partial y_0}{\partial x_0} & \frac{\partial y_0}{\partial x_1} & \cdots & \frac{\partial y_0}{\partial x_{n-1}} \\ \frac{\partial y_2}{\partial x_0} & \frac{\partial y_2}{\partial x_1} & \cdots & \frac{\partial y_2}{\partial x_{n-1}} \\ \vdots & \vdots & \cdots & \vdots \\ \frac{\partial y_{m-1}}{\partial x_0} & \frac{\partial y_{m-1}}{\partial x_1} & \cdots & \frac{\partial y_{m-1}}{\partial x_{n-1}} \end{bmatrix}$$

The intuition behind the Jacobian is similar to that of the gradient: at a particular point in the domain of the target function, the Jacobian shows how the output vector changes given a small change in the input vector.

One application of the Jacobian is in analysing dynamical systems. In such a system, $\mathbf{y} = f(\mathbf{x})$, suppose $f : \mathbf{R}^n \rightarrow \mathbf{R}^m$ is differentiable and its jacobian is $\mathbf{J}$.

According to the Hartman-Grobman theorem [19], the stability of a dynamical system near a stationary point is determined by the eigenvalues of $\mathbf{J}$: the system is stable if all the eigenvalues have negative real parts and unstable otherwise, unless the largest real part of the eigenvalues is zero in which case stability cannot be determined from the eigenvalues.

Let us revisit the two-body problem from Chap. 6 and examine the system's stability. This dynamical system is described by a group of differential equations:

$$y_0' = y_2$$
$$y_1' = y_3$$
$$y_2' = -\frac{y_0}{r^3}$$
$$y_3' = -\frac{y_1}{r^3}$$

Here $r = \sqrt{y_0^2 + y_1^2}$. We can express this system as follows:

```
open Algodiff.D

let f y =
  let y0 = Mat.get y 0 0 in
  let y1 = Mat.get y 0 1 in
  let y2 = Mat.get y 0 2 in
  let y3 = Mat.get y 0 3 in
```

```
let r = Maths.(sqrt ((sqr y0) + (sqr y1))) in
let y0' = y2 in
let y1' = y3 in
let y2' = Maths.( neg y0 / pow r (F 3.)) in
let y3' = Maths.( neg y1 / pow r (F 3.)) in

let y' = Mat.ones 1 4 in
let y' = Mat.set y' 0 0 y0' in
let y' = Mat.set y' 0 1 y1' in
let y' = Mat.set y' 0 2 y2' in
let y' = Mat.set y' 0 3 y3' in
y'
```

We can then calculate the Jacobian of f at the point $(1, 1, 1, 1)$ using the `Algodiff.D.jacobian` function:

```
let y = Mat.ones 1 4;;
let result = jacobian f y;;

let j = unpack_arr result;;
- : A.arr =
          C0          C1 C2 C3
R0         0           0  1  0
R1         0           0  0  1
R2  0.176777  0.53033    0  0
R3   0.53033  0.176777   0  0
```

Finally, we compute the eigenvalues of this Jacobian using Owl's linear algebra module, introduced previously in Sec. 4.5 of the Linear Algebra chapter.

```
let eigenvalues = Owl_linalg.D.eigvals j;;

val eigenvaluess : Owl_dense_matrix_z.mat =
                C0              C1              C2              C3
R0 (0.840896, 0i) (-0.840896, 0i) (0, 0.5946i) (0, -0.5946i)
```

We see that one of the eigenvalue is both real and positive and so we know that the system is unstable at the chosen point, $(1, 1, 1, 1)$.

8.6.3 Hessian and Laplacian

Given a multivariate function that maps n input variables to a scalar, we can examine its second order derivatives as a matrix called the *Hessian*.

$$\mathbf{H}(y) = \begin{bmatrix} \dfrac{\partial^2 y_0}{\partial x_0^2} & \dfrac{\partial^2 y_0}{\partial x_0 x_1} & \cdots & \dfrac{\partial^2 y_0}{\partial x_0 x_{n-1}} \\[2mm] \dfrac{\partial^2 y_1}{\partial x_1 x_0} & \dfrac{\partial^2 y_1}{\partial x_1^2} & \cdots & \dfrac{\partial^2 y_1}{\partial x_1 x_{n-1}} \\[2mm] \vdots & \vdots & \cdots & \vdots \\[2mm] \dfrac{\partial^2 y_{m-1}}{\partial x_{n-1} x_0} & \dfrac{\partial^2 y_{m-1}}{\partial x_{n-1} x_1} & \cdots & \dfrac{\partial^2 y_{m-1}}{\partial x_{n-1}^2} \end{bmatrix}$$

The Hessian can be calculated using the `Algodiff.D.hessian` function. An example of its use is in *Newton's method* for finding the minimum value of a function.

$$x_{n+1} = x_n - \alpha \mathbf{H}^{-1} \nabla f(x_n) \tag{8.7}$$

Rather than moving only in the direction of the gradient, Newton's method combines the gradient with the second order gradients of a function, $\dfrac{\nabla f(x_n)}{\nabla^2 f(x_n)}$, starting from a random position, and iterating until convergence, per Eq. 8.7.

```
open Algodiff.D

let rec newton ?(eta=F 0.01) ?(eps=1e-6) f x =
  let g = grad f x in
  let h = hessian f x in
  if (Maths.l2norm' g |> unpack_flt) < eps then x
  else newton ~eta ~eps f Maths.(x - eta * g *@ (inv h))
```

We can then apply this method to a two dimensional triangular function, starting from a random initial point, to find a local minimum. Note that the `newton` function takes a vector as input and outputs a scalar.

```
let _ =
  let f x = Maths.(cos x |> sum') in
  newton f (Mat.uniform 1 2)
```

Another useful related function is the `laplacian` which calculates the *Laplacian operator* $\nabla^2 f$, the trace of a Hessian matrix:

$$\nabla^2 f = \text{trace}(H_f) = \sum_{i=0}^{n-1} \frac{\partial^2 f}{\partial x_i^2}$$

The Laplacian is commonly used to represent the flux density of the gradient flow of a function, e.g. the moving rate of a chemical in a fluid. Therefore, differential equations which contain the Laplacian are frequently used in many fields to describe physical systems, including gravitational potentials, fluid flow, wave propagation, electric fields, and many more.

Table 8.5: List of other APIs in the AD module of Owl

Functions	Derivatives
diff'	as diff, but returns (f x, diff f x)
grad'	as grad, but returns (f x, grad f x)
jacobian'	as jacobian, but returns (f x, jacobian f x)
jacobianv	Jacobian vector product of f : vector -> vector at x along v, i.e., (jacobian x) v
jacobianv'	as jacobianv, but returns (f x, jacobianv f x)
jacobianTv	equivalent to transpose ((jacobianv f x v))
jacobianTv'	as jacobianTv, but returns (f x, transpose (jacobianv f x v))
hessian'	hessian vector product of f : (scalar -> scalar) at x along v, i.e., (hessian x) v
hessianv'	as hessianv, but returns (f x, hessianv f x v)
laplacian'	as laplacian, but returns (f x, laplacian f x)
gradhessian	(grad f x, hessian f x) for f : scalar -> scalar
gradhessian'	(f x, grad f x, hessian f x)
gradhessianv	(grad f x v, hessian f x v)
gradhessianv'	(f x, grad f x v, hessian f x v)

8.6.4 Other APIs

There are many other convenience functions in the AD module, such as jacobianv for calculating Jacobian vector product, and diff' for calculating both f x and diff f x at once. A complete list of functions is shown in Tbl. 8.5.

8.7 Internals of Algorithmic Differentiation

Having seen how to implement both forward and reverse mode differentiation, you are already an expert in AD. Moreover, you have also learnt how to use the high level APIs that in Owl. But as a software designer, you might be wondering how these pieces are put together in the real world.

It turns out that the simple implementations given above are not very far away from the industry-level implementations in Owl. While there are numerous details that must be dealt with, the gist is the same. While avoiding diving too deeply into the code, in this section we give an overview of the AD module to see how it differs from the simple versions above.

8.7.1 Architecture and Components

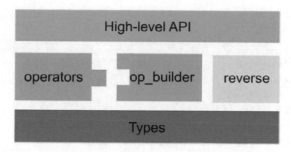

Fig. 8.4: Architecture of the AD module

Fig. 8.4 shows the structure of the AD module, and we will introduce its components one by one, starting with the **type definition**.

```
type t =
  | F   of A.elt
  | Arr of A.arr
  (* primal, tangent, tag *)
  | DF  of t * t * int
  (* primal, adjoint, op, fanout, tag, tracker *)
  | DR  of t * t ref * op * int ref * int * int ref
and adjoint = t -> t ref -> (t * t) list -> (t * t) list
and register = t list -> t list
and label = string * t list
and op = adjoint * register * label
```

There are several obvious differences to our earlier simple implementation. First, besides DF and DR, the type definition also contains two other constructors, F and Arr. F allows us to pack float numbers for scalar functions, whilst Arr can pack input and output for the functions handle ndarrys. You may also notice some extra fields in the DF and DR constructors. One important field of them is tag of type int, which is used to solve a particular problem when calculating higher order derivatives with nested forward and backward modes. This problem is referred to as *perturbation confusion*. Properly handling perturbation confusion is important for any AD implementation; we only scratch the surface of this problem here.

Consider: how should we compute the derivative of

$$f(x) = x \frac{d(x+y)}{dy}$$

i.e. a function that contains another derivative function. It initially seems straightforward: as $\frac{d(x+y)}{dy} = 1$ so $f'(x) = x' = 1$. Unfortunately, applying our previous simple implementation does not work.

```
# let diff f x =
    match x with
    | DF (_, _)      ->
      f x |> tangent
    | DR (_, _, _) ->
      let r = f x in
      reverse_push [(1., r)];
      !(adjoint x);;
val diff : (t -> t) -> t -> float = <fun>

# let f x =
    let g = diff (fun y -> add_ad x y) in
    mul_ad x (make_forward (g (make_forward 2. 1.)) 1.);;
val f : t -> t = <fun>

# diff f (make_forward 2. 1.);;
- : float = 4.
```

The result is 4 at point $(2, 2)$ but we have previously calculated it should be 1 at any point, so what has gone wrong?

Did you notice that x=DF(2,1)? The tangent value equals to 1, which means that $\frac{dx}{dx} = 1$. Now if we continue to use this same x value in function g, whose variable is y, the same x=DF(2,1) can be incorrectly translated by the AD engine as $\frac{dx}{dy} = 1$. Therefore, when used within function g, x should actually be treated as DF(2,0).

Tagging solves this nested derivative problem by distinguishing derivative calculations and their associated attached values using a unique tag for each application of the derivative operator. In that way, different inputs will be recognised properly by the AD engine. More details of this method is explained in [20].

Moving up we have the **builder** module that abstracts over the general process of forward and reverse modes, and the **operators** module that contains the specific calculation methods for each operation. Our simple implementation demonstrated how to construct differentiation for unary and binary operations using two builder modules, but of course there are many other operators with different types and so Owl abstracts according to their number of input variables and output variables. For example, the qr operations calculates QR decomposition of an input matrix. This operation uses the SIPO (single-input-pair-output) builder template.

Each operation in ops specifies three functions as in our simple example: (*i*) ff for calculating the primal value, (*ii*) df the tangent value, and (*iii*) dr the adjoint value). However, in our simple example all the constants are either DF or DR and so we must define both f_forward and f_reverse even though only the definition of constants differs. But now that the float number is included in the data type t, we can define just one computation function for both modes:

```
let f_forward x =
  let x0, x1 = x in
  let v2 = sin_ad x0 in
  let v3 = mul_ad x0 x1 in
  let v4 = add_ad v2 v3 in
  let v5 = F 1. in
  let v6 = exp_ad v4 in
  let v7 = F 1. in
  let v8 = add_ad v5 v6 in
  let v9 = div_ad v7 v8 in
  v9
```

We next consider how to compute DR and F data types together, by considering more cases in an operation than in our simple implementation which used just three functions (ff, df, and dr). For ff a b, we must consider cases such as a is scalar while b is an Ndarray. For df and dr, we must consider what happens if one input is constant, either F or Arr. Finally, above these parts sits the high level APIs introduced earlier.

8.7.2 Extending AD

A significant benefit of the module design described above is that it can be easily extended by providing modules representing new functions. For example, suppose that the AD system did not support the natural logarithm, $\sin x$, whose derivative is $\sin' x = \cos x$. Including this function is a simple matter of defining the necessary functions for calculating primal, tangent, and adjoint values in a module, and applying the relevant function from the Builder module – in this case, build_siso for building "single input, single output" functions.

```
open Algodiff.D

module Sin = struct
  let label = "sin"
  let ff_f a = F A.Scalar.(sin a)
  let ff_arr a = Arr A.(sin a)
  let df _cp ap at = Maths.(at * cos ap)
  let dr a _cp ca = Maths.(!ca * cos (primal a))
end

let new_sin_ad = Builder.build_siso (module Sin : Builder.Siso)
```

We can now use this function as if it were a native operation from the AD module. For example, we can use this new operator to build the calculation from our earlier example, and obtain the derivatives.

```
# let f x =
    let x1 = Mat.get x 0 0 in
    let x2 = Mat.get x 0 1 in
    Maths.(div (F 1.) (F 1. + exp (x1 * x2 + (new_sin_ad x1))));;
val f : t -> t = <fun>

# let x = Mat.ones 1 2;;
val x : t = [Arr(1,2)]

# let _ = grad f x |> unpack_arr;;
- : A.arr =
           C0          C1
R0 -0.181974 -0.118142
```

8.7.3 Lazy Evaluation

The Builder module makes it convenient to add new operations, simplifying code and improving productivity but not without cost: efficiency. Imagine a large problem that consists of thousands of operations, with one function occurring many times among those operations.[3] With the Builder approach described above, the operation will be recreated every time it is used which is rather inefficient. Fortunately, the simple caching provided by OCaml's *lazy evaluation* comes to our rescue.

OCaml provides a built-in function lazy that accepts an input of type 'a and returns a value of type 'a lazy_t where the computation of the value of type 'a has been delayed. This lazy expression won't be evaluated until it is called by Lazy.force, and the first time it is so called the expression is evaluated and the result cached. Subsequent applications of Lazy.force will simply return the cached result without further re-evaluation.

For example, imagine again that we need to add support for the sin operation. The definition of sin remains the same:

```
open Algodiff.D

module Sin = struct
  let label = "sin"
  let ff_f a = F A.Scalar.(sin a)
  let ff_arr a = Arr A.(sin a)
  let df _cp ap at = Maths.(at * cos ap)
  let dr a _cp ca = Maths.(!ca * cos (primal a))
end
```

---

[3] Such situations are actually quite common when using AD with neural networks where large computation graphs are created that use functions such as add and mul many hundreds of times.

However, we can instead use lazy evaluation to actually build the implementation, and benefit from the efficiency gains of the caching it provides.

```
let _sin_ad = lazy Builder.build_siso (module Sin : Builder.Siso);;
let new_sin_ad = Lazy.force _sin_ad;;
```

In this way, regardless of how many times this `sin` function is called in a massive computation graph, the `Builder.build_siso` process is only evaluated once.

8.8 Summary

In this chapter we have discussed algorithmic differentiation (AD) and its underlying theory, starting from the classic chain rule in differentiation, introducing the basic forward and reverse modes of AD, followed by a step-by-step illustration of how to implement a simple AD module. We finished with a more in-depth discussion of Owl's AD module, and how to extend and optimise it. After finishing this chapter, you should have a solid understanding of both AD theory and its implementation.

References

1. Baydin AG, Pearlmutter BA, Radul AA, Siskind JM. Automatic differentiation in machine learning: a survey. Journal of machine learning research. 2018 Apr 1;18.
2. Hecht-Nielsen R. Theory of the backpropagation neural network. Neural networks for perception 1992 Jan 1 (pp. 65-93). Academic Press.
3. Feynman RP, Leighton RB, Sands M. The feynman lectures on physics; vol. i. American Journal of Physics. 1965 Sep;33(9): 750-2.
4. Courant R, John F. Introduction to calculus and analysis I. Springer Science & Business Media; 2012 Dec 6.
5. Osler TJ. The fractional derivative of a composite function. SIAM Journal on Mathematical Analysis. 1970 May;1(2):288-93.
6. Atkinson KE. An introduction to numerical analysis. John wiley & sons; 2008.
7. Fischer H, Warsitz H. Complexity of derivatives generated by symbolic differentiation. In-Computer Algebra in Scientific Computing 2000 (pp. 129-144). Springer, Berlin, Heidelberg.
8. Margossian CC. A review of automatic differentiation and its efficient implementation. Wiley interdisciplinary reviews: data mining and knowledge discovery. 2019 Jul;9(4):1305.
9. Paszke A, Gross S, Chintala S, Chanan G, Yang E, DeVito Z, Lin Z, Desmaison A, Antiga L, Lerer A. Automatic differentiation in pytorch.
10. Naumann U. The art of differentiating computer programs: an introduction to algorithmic differentiation. Society for Industrial and Applied Mathematics. 2011 Jan.
11. Griewank A. Mathematical programming: Recent developments and applications. InOn automatic differentiation 1989 (pp. 83-108). Kluwer Academic Publishers.
12. Python AD [Internet]. Welcome to the ad package - ad Python package 1.3.2 documentation. [cited 2021 Dec 5]. Available from: https://pythonhosted.org/ad/.
13. Juliadiff [Internet]. JuliaDiff. [cited 2021 Dec 5]. Available from: https://www.juliadiff.org/.
14. MATLAB AD [Internet]. Cayugaresearch.com. [cited 2021 Dec 5]. Available from: http://www.cayugaresearch.com/admat.html.

15. Vishwanathan SV, Schraudolph NN, Schmidt MW, Murphy KP. Accelerated training of conditional random fields with stochastic gradient methods. Proceedings of the 23rd international conference on Machine learning 2006 Jun 25 (pp. 969-976).
16. Homescu C. Adjoints and automatic (algorithmic) differentiation in computational finance. Available at SSRN 1828503. 2011 Sep 12.
17. Ruder S. An overview of gradient descent optimization algorithms. arXiv preprint arXiv:1609.04747. 2016 Sep 15.
18. Curry HB. The method of steepest descent for non-linear minimization problems. Quarterly of Applied Mathematics. 1944. 2(3):258-61.
19. Hartman-Grobman theorem [Internet]. Wikipedia. Wikimedia Foundation; 2021 [cited 2021 Dec 5]. Available from: `https://en.wikipedia.org/wiki/Hartman-Grobman_theorem`.
20. Siskind JM, Pearlmutter BA. Perturbation confusion and referential transparency: Correct functional implementation of forward-mode AD.

Chapter 9
Optimisation

Abstract Optimisation is one of the most fundamental areas of numerical computing. From simple root finding to advanced machine learning, optimisation is everywhere. In this chapter, we will give you a brief overview of this topic and how the OCaml numerical library, Owl, supports basic optimisation methods.

9.1 Objective Functions

Mathematical optimisation is the problem of finding the minima (or maxima) of a function, either as a closed-form solution if one exists, or numerically. [1] Formally, an optimisation problem can be expressed in the form:

$$\text{minimise} \quad f_0(\mathbf{x}),$$
$$\text{subject to } f_i(\mathbf{x}) \le b_i, i = 1, 2, \ldots, n. \tag{9.1}$$

The function $f_0 : \mathbf{R}^n \to \mathbf{R}$ is the optimisation target, called the *objective function* or *cost function*. $\mathbf{x}$ is a vector of the *optimisation variables*, i.e. $\mathbf{x} = [x_1, x_2, \ldots x_n]$. An optimisation problem bounds the objective function zero or more *constraints* expressed as *constraint functions* $f_i : \mathbf{R}^n \to \mathbf{R}$ bounded by the b_i. The goal is to find the optimal values of $\mathbf{x} = \mathbf{x}^*$ where f_0 obtains the minimum value.

Solving an optimisation problem embodies the idea "maximise benefit/minimise cost with given constraints", which applies to many real world problems including task scheduling, resource management, investment portfolio optimisation, logistics, aeronautical engineering, and competitive games [2]. Optimisation problems can be categorised into different types. The simplest are called *linear optimisation* (or *linear programming*) where both the objective and the constraint functions are linear. Mathematically, this means the functions satisfy the conditions in Eq. 9.1.

$$f_i(\alpha x + \beta y) = \alpha f_i(x) + \beta f_i(y), \tag{9.2}$$

L. Wang et al., *OCaml Scientific Computing*, Undergraduate Topics in Computer Science, https://doi.org/10.1007/978-3-030-97645-3_9

Linear optimisation is important because non-negativity is a common constraint on real world quantities such as resource planning, and people are often interested in additive bounds. Additionally, many problems can be approximated by a linear model, e.g., financial portfolio optimisation, supermarket production planning, and delivery route optimisation [3]. Though practically often limited by the problem size, most linear optimisation problems are already solved and are provided for by off-the-shelf software tools.

Linear optimisation can be extended in various ways. For example, solving for a quadratic objective function under linear constraints is called *quadratic programming*. If we replace the "=" with "≤" in Eq. 9.2, it would make all the constraint function *convex* and the problem becomes *convex optimisation*, a generalised form of linear optimisation. In the optimisation world, convexity is considered a watershed between "easy" and "difficult" problems: convex problems have efficient algorithmic solutions, whereas most non-convex problems do not. The textbook by Boyd [4] focuses exclusively on the topic of convex optimisation.

Alternatively, if we remove all the constraints from Eq. 9.1, it becomes an *unconstrained optimisation* problem. If f is both convex and differentiable, this problem can be solved by finding the root x^* of the derivative of f so that $f'(x^*) = 0$. One particular type of constrained optimisation is the *equality constrained optimisation*, where all the constraints are expressed in the form of equality $Ax = b$. Such problems can be simplified into the corresponding unconstrained problems.

Optimisation covers a very wide range of topics and techniques, and this chapter gives only a very brief introduction. The rest of this chapter focuses mainly on the unconstrained and local optimisation. We refer readers to Boyd et al. [4] and Fletcher [5] for a more in-depth discussion. [1].

9.2 Root Finding

Root finding is a process searching for the variable values which make a continuous function equal to zero. These variable values are the *roots* of the function. We have already seen examples of root finding in Chapter 2. Actually, it is not an optimisation problem, but the two topics are closely related, and understanding the techniques used in root finding is also helpful for us to understand some of the the the optimisation methods.

The *bisection method* is a simple iterative method to find roots. Let's use $f(x) = x^2 = 2$ as an example. The root of this equation is $x = \sqrt{2} = 1.4142135623\ldots$. To find this solution, the bisection method proceeds as follows. We know the solution must be between 1 and 2, so first we set x to the mid-point $x = 1.5$. As $x^2 = 2.25$ is larger than 2, x is too large so we try $x = 1\frac{1}{4}$ next, which turns out to be too small. We then try $1\frac{3}{8}$, and so on, until a number is found, the square of which is close enough to

---

[1] When calculation of derivatives is required, we use here algorithmic differentiation as introduced in Chapter 8, due to its many benefits; but you may surely opt to use numerical differentiation if you prefer to do so.

2. The bisection method is often slow to converge, taking about 50 iterations to reach a precision of 14 digits after the decimal point, for example. However, bisection is a solid and reliable method to find roots as long as the function is continuous in the given region. Owl provides the `Owl_maths_root.bisec` function that implements this method.

The *Newton method* (also known as the *Newton-Raphson method*) is more complex, using the derivative of objective function f, but has faster convergence. It starts with an initial value x_0, and iteratively updates the variable as follows until the change in consecutive values is deemed too small:

$$x_{n+1} = x_n - \frac{f(x_n)}{f'(x_n)}. \tag{9.3}$$

We can use the Algorithm Differentiation module in Owl to perform the derivative calculation. Again, let's find the root of $x^2 - 2 = 0$ with this method using the following code, a direct translation of Eq. 9.3.

```
open Algodiff.D

let f x = Maths.(x ** (F 2.) - (F 2.))

let _ =
  let x = ref 1. in
  for _ = 0 to 6 do
    let g = diff f (F !x) |> unpack_elt in
    let v = f (F !x) |> unpack_elt in
    x := !x -. v /. g;
    Printf.printf "%.15f\n" !x
  done;;
```

The resulting sequence is very short compared to the bisection method, taking only 5 iterations to find the root:

```
1.500000000000000
1.416666666666667
1.414215686274510
1.414213562374690
1.414213562373095
1.414213562373095
1.414213562373095
```

The Newton method is very efficient due to it having *quadratic convergence*: the square of the error at one iteration is proportional to the error at the next iteration. As a result, it is the basis of many powerful numerical methods, such as optimisation, multiplicative inverses of numbers and power series, and solving transcendental equations.

However, the Newton method requires a function to be differentiable. If f is not smooth or computing the derivative is not possible, we need to approximate the tangent at one point with a secant through two points – the *Secant Method*:

$$f'(x) \approx \frac{f(x_n) - f(x_{n-1})}{x_n - x_{n-1}}. \tag{9.4}$$

This method does not need to compute derivatives but has similar convergence property as the Newton method. With the Secant method, two points are used at each iteration in approximating the derivative. An improved version, the *Inverse Quadratic Interpolation* (IQI) method, uses three points to perform the interpolation. This has the benefit that it's fast at the later phase of the process at the cost of intermediate results being potentially very unstable.

These methods can be combined to make the even more powerful *Brent's method* which is arguably the best amongst various root-finding methods. It combines the robustness of the Bisection method with the fast iteration speed of Secant and IQI methods. The core idea is to use the fast algorithm whenever possible, but switch to the robust one when intermediate results becomes unstable. `Owl_maths_root.brent` implements the Brent's method. The above example can then be simply solved with a one-liner.

```
# let f x = x *. x -. 2.;;
val f : float -> float = <fun>
# Owl_maths_root.brent f 0. 2.;;
- : float = 1.41421331999550248
```

9.3 Univariate Function Optimisation

Having briefly introduced several classic root-finding methods, we now turn to the main topic of this chapter: unconstrained optimisation problems. Let's consider first a simple case where there is only one variable in the objective function. We will see in the next section that the same optimisation methods used for multivariate functions can be applied to the univariate case, but they are much slower than the specific algorithm covered in this section. Understanding the optimisation of univariate functions is also an indispensable step towards understanding optimisation of multivariate ones.

9.3.1 Use Derivatives

If a function is continuous and differentiable, an obvious solution for finding extreme values is to locate where its first-order derivatives equals 0:

$$f'(x) = 0$$

This takes us back to our root finding solutions. If we already know the analytical form of $f'(x)$, life becomes very easy. For example, if $f(x) = x^2 - x$, then you can directly calculate root with $f'(x) = 2x - 1$. Otherwise, you can use the differentiation functions in Owl. Let's look at an example of optimising an objective function with a hump shape:

$$f(x) = \frac{1}{(x - 0.3)^2 + 0.01} + \frac{1}{(x - 0.9)^2 + 0.04} - 6$$

```
open Algodiff.D

let f x = Maths.(
  (F 1.) / ((x - F 0.3) ** (F 2.) + F 0.01) +
  (F 1.) / ((x - F 0.9) ** (F 2.) + F 0.04) - F 6.)

let g = diff f
let f' x = f (F x) |> unpack_flt
let g' x = g (F x) |> unpack_flt;;
```

To better understand the optimisation process, we can visualise the function and its derivative in Fig. 9.1.

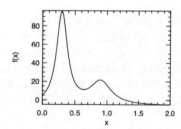 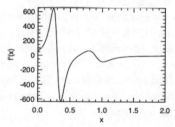

Fig. 9.1: The hump function and its derivative function

You can then obtain the extreme values by calling the Owl root finding functions, including Brent's:

```
# Owl_maths_root.brent g' 0. 0.4;;
- : float = 0.30037562625819042
# Owl_maths_root.brent g' 0.4 0.7;;
- : float = 0.63700940626897
# Owl_maths_root.brent g' 0.7 1.0;;
- : float = 0.892716303287079405
```

We can see that Brent's method works fine in this case, it successfully calculates extreme values in the given range. However, one problem of this method is that we cannot be certain which are maxima and which are minima.

9.3.2 Golden Section Search

We are again confronted by the question: what if calculating the derivatives of a function is difficult, or analytical solutions are not available? That leads us to some search-based approaches. In the previous section, we saw how we can quickly reach the solution by continuously shrinking an interval where the root locates. The *Golden Section Search* is built on the same idea, and for a function whose derivative is discontinuous, this numerical algorithm is an ideal choice since it does not require calculating derivatives [6].

As in the bisection method, Golden Section Search keeps shrinking intervals until they are "small enough". The bisection method uses a tuple [a, b], and iteratively inserts a third point c to improve this tuple. Golden section search this method uses three numbers rather than two: [a, b, c], and seeks to iteratively improve this tuple by inserting a forth number. It contains two ranges: [a, b] and [b, c]. For every iteration, we need to find a new number d within one of the two ranges. For example, if we choose the d within [b, c], and if $f(b) > f(d)$, then the new triplet becomes [b, d, c], otherwise the new triplet is chosen as [a, d, b]. With this approach, the range of this triplet keeps reducing until it is small enough and the minimum value can thus be found.

The only question left is to choose a suitable d point at each step. This approach first chooses the larger of the two ranges, either [a, b] or [b, c]. And then, instead of choosing the middle point in that range, it uses the fractional distance 0.38197 from the central point of the triplet. Indeed, the name of this method comes from the Golden Ratio used in this step. In general, this method is rather robust. It guarantees that each new iteration will bracket the minimum to a range just 0.61803 times the size of the previous one [6].

9.4 Multivariate Function Optimisation

The methods for univariate scenarios can be extended to f multivariate optimisation problems. The equivalent of derivative in this multi-dimensional space is the *gradient*, already seen in Chap. 8. To find the extreme values of a function with multiple independent values, we will introduce two methods, with and without using gradient.

9.4.1 Nelder-Mead Simplex Method

As with the Golden Section Search or Brent's, a non-gradient method is often slow but quite robust. One such method is the *Nelder-Mead Simplex Method* [7]. As the name suggested, it is probably the simplest way to minimise a well-behaved multivariate function. It simply goes "downhill" in a straightforward way, without any special assumptions about the objective function. [8]

The method utilises the concept of a *simplex*, namely a polytope of $n + 1$ vertices in an n-dimensional space, such as a triangle on a 2D plane. The basic idea of this method is to move a simplex gradually towards the lowest point on the function. In an N-dimensional space (where the function contains N variables), this simplex consists of $N + 1$ points. For example, for a function that has 2 variables, the simplex is a triangle; for 3 variables, a tetrahedron is used, etc.

From a starting point, the simplex move downwards step by step. At each step, the "highest" point p, having the largest value on the function, is found followed by moving p in three possible ways:

1. through the opposite face of the simplex to a new point; if this *reflected* point is not the "worst" point (point that leads to largest values on the function) among the $N + 1$ simplex points, accept it;
2. if the reflected point is the worst, then try to *contract* it towards the remaining points; if it is not the worst point, accept it;
3. if the contracted point is still the worst, then we have to *shrink* the simplex.

Repeat this process until it reaches a "valley", where the method starts to contract itself in the transverse direction and moves down the valley. The three different methods are illustrated in Fig. 9.2 where there are two variables and the simplex is a triangle.

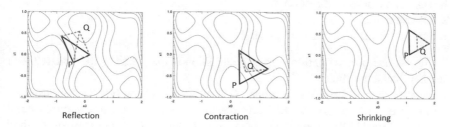

Fig. 9.2: Different movement of simplex in Nelder-Mead optimisation method

There are some other methods which do not rely on computing gradients such as Powell's method [12]. If the function is smooth, this method can find the direction it is going downhill, but using a one-dimensional optimisation method rather than computing gradients, and therefore faster than the simplex method. But this method is always a robust and cost-effective way to try solving an optimisation problem at the beginning.

9.4.2 Gradient Descent Methods

A *descent method* is an iterative optimisation process. [9] The idea is to start from an initial value, and then find a certain *search direction* along a function to decrease the value by certain *step size* until it converges to a local minimum. This process can be illustrated in Fig. 9.3.

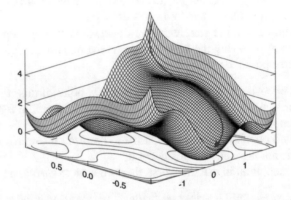

Fig. 9.3: Reach the local minimum by iteratively moving downhill

We can thus describe the *n*-th iteration of descent method as:

1. calculate a descent direction d;
2. choose a step size α;
3. update the location: $x_{n+1} = x_n + \alpha\, d$.

Repeat this process until a stopping condition is met, such as the update being smaller than a threshold.

Among descent methods, *Gradient Descent* is one of the most widely used algorithms to perform optimisation and the most common way to optimise neural networks. [10, 11] Based on the descent process above, Gradient Descent method uses the function gradient to decide its direction d, and can be described as:

1. calculate a descent direction $-\nabla f(x_n)$;
2. choose a step size α;
3. update the location: $x_{n+1} = x_n + \alpha\, \nabla f(x_n)$.

Here ∇ denotes the gradient. The distance α along a certain direction is also called *learning rate*. In a gradient descent process, when searching for the minimum, the point always follows the direction that is against the direction that is represented by the negative gradient.

Again, it is straightforward to implement this process with the algorithmic differentiation module in Owl. Let's look at one example. Here we use the Rosenbrock function [13] which is usually used as a performance test for optimisation problems. The function is defined as:

$$f(x, y) = (a - x)^2 + b(y - x^2)^2. \qquad (9.5)$$

The parameters are usually set as $a = 1$ and $b = 100$.

```
open Algodiff.D
module N = Dense.Ndarray.D

let rosenbrock a =
  let x = Mat.get a 0 0 in
  let y = Mat.get a 0 1 in
  Maths.( (F 100.) * (y - (x ** (F 2.))) ** (F 2.) +
    (F 1. - x) ** (F 2.) |> sum');;
```

Let's apply the gradient descent method and observe the optimisation trajectory:

```
let v = N.of_array [|2.; -0.5|] [|1; 2|]
let traj = ref (N.copy v)
let a = ref v
let eta = 0.0001
let n = 200;;
```

As preparation, we use an initial starting point [2, -0.5]. The step size eta is set to 0.0001, and the iteration number is 100. Then we can perform the iterative descent process. You can also run this process in a recursive manner.

```
let _ =
  for i = 1 to n - 1 do
  let u = grad rosenbrock (Arr !a) |> unpack_arr in
  a := N.(sub !a (scalar_mul eta u));
  traj := N.concatenate [|!traj; (N.copy !a)|]
  done;;
```

We apply the grad method to the Rosenbrock function iteratively, and the updated data a is stored in the traj array. Finally, let's visualise the trajectory of the optimisation process.

```
let plot () =
  let a, b = Dense.Matrix.D.meshgrid (-2.) 2. (-1.) 3. 50 50 in
  let c = N.(scalar_mul 100.
    (pow_scalar (sub b (pow_scalar a 2.)) 2.) +
    (pow_scalar (scalar_sub 1. a) 2.)) in
```

```
let h = Plot.create ~m:1 ~n:2 "plot_gradients.png" in
Plot.subplot h 0 0;
Plot.(mesh ~h ~spec:[ NoMagColor ] a b c);

Plot.subplot h 0 1;
Plot.contour ~h a b c;
let vx = N.get_slice [[]; [0]] !traj in
let vy = N.get_slice [[]; [1]] !traj in
Plot.plot ~h vx vy;
Plot.output h;;
```

We first create a mesh grid based on the Rosenbrock function to visualise the 3D image, then on the 2D contour image of the same function we plot how the result of the optimisation is updated, from the initial starting point towards a local minimum point. The visualisation results are shown in Fig. 9.4. On the right figure the black line shows the moving trajectory. You can see how it moves downwards along the slope to its minimum.

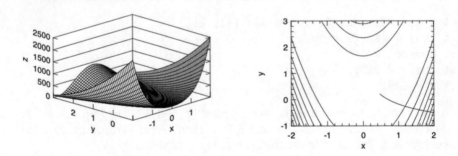

Fig. 9.4: Optimisation process of gradient descent on multivariate function

Optimisation is the foundation of machine learning and neural network training. In the `Owl.Optimise` module, we provide a `minimise_fun` function to perform this task. This function is actually an internal function that aims mainly to serve the Neural Network module, but nevertheless we can still use this function to solve an optimisation problem with gradient descent method. This function is based on the Algorithm Differentiation module. It minimises a function in the form of f: x → y with regard to x, an AD ndarray, and y is an AD scalar value. This function is implemented following the iterative descent approach. Let's use the previous Rosenbrock function example to demonstrate how it works.

```
let p = Owl_optimise.D.Params.default ()
let _ = p.epochs <- 10.
let _ = p.gradient <- Owl_optimise.D.Gradient.GD;;
```

First, we set the optimisation parameters. The `Owl_optimise.D.Params` sub-module contains several categories of parameters, including the gradient method, learning rate, loss functions, regularisation method, momentum method, epoch and batch etc. We will introduce these different parts in Chapter 10. For now it suffices simply to set the iteration limit epochs to something like 10 or 20 iterations. Then we set the gradient method to be the gradient descent and simply execute the code, starting from the same initial values:

```
let init_value = N.of_array [|2.;-0.5|] [|1;2|] |> pack_arr
let _ = Owl_optimise.D.minimise_fun p rosenbrock init_value;;
```

This function outputs detailed logs to track the intermediate results, looking in part like below. It shows how the function value, starting from 2926 at the initial point, is quickly reduced to about 2.5 within only 10 steps using gradient descent.

```
...
10:46:49.805 INFO : T: 00s | E: 1.0/10 | B: 1/10 | L: 2026.000
10:46:49.806 INFO : T: 00s | E: 2.0/10 | B: 2/10 | L: 476.1010
10:46:49.807 INFO : T: 00s | E: 3.0/10 | B: 3/10 | L: 63.83614
10:46:49.807 INFO : T: 00s | E: 4.0/10 | B: 4/10 | L: 37.77679
10:46:49.808 INFO : T: 00s | E: 5.0/10 | B: 5/10 | L: 21.39686
10:46:49.809 INFO : T: 00s | E: 6.0/10 | B: 6/10 | L: 11.74234
10:46:49.809 INFO : T: 00s | E: 7.0/10 | B: 7/10 | L: 6.567733
10:46:49.809 INFO : T: 00s | E: 8.0/10 | B: 8/10 | L: 4.085909
10:46:49.810 INFO : T: 00s | E: 9.0/10 | B: 9/10 | L: 3.016714
10:46:49.810 INFO : T: 00s | E: 10.0/10 | B: 10/10 | L: 2.5943
...
```

9.4.3 Conjugate Gradient Method

Unfortunately, a problem with Gradient Descent is that it may perform badly on certain types of functions. For example, if a function is steep and narrow then gradient descent will take very many small steps to reach the minimum, bouncing back and forth, even if the function is in quadratic form.

This can be fixed by the *Conjugate Gradient* (CG) method, first proposed by Hestenes and Stiefel [14]. The CG method is one of the most popular iterative method to solve large – and particularly *sparse* – linear system in the form of $Ax = b$ where system matrix A should be symmetric and positive-definite. The CG method is similar to gradient descent, but the new direction at each step does not completely follow the new gradient but is somehow *conjugated* to the old gradients and to all previous directions traversed.

For example, Fig. 9.5 compares the descent efficiency of the conjugate gradient against gradient descent. Both methods start from the same position and go for the same direction. At the next point, gradient descent follows the direction of the

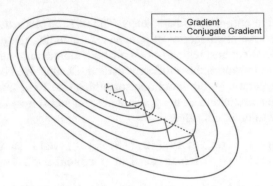

Fig. 9.5: Compare conjugate gradient and gradient descent

descent, which is a blunt one since this function is steep. But the conjugate method thinks, "hmm, this seems like a steep turn of direction, and I would prefer following the previous momentum a little bit". As a result, the conjugate method follows a direction in between (the red dotted line), and it turns out that the new direction avoids lots of unnecessary bounces and finds the minimum more efficiently than the gradient descent method.

When calculating, rather than using $-\nabla f(x_n)$, the conjugate gradient method chooses another way to calculate the descent direction. It maintains two sequences of updates:

$$x_{n+1} = x_n - \alpha_n \, A \, y_n$$
$$y_{n+1} = x_{n+1} + \beta_n \, y_n$$

where $\alpha_n = \frac{x_n^T \, y_n}{y_n^T \, A y_n}$, and $\beta_n = \frac{x_{n+1}^T \, x_{n+1}}{x_n^T \, x_n}$. Here x_n is the function variable to be optimised. As with the gradient descent method, the conjugate gradient is supported in Owl optimisation module as `Owl_optimise.D.Gradient.CG`.

9.4.4 Newton and Quasi-Newton Methods

Not to be confused with the newton method used in root-finding, there is also a Newton Method in optimisation [15]. Still following the basic process of descent method, the Newton method starts from an initial point and then repeats the process:

1. compute the descent direction: $d = -\frac{\nabla f(x_n)}{\nabla^2 f(x_n)}$
2. choose a step size α;
3. update the location: $x_{n+1} = x_n + \alpha d$.

We denote $\nabla^2 f(x_n)$ as the second-order derivatives of function f. For a scalar-valued function, its 2nd order derivatives can be represented by its Hessian matrix, which is introduced in Chap. 8. With the hessian matrix $\mathbf{H}$, the update process of newton method can be expressed as:

$$x_{n+1} = x_n - \alpha \mathbf{H_n}^{-1} \nabla f(x_n).$$

This equation can be implemented using the AD module as:

```
open Owl
open Algodiff.D


let rec newton ?(eta=F 0.01) ?(eps=1e-6) f x =
  let g, h = (gradhessian f) x in
  if (Maths.l2norm' g |> unpack_flt) < eps then x
  else newton ~eta ~eps f Maths.(x - eta * g *@ (inv h))

let _ =
  let f x = Maths.(cos x |> sum') in
  let y = newton f (Mat.uniform 1 2) in
  Mat.print y;;
```

One nice property about the Newton's method is that it converges *quadratically*. However, a big problem with the Newton method is its scalability, the method is suitable for small or medium problem size. In many real world applications, it is common to see optimisation problems with millions or more variables. In these cases, the complexity of computing the Hessian matrix explodes, as does computing its inverse.

To get around the scalability issue, *quasi newton* methods are proposed. The basic idea is to iteratively build up an approximation of the inverse of the Hessian matrix. Their convergence is fast, but not as fast as Newton's method. It takes about n quasi-newton iterations to progress similarly as the Newton's method. The most important method in this category is BFGS (Broyden-Fletcher-Goldfarb-Shanno), named after its four authors [5]. In the BFGS algorithm, the search direction d_i at each iteration i is calculated by

$$A_i p_i = -\nabla f(x_i).$$

In the formula, A_i is the approximation of Hessian matrix which is of the same shape $m \times m$. It is iteratively updated at each step. As a practical enhancement to the algorithm, the Limited-BFGS (L-BFGS) method [16] address the memory usage issue in BFGS. Instead of the large approximate matrix A_i, this method only stores a small number of vectors to represent this matrix, and also keeps updates from the last several iterations.

9.5 Global Optimisation and Constrained Optimisation

So far we have focused on unconstrained optimisation, where we are trying to find local optima. In the rest of this chapter we will give a very brief introduction to global optimisation and constrained optimisation.

The basic idea of global optimisation is to provide search methods and heuristics to traverse the search space effectively. One method is to start from a sufficient number of initial points and find the local optima, then choose the smallest/largest value from them. Another heuristic is to try stepping away from a local optimal value by taking a finite amplitude step away from it, performing the optimisation method, and verifying if it leads to a better solution.

An example of a global optimisation algorithm is the *Simulated Annealing Method* [17], inspired by the thermodynamics of metals cooling and annealing after being at a high temperature. During this process, the atoms often find the minimum energy state of the system automatically, just like finding the global optimum. Simulated annealing has already been used to solve the famous NP-hard travelling salesman problem to find the shortest path.

A suitable system to apply Simulated Annealing has several elements, starting with a finite set S, and a cost function f that is defined on the set. There is also a non-increasing function T that projects the set of positive integers to real positive values. $T(t)$ is called the *temperature* at time t. Suppose at time t, the current state is i in S. It randomly selects one of its neighbours j. Next, if $f(i) < f(j)$ then j is used as the next state. If not so, then j is chosen as the next state with a probability of $e^{-\frac{f(j)-f(i)}{T(t)}}$, otherwise the state remains i. Starting from an initial state x_0, this process is repeated for a finite number of steps to find the optimum.

Constrained Optimisation is another large optimisation topic. Unlike the optimisation problems we have seen so far, in a real-world problem it is hardly the case that you can freely explore the solution space with zero restriction, local or global. More often than not there will be constraints on the variables. The Linear Programming, Convex Programming, and Quadratic Programming we have briefly introduced in Sec. 9.1 all belong to this type of optimisation. Sometimes the variables are only allowed to take integer values, then it is called Integer Linear Programming (ILP).

With the help of algorithms such as the simplex algorithm and interior point methods, linear programming problems can often be efficiently solved with existing method and off-the-shelf tools. The tools such as LINDO [18] provide functionality to solve various sorts of optimisation problems. CVXOPT is a Python-based software package for convex optimization [19]. General scientific computing libraries and environments such as SciPy and MATLAB also provides optimisation tools.

Compared to linear optimisation, solving *non-linear optimisation* problems is even more challenging, especially the non-convex and non-quadratic problems. If both objective function and constrains are continuous, then the Lagrangian optimisation with or without Karush-Kuhn-Tucker (KKT) condition, can be used [20] [21]. If not, we can only rely on heuristics such as the simulated annealing methods we have just mentioned.

Finding a global solution that maximises or minimises a non-linear objective function is often time-consuming, even for only a small set of variables. Therefore, global optimisation of a non-linear problem is normally only used when it is absolutely necessary. For example, suppose a system pressure test is modelled as an optimisation problem. Given a small number of variants in the system, a global extreme value has to be found to test if the system is robust enough. Otherwise, a local maximum or minimum is often used instead as an approximation. In most engineering applications, a local extreme value is good enough. Even though optimisation cannot promise a true extreme, and can be easily affected by algorithm parameters and initial guess in iterative algorithms, as a trade-off, local optimum is much faster and thus still deemed acceptable.

9.6 Summary

In this chapter we have given an overview of the mathematical optimisation field. Optimisation problems can be viewed in two dimensions: local or global, unconstrained or constrained. We primarily focused on the local unconstrained problem. In the beginning, we introduce the techniques to find roots of function, since the basic idea of root finding is similar to that of optimisation. Next, the local unconstrained problem is explained in two parts: univariate and multivariate optimisation, depending on the number of target optimisation variant(s). An optimisation problem can be solved with or without calculating derivatives. We thus introduced methods in both aspects in these parts. Such methods including golden section search, simplex, newton method, gradient descent, etc. Of all these methods, gradient descent is especially important and we will encounter it again in the regression and neural network chapters. Finally, we give a brief introduction on the topic of global and constrained optimisation problems.

References

1. Jamil M, Yang XS. A literature survey of benchmark functions for global optimisation problems. International Journal of Mathematical Modelling and Numerical Optimisation. 2013 Jan 1;4(2):150-94.
2. Floudas CA, Pardalos PM, editors. Encyclopedia of optimization. Springer Science & Business Media; 2008 Sep 4.
3. Dantzig GB. Reminiscences about the origins of linear programming. InMathematical Programming The State of the Art 1983 (pp. 78-86). Springer, Berlin, Heidelberg.
4. Boyd S, Boyd SP, Vandenberghe L. Convex optimization. Cambridge university press; 2004 Mar 8.
5. Fletcher R. Practical methods of optimization. John Wiley & Sons; 2013 Jun 6.
6. Kiefer J. Sequential minimax search for a maximum. Proceedings of the American mathematical society. 1953 Jun 1;4(3):502-6.
7. Nelder JA, Mead R. A simplex method for function minimization. The computer journal. 1965 Jan 1;7(4):308-13.

8. Powell MJ. On search directions for minimization algorithms. Mathematical programming. 1973 Dec;4(1):193-201.
9. Courant R. Variational methods for the solution of problems of equilibrium and vibrations. Lecture notes in pure and applied mathematics. 1994 Aug 10.
10. Chong EK, Zak SH. An introduction to optimization. John Wiley & Sons; 2004 Apr 5.
11. Ruder S. An overview of gradient descent optimization algorithms. arXiv preprint arXiv:1609.04747. 2016 Sep 15.
12. Powell MJ. An efficient method for finding the minimum of a function of several variables without calculating derivatives. The computer journal. 1964 Jan 1;7(2):155-62.
13. Rosenbrock H. An automatic method for finding the greatest or least value of a function. The Computer Journal. 1960 Jan 1;3(3):175-84.
14. Hestenes MR, Stiefel E. Methods of conjugate gradients for solving linear systems. Washington, DC: NBS; 1952 Dec 6.
15. Avriel M. Nonlinear programming: analysis and methods. Courier Corporation; 2003.
16. Liu DC, Nocedal J. On the limited memory BFGS method for large scale optimization. Mathematical programming. 1989 Aug;45(1):503-28.
17. Pincus M. Letter to the editor—a Monte Carlo method for the approximate solution of certain types of constrained optimization problems. Operations research. 1970 Dec; 18(6):1225-8.
18. LINDO API for custom optimization application [Internet]. LINDO System, Inc. [cited 2021 Dec 5]. Available from: https://www.lindo.com/.
19. Andersen M, Dahl J, Vandenberghe L. CVXOPT: Convex Optimization. Astrophysics Source Code Library. 2020 Aug.
20. Kuhn HW, Tucker AW. Nonlinear programming. Berkeley. University of California Press. 1951; 13:54.
21. Karush W. Minima of functions of several variables with inequalities as side constraints. M. Sc. Dissertation. Dept. of Mathematics, University of Chicago. 1939.

Chapter 10
Regression

Abstract Regression is an important topic in statistical modelling and machine learning. It's about modelling problems which include one or more variables (also called "features" or "predictors") and require us to make predictions of another variable ("output variable") based on previous values of the predictors. Regression analysis includes a wide range of models, from linear regression to isotonic regression, each with different theoretical backgrounds and applications – explaining all these models is beyond the scope of this book. Here, we focus on several common forms of regression, particularly linear and logistic regression. We introduce their basic ideas, how they are supported in the numerical library, and how to use them to solve real problems.

10.1 Linear Regression

Linear regression models the relationship between input features and the output variable with a linear model. It is the most widely used regression model and the easiest to understand, so it is a good starting point for us to understand regression. Let's start with a simple problem where only one feature needs to be considered.

10.1.1 Problem: Where to open a new McDonald's restaurant?

McDonald's is undoubtedly one of the most successful fast food chains in the world. By 2018, it had opened more than 37,000 stores worldwide, and more are surely being built as you read this book. One question you might be interested in then is: where to open a new McDonald's restaurant?

© The Author(s), under exclusive license to Springer Nature Switzerland AG 2022
L. Wang et al., *OCaml Scientific Computing*, Undergraduate Topics in Computer Science,
https://doi.org/10.1007/978-3-030-97645-3_10

According to the website of McDonald's website[1], a lot of factors are considered: area population, existing stores in the area, proximity to retail parks, shopping centres, etc. Let's simplify the problem by assuming that expected profit is related only to area population. Suppose you are the decision maker in McDonald's, and also have access to the data about each branch store's profit and local population. How would you decide where to locate your next branch store? One way would be to use linear regression.

Since we don't really have the McDonald data at hand, we have prepared some synthesised data (in file `data.csv`, included in the code repository of this chapter) to use in our example. It consists of two columns, the first column represents the profit and the second one represents the population. They can be visualised using the code below. It extracts the two columns from the data file, converts them to a dense matrix, and get its two columns. They can be further visualises in a scatter plot as shown in Fig. 10.1. As expected, there is a clear trend that larger population and larger profit are co-related with each other. But precisely how?

```
let extract_data csv_file =
  let data = Owl_io.read_csv ~sep:',' csv_file in
  let data = Array.map (fun x ->
    Array.map float_of_string x) data
    |> Mat.of_arrays in
  let x = Mat.get_slice [[];[1]] data in
  let y = Mat.get_slice [[];[0]] data in
  x, y;;
```

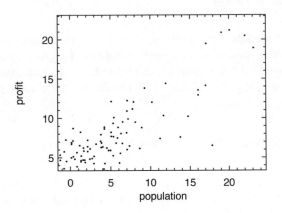

Fig. 10.1: Visualise data for regression problem

10.1.2 Cost Function

Let's start with a linear model that assumes the relationship between population and profit can be written as:

$$y = \theta_0 + \theta_1 x + \epsilon \tag{10.1}$$

where y denotes the profit we want to predict, and input variable x is the population number in this example. Since modelling can hardly make a perfect fit with the real data, we use ϵ to denote the error between our prediction and the data. Specifically, we represent the prediction part as:

$$h_{\theta_0,\theta_1}(x) = \theta_0 + \theta_1 x \tag{10.2}$$

The θ_0 and θ_1 are the parameters of this model. To be more concise, we can express $h_{\theta_0,\theta_1}()$ as $h_\theta()$. Mathematically they define a line on a plane. We first randomly choose some values for these parameters, to see how good the result looks, shown in Fig. 10.2. Our aim therefore is to find suitable values for these parameters so that the line is *close* to data we observed.

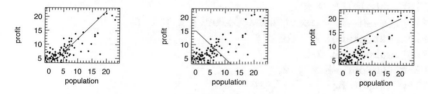

Fig. 10.2: Find possible regression line for given data

How do we define the "closeness" between the prediction line and the observed data? One frequently used method is to use the *ordinary least squares* method to minimise the sum of squared distances between the data and line [1]. We have shown the "(x, y)" data pairs above. We represent the total number of data pairs with n, and thus the i'th pair of data can be represented with $x^{(i)}$ and $y^{(i)}$. With these notations, we can represent a metric to measure the *closeness* as:

$$C_\theta(\boldsymbol{x}, \boldsymbol{y}) = \frac{1}{2n} \sum_{i=1}^{n} (h_\theta(x^{(i)}) - y^{(i)})^2 \tag{10.3}$$

Here $\boldsymbol{x}$ and $\boldsymbol{y}$ are both vectors of length n. In regression, we call this function the *cost function*. It measures how close the models are to an ideal one, and our target is thus clear: find suitable θ parameters to minimise the cost function.

Why do we use least square in the cost function? Physically, the cost function C represents the average distance of each data point to the line. By "distance" we mean the Euclidean distance between a data point and the point on the line with the same x-axis. A reasonable solution can thus be achieved by minimising this average distance.

10.1.3 Solving Problem with Gradient Descent

To give a clearer view, we can visualise the cost function with a contour graph. According to Eq. 10.3, the cost function is implemented as below:

```
let c x_data y_data theta0 theta1 =
  let f x = x *. theta1 +. theta0 in
  Mat.(pow_scalar (map f x_data - y_data) 2. |> mean') *. 0.5;;
```

Here x_data and y_data are the data introduced in Sec. 10.1.1. We can then visualise this cost function within a certain range using surface and contour graphs:

```
let plot_surface x_data y_data =
  let x, y = Mat.meshgrid (-20.) 10. (-20.) 10. 100 100 in
  let z = Mat.(map2 (c x_data y_data) x y) in
  let h = Plot.create ~m:1 ~n:2 "reg_cost.png" in
  Plot.subplot h 0 0;
  Plot.(mesh ~h ~spec:[ NoMagColor ] x y z);
  Plot.set_zlabel h "cost";
  Plot.subplot h 0 1;
  Plot.contour ~h x y z;
  Plot.output h;;
```

In Fig. 10.3 we can see that the cost function varies with parameters θ_0 and θ_1 with a bowl-like shape curve surface. The minimum point lies somewhere at the bottom of the "valley". It is thus natural to recall gradient descent [2], introduced in the optimisation chapter, and use it to find the minimal point in this bowl-shape surface.

Recall from the previous chapter that gradient descent works by starting at one point on the surface, and moves in the *direction* of steepest descent at some *step size*, then gradually approaches to a local minimum, hopefully as fast as possible. Let's use a fixed step size α, and the direction at a point on the surface can be obtained by using partial derivatives on the surface. Therefore, what we need to do is to apply this update process iteratively for both θ_i parameters:

$$\theta_i \leftarrow \theta_i - \alpha \frac{\partial}{\partial \theta_i} C_\theta(x, y), \tag{10.4}$$

for $i = 1, 2$. This process may seem complex at first sight, but by solving the partial derivative we can calculate it in two parts:

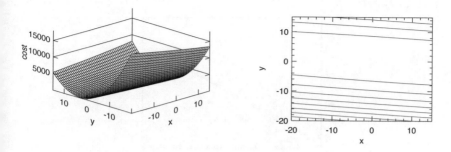

Fig. 10.3: Visualise the cost function in linear regression problem

$$\theta_0 \leftarrow \theta_0 - \frac{\alpha}{n} \sum_{i=1}^{n} (h_\theta(x_0^{(i)}) - y^{(i)}) x_0^{(i)}, \tag{10.5}$$

and

$$\theta_1 \leftarrow \theta_1 - \frac{\alpha}{n} \sum_{i=1}^{n} (h_\theta(x_1^{(i)}) - y^{(i)}) x_1^{(i)}. \tag{10.6}$$

Here the $x_0^{(i)}$ and $x_0^{(i)}$ are just different input features of the i-th row in data. Since we currently focus on only one feature x in our problem, $x_0^{(i)} = 1$ and $x_0^{(i)} = x^{(i)}$. Following these equations, we can perform gradient descent manually until it converges:

```
let gradient_desc x y =
  let alpha = 0.01 in
  let theta0 = ref 10. in
  let theta1 = ref 10. in
  for i = 0 to 500 do
    let f x = x *. !theta1 +. !theta0 in
    theta0 := !theta0 -. Mat.(map f x - y |> mean') *. alpha;
    theta1 := !theta1 -. Mat.((map f x - y) * x
      |> mean') *. alpha
  done;
  theta0, theta1;;
```

In the code above, we set the step size $\alpha = 0.01$, and start from a set of initial parameters: $\theta_0 = \theta_1 = 10$, aiming to gradually improve them. We then iteratively update the parameters over 500 iterations. Note that instead of manual summation in the equations, we use the vectorised ndarray operations. The computing result is $\theta_0 = 5.01$ and $\theta_1 = 0.57$. To validate whether these are indeed suitable parameters, we can visualise them against the input data. The resulting Fig. 10.4 shows a line that visually aligns well with the input data.

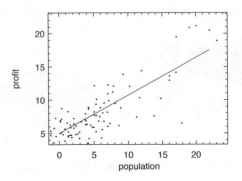

Fig. 10.4: Validate regression result with original dataset

Of course, there is no need to manually solve linear regression problems in Owl as it provides a number of high-level regression functions. For example, the `ols` function in the `Regression` module uses the ordinary least square method we have introduced to perform linear regression.

```
val ols : ?i:bool -> arr -> arr -> arr array
```

Here the boolean parameter `i` denotes if the constant parameter θ_0 is used or not. By default it is set to `false`. We can use this function to directly solve the problem, and the resulting parameters are close to what we have calculated manually:

```
# let theta = Regression.D.ols ~i:true x y;;
val theta : Owl_algodiff_primal_ops.D.arr array =
  [|     C0
R0 0.603401
;       C0
R0 4.63658 |]
```

The API is not limited to the regression module. The `linreg` function in the Linear Algebra module `Linalg` can also be used to perform the same task. You can try to run this function on the input data `x` and `y` to see if its output agrees with that of `ols` function.

Since the process of finding suitable parameters can be performed using gradient descent methods, we can use the gradient descent optimisation methods introduced in Chap. 9 and apply them on the cost function Eq. 10.3 to get the `theta` parameters. In fact, the regression functions in the `Regression` module are mostly implemented using the `minimise_weight` function from the `Optimisation` module.

Table 10.1: Sample of input data: multiple features

x_0	x_1	y
2369	3.78	450258
2136	4.39	286305
1261	3.16	199164
1607	2.34	368848
...	...	...

10.2 Multiple Regression

Back to our McDonald's problem. We have seen how a new store's profit can be related to the population of its surrounding, and we can predict it given past data. However, in the real world population is not the only feature that affects a store's profit – other factors such as pre-existing stores in the area, proximity to retail parks, shopping centres, etc. also play a role. How can we extend our single-variable linear regression to handle multiple features? For example, we can extend our previous prediction function to using multiple parameters as in Eq. 10.7.

$$h_{\theta_0,\theta_1,\theta_2,\theta_3,\dots}(x_1,x_2,x_3,\dots) = \theta_0 + \theta_1 x_1 + \theta_2 x_2 + \theta_3 x_3 \dots \quad (10.7)$$

We typically do not wish to list all parameters explicitly as many real-world problems may involve thousands (or even more) features. Instead we use the vectorised notation:

$$h_{\theta}(\mathbf{x}^{(i)}) = \theta \mathbf{x}^{(i)}, \quad (10.8)$$

where $\theta = [\theta_0, \theta_1, \theta_2, \theta_3, \dots]$, and $\mathbf{x}^{(i)} = [1, x_1^{(i)}, x_2^{(i)}, x_3^{(i)}, \dots]^T$ contains all the features from the i-th row in data. Accordingly, the cost function can be represented as Eq. 10.9, where $y^{(i)}$ is the output variable value on the i-th row of input data.

$$C_{\theta}(\mathbf{x}, \mathbf{y}) = \frac{1}{2n} \sum_{i=1}^{n} (\theta \mathbf{x}^{(i)} - y^{(i)})^2, \quad (10.9)$$

The derivative and manual gradient descent method can be similarly applied to solve this regression problem. But here we only show an example of using the regression function that the `Regression` module has provided. As in the previous problem, we provide data for this multiple variable problem in the code repository. It consists of several rows, each row contains three values: x_0, x_1, and y. Part of the data is listed in Tbl. 10.1.

Similar to the single-variable regression problem in the previous section, by using the `ols` regression function, we can easily get the multi-variable linear model. The data loading method is exactly the same as before.

```
let multi_regression csv_file =
  let data = Owl_io.read_csv ~sep:',' csv_file in
  let data = Array.map
    (fun x -> Array.map float_of_string x) data
    |> Mat.of_arrays in
  let x = Mat.get_slice [[];[0; 1]] data in
  let y = Mat.get_slice [[];[2]] data in
  Regression.D.ols ~i:true x y;;
```

The result is $\theta_0 = 57.52; \theta_1 = 57.47; \theta_2 = 57.0$, giving us a linear model that is approximately:

$$y = 57 + 57x_0 + 57x_1$$

This might be right – but something feels wrong. A simple validation check is to apply this model to any line of data from Tbl. 10.1, whereupon we see that the prediction result deviates too much from the true y value. So what went wrong? The answer requires us to address an important issue: normalisation.

10.2.1 Feature Normalisation

Crunching some numbers doesn't mean our work is properly done. Let's look at the multi-variable data in Tbl. 10.1 again. We see that the first feature is orders of magnitude larger than the second feature. That means the model and cost function are dominated by the first feature, and a minor change of the value in this column will have a disproportionately large impact on the model – and that's why the model we obtained above is wrong.

To overcome this problem, we must pre-process the data before the regression, normalising features to a common interval, say $[-1, 1]$. This step is also called *feature scaling* [3]. There are many ways to do this, one of which is *mean normalisation*: for a column of features, calculate its mean, and divide by the difference between the largest value and smallest value, as shown in the code below:

```
let norm_ols data =
  let m = Arr.mean ~axis:0 data in
  let r = Arr.(sub (max ~axis:0 data) (min ~axis:0 data)) in
  let data' = Arr.((data - m) / r) in
  let x' = Mat.get_slice [[];[0; 1]] data' in
  let y' = Mat.get_slice [[];[2]] data' in
  Regression.D.ols ~i:true x' y';;
```

Here data is the matrix we get from loading the CSV file from the previous section. This time we get a new set of parameters for the normalised data. The output is an array of size two, the first being an ndarray that contains two parameters: 0.93, −0.02, and the second being and ndarray that contains only one value: 0. These parameters set the model as $y = 0.93x_0 - 0.02x_1$. This result can be cross-validated with the analytical solution shown in the next section. You can also manually check this result with the normalised data:

```
val data' : (float, Bigarray.float64_elt)
  Owl_dense_ndarray_generic.t =
          C0      C1      C2
R0     0.094   0.154   0.208
R1     0.031   0.324  -0.099
...
```

Another benefit of performing data normalisation is to accelerate gradient descent [4]. We have already seen that Gradient Descent, which always tries to find the steepest downward path, may perform poorly on a slight slope. Normalisation effectively reshapes the slope to a better shape to avoid unnecessary zigzags. The illustration in Fig. 10.5 shows this.

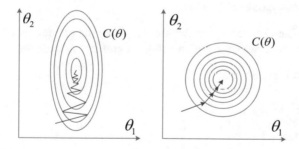

Fig. 10.5: Compare gradient descent efficiency with and without data normalisation

Normalisation is used in many other data analysis and machine learning tasks, not just regression. For example, in computer vision tasks, an image is represented as an ndarray with three dimensions. Each element represents a pixel in the image, with a value between 0 and 255. More often than not, this ndarray needs to be normalised in data pre-processed for the next step processing such as image classification [5].

10.2.2 Analytical Solution

Before taking a look at some other forms of regression, let's discuss solution to the linear regression besides gradient descent. It turns out that there is actually one closed form solution to linear regression problems, as shown in Eq. 10.10. Chap.3 of [12] gives the details of the derivation of this solution if you are interested.

$$\boldsymbol{\theta} = (X^T X)^{-1} X^T \mathbf{y} \qquad (10.10)$$

Suppose the linear model contains m features, and the input data contains n rows, then here X is an $n \times (m + 1)$ matrix representing feature data, and the output data $\mathbf{y}$ is an $n \times 1$ matrix. The reason there are $m + 1$ columns in X is that we need an extra constant feature for each data, equal to one for each data point.

With this method, we don't need to iterate over and over until we reach convergence – we can compute the result with one pass over the given input data. This calculation can be efficiently performed in Owl using its Linear Algebra module. Let's use the dataset from multi-variable regression again and perform the computation.

```
let o = Arr.ones [|(Arr.shape x).(0); 1|];;
let z = Arr.concatenate ~axis:1 [|o; x'|];;
let solution = Mat.dot (Mat.dot
    (Linalg.D.inv Mat.(dot (transpose z) z))
    (Mat.transpose z)) y';;
```

Here the x' and y' are the normalised data from the previous section. The result is very close to what we have gotten using the gradient descent based method:

```
val solution : Mat.mat =
            C0
R0 -4.5176E-17
R1    0.932052
R2   -0.0181452
```

Compared to the previous iterative methods, this method does not require multiple iterations, and you don't need to worry about setting hyper-parameters such as the choice of learning rate. However this approach does have its own problems. When the size of X, the input data, becomes very large the computation of large linear algebra operations such as matrix multiplication and inversion can become very slow. Or even worse: your computer may not have enough memory to perform the computation at all. In contrast, gradient descent works well even when the dataset is large.

Besides, there could be no solution at all using this method when the $X^T X$ matrix is non-invertible, e.g. it's a singular matrix. This could be caused by multiple reasons: perhaps some features are linearly dependent, or many features are redundant. Then techniques such as feature selection or regularisation are required. Most importantly, there will not always be a closed-form solution for you to use in other regression or machine learning problems. Gradient descent is both flexible and robust.

10.3 Non-Linear Regressions

As powerful as it is, not all regression problems can be solved with the linear model above. A lot of data can follow other patterns than a linear one. We can show this point with data from the Boston Housing Dataset [7] which contains information collected by the U.S Census Service concerning housing in the area of Boston, Massachusetts. It contains 506 cases, each containing 14 features, such as crime rate, nitric oxides concentration, average number of rooms per dwelling, etc. For this example, we observe that the relationship between percentage of lower status of the population ("LSTAT") and the median value of owner-occupied homes is in $1000's ("MDEV").

```
let f ?(csv_file="boston.csv") () =
  let data = Owl_io.read_csv ~sep:' ' csv_file in
  let data = Array.map
    (fun x -> Array.map float_of_string x) data
    |> Mat.of_arrays in
  let lstat = Mat.get_slice [[];[12]] data in
  let medv = Mat.get_slice [[];[13]] data in
  lstat, medv;;
```

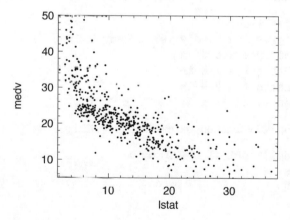

Fig. 10.6: Visualise part of the boston housing dataset

We can then visualise the data to see the trend, in Fig. 10.6. The relationship basically follows a convex curve – so while you could try to fit a straight line to these data, it's quite likely that the result would not be a good fit. Instead, we might use a non-linear model. In this section, we use examples to present a type of non-linear regression, the *polynomial regression* [8].

In polynomial regression, the relationship between the feature x and the output variable is modelled as an n-th degree polynomial with regards to feature x:

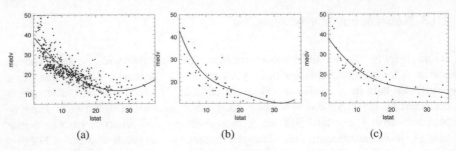

(a)   (b)   (c)

Fig. 10.7: (a) Polynomial regression based on Boston housing dataset; (b) polynomial regression with higher order; (c) revised polynomial model by applying regularisation in regression.

$$h_\theta(x) = \theta_0 + \theta_1 x + \theta_2 x^2 + \theta_3 x^3 \ldots \qquad (10.11)$$

Owl provides a function to perform this type of regression:

```
val poly : arr -> arr -> int -> arr
```

In the code, we use the `poly` function in the `Regression` module to get the model parameter. We limit the model to be second order.

```
let poly lstat medv =
  let a = Regression.D.poly lstat medv 2 in
  let a0 = Mat.get a 0 0 in
  let a1 = Mat.get a 1 0 in
  let a2 = Mat.get a 2 0 in
  fun x -> a0 +. a1 *. x +. a2 *. x *. x;;
```

By executing the `poly` function, the parameters we can get are:

```
- : Owl_algodiff_primal_ops.D.arr =
          C0
R0     42.862
R1   -2.33282
R2 0.0435469
```

That gives us the polynomial model:

$$f(x) = 42.8 - 2.3x + 0.04x^2 + \epsilon$$

We can visualise this model to see how well it fits the data in Fig. 10.7a.

10.4 Regularisation

Regularisation is an important issue in regression, and is widely used in many regression models. The motivation arises from the problem of *over-fitting*, where a model is tuned too closely to a particular data set and so may fail to predict future observations well [9].

Let's use polynomial regression as an example. Instead of limiting to second order, we'll use a fourth order polynomial, and take only a subset of 50 rows from the full data set to better visualise the over-fitting problem:

```
let subdata, _ = Mat.draw_rows ~replacement:false data 50;;
```

We then obtain the new fourth order model:

$$f(x) = 63 - 10.4x + 0.9x^2 - 0.03x^3 + 0.0004x^4.$$

This is visualised in Fig. 10.7b where we can see that it over-fits to the given data, capturing even the outliers – points that differs significantly from other observation. The result is that it is not good and predicting future, as-yet unknown, values.

To reduce the effect of higher order parameters, we can penalise these parameters in the cost function, designing the cost function so that the large parameter values lead to higher cost, and therefore by minimising the cost function we keep the parameters relatively small. This is actually a fairly simple change to the cost functions: all we need do is add an extra term at the end, e.g.,

$$C_\theta(x, y) = \frac{1}{2n} \left[\sum_{i=1}^{n} \left(h_\theta(x^{(i)}) - y^{(i)} \right)^2 + \lambda \sum_{l=1}^{m} \theta_i^2 \right]. \tag{10.12}$$

Here the sum of squared parameter values is the *penalty* [10] we add to the original cost function, and λ is a regularisation control parameter. That leads to a minor change in the derivative of $C(\theta)$ when using gradient descent:

$$\theta_j \leftarrow \theta_j - \frac{\alpha}{n} \left[\sum_{i=1}^{n} (h_\theta(x^{(i)}) - y^{(i)})x_j^{(i)} - \lambda\theta_j \right]. \tag{10.13}$$

We can now apply the new update procedure in the gradient descent code, using a polynomial model with 4th order. Currently not all the regression methods in Owl support regularisation. But we can implement one easily for the polynomial regression:

```
open Optimise.D
open Optimise.D.Algodiff
let poly_ridge ~alpha x y n =
  let z =
    Array.init (n + 1) (fun i ->
      A.(pow_scalar x (float_of_int i |> float_to_elt)))
  in
```

```
let x = A.concatenate ~axis:1 z in
let params =
  Params.config
    ~batch:Batch.Full
    ~learning_rate:(Learning_Rate.Const 1.)
    ~gradient:Gradient.Newton
    ~loss:Loss.Quadratic
    ~regularisation:(Regularisation.L2norm alpha)
    ~verbosity:false
    ~stopping:(Stopping.Const 1e-16)
    100.
in
(Regression.D._linear_reg false params x y).(0);;
```

The implementation is based on the `Optimisation` module and its general low level `_linear_reg` function. It uses the optimisation parameter module `Params`, which will be explained in detail in the neural network chapter. For now, it suffices to know that the key point is to use the L2-norm function as regularisation method.

By using this regularised version of polynomial regression, we obtain the updated model shown in Fig. 10.7c. Here we choose $\alpha = 20$. We can see that by using regularisation the model is less prone to over-fitting compared to Fig. 10.7b. Note that linear regression is used as an example in the equation, but regularisation is widely used in all kinds of regressions.

10.4.1 Ols, Ridge, Lasso, and Elastic_net

You might notice that Owl provides a series of functions other than `ols` in the regression module, such as:

```
val ridge : ?i:bool -> ?alpha:float -> arr -> arr -> arr array
val lasso : ?i:bool -> ?alpha:float -> arr -> arr -> arr array
val elastic_net : ?i:bool -> ?alpha:float -> ?l1_ratio:float
  -> arr -> arr -> arr array
```

What are they? The short answer is that they provide regularisation in regression using different methods. The `ridge` cost function [11] adds the L2 norm of θ as the penalty term: $\lambda \sum \theta^2$, which is what we have just introduced. The `lasso` cost function is similar, adding the *L1 norm*, or absolute value of the parameter as penalty: $\lambda \sum |\theta|$. This difference means that `lasso` permits coefficients to be zero, very useful for feature selection [13] [14]. In [21] the authors propose `elastic_net` which combines the penalties of the previous two:

$$\lambda(\frac{1-a}{2} \sum \theta^2 + a \sum |\theta|),$$

where a is a hyper parameter balancing between `ridge` and `lasso`. This method aims to make feature selection less dependent on the input data. We can thus choose one of these functions to perform regression with regularisation on the dataset in the previous chapter.

10.5 Logistic Regression

Whether we have used linear or polynomial regression, we have so far been interested in predicting a particular value from our problem. But what if we care about, not the particular value, but rather a *classification*? For example, we might have some historical medical data and need to use it to decide if a tumour signifies cancer or not, based on several features. Such variables are called *categorical variables*, and they represent data that can be divided into groups. Such variables might include "age group", "religion", "nationality", etc.

To predict which group a given data point belongs to, we use *Logistic Regression* which seeks to determine "boundaries" in the the data space so that different data is divided into corresponding variable groups. Let's start with investigating how it works.

10.5.1 Sigmoid Function

As a first naive solution, we might still try to use linear regression, interpreting the model as the possibility of one or other category. However, the prediction could easily fall outside the bounds $[0, 1]$ – so how might we normalise the result to fall in this interval? The solution is to use the sigmoid function (or logistic function)

$$\sigma(x) = \frac{1}{1 + e^{-x}}. \tag{10.14}$$

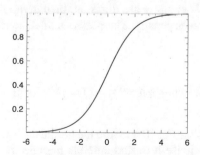

Fig. 10.8: The logistic function curve

As shown in Fig. 10.8, this function projects value within the interval $[0, 1]$. Applying this function on the returned value of a regression, we can get a model that returns value within $[0, 1]$.

$$h_\theta(x) = \sigma(\theta^T x) = \frac{1}{1 + e^{-\theta^T x}}. \tag{10.15}$$

This model is quite straightforward to interpret. The function value can be viewed as a possibility. If it is larger than 0.5, then the classification result is 0, otherwise it returns 1. Remember that in logistic regression we only care about the classification. So for a 2-class classification, returning 0 and 1 is enough.

10.5.2 Cost Function

With the new model comes a new cost function. Recall that we can find suitable parameters by minimising the cost function with training data. The cost function of linear regression in Eq. 10.3 indicates the sum of squared distances between the data and the model line. We can continue to use it here, with the new $h(x)$ function defined using sigmoid function. But the problem in this case is that it will end up being non-convex, so gradient descent can only give us one of many local minima. Therefore, we define the cost function in logistic regression as:

$$C_\theta(x, y) = \frac{1}{m} \sum_{i=1}^{m} g(h_\theta(x^{(i)}), y^{(i)}), \tag{10.16}$$

where the function g is defined as:

$$g(h_\theta(x), y) = \begin{cases} -log(h_\theta(x)), & \text{if } y = 1 \\ -log(1 - h_\theta(x)), & \text{if } y = 0 \end{cases} \tag{10.17}$$

Both forms of function $g()$ capture the same idea. As the h function is in the range $[0, 1]$, the range of g is $[0, \infty])$. When the values of $h(x)$ and y are close, the item within the summation in Eq. 10.16 $g(h(x)) - y$ will close to 0; on the other hand, if the prediction result $h(x)$ and y are different, then $g(h(x)) - y$ will incur a large value to the cost function as penalty. The previous three equations can be combined into Eq. 10.18.

$$C_\theta(x, y) = \frac{1}{m} \sum_{i=1}^{m} (y^{(i)} \log(h_\theta(x^{(i)})) + (1 - y^{(i)}) \log(1 - h_\theta(x^{(i)}))) \tag{10.18}$$

The next step is to follow Eq. 10.4 to find the partial derivative of this cost function and then iteratively minimise it to find suitable parameters θ. It turns out that the partial derivative of this cost function is similar as that in linear regression:

$$\frac{\partial C_\theta(\boldsymbol{x}, \boldsymbol{y})}{\partial \theta_j} = \frac{1}{m} \sum_{i=1}^{m} (\sigma_\theta(x^{(i)}) - y^{(i)})^2 x_j^{(i)} \tag{10.19}$$

This solution benefits from the fact that the sigmoid function has a simple derivative: $\sigma'(x) = \sigma(x)(1 - \sigma(x))$. With this derivative at hand, the remainder follow what we did with linear regression: follow Eq. 10.4 and repeat this gradient descent step until it converges. Owl also provides a `logistic` function in the `Regression` module. In the next section, we will show an example of binary categorisation with logistic regression.

10.5.3 Example

To perform the logistic regression, let's first prepare some data. We can generate the data with the code below:

```
let generate_data () =
  let open Mat in
  let c = 500 in
  let x1 = (gaussian c 2 *$ 2.) in
  let a, b = float_of_int (Random.int 5),
    float_of_int (Random.int 5) in
  let x1 = map_at_col (fun x -> x +. a) x1 0 in
  let x1 = map_at_col (fun x -> x +. b) x1 1 in
  let x2 = (gaussian c 2 *$ 1.) in
  let a, b = float_of_int (Random.int 5),
    float_of_int (Random.int 5) in
  let x2 = map_at_col (fun x -> x -. a) x2 0 in
  let x2 = map_at_col (fun x -> x -. b) x2 1 in
  let y1 = create c 1 ( 1.) in
  let y2 = create c 1 ( 0.)in
  let x = concat_vertical x1 x2 in
  let y = concat_vertical y1 y2 in
  x, y

let x, y = generate_data ();;
```

Basically this code creates two groups of random data with `gaussian` function. Data x is of shape `[|1000; 2|]` and is equally divided into two groups, the first at a higher position with the corresponding y label positive, and the second in a lower group labelled as negative. Our task is to try to divide a given data point into one of these two categories.

With the `logistic` function, we train a model:

$$h_\theta(x_0, x_1) = \sigma(\theta_0 x_0 + \theta_1 x_1 + \theta_2).$$

In the linear model within the sigmoid function, we have two parameters θ_0 and θ_1 for the two variables that represent the two coordinates of a data point. The logistic functions takes an i argument. If i is set to true, the linear model contains an extra parameter θ_2. Based on the data, we can get the parameters by simply executing:

```
# let theta =
    Owl.Regression.D.logistic ~i:true x y;;
val theta : Owl_algodiff_primal_ops.D.arr array =
    [|     C0
R0 16.4331
R1 12.4031
;       C0
R0 20.7909
|]
```

Therefore, the model we get is:

$$h(x_0, x_1) = \sigma\ (16\ x_0 + 12\ x_1 + 20). \tag{10.20}$$

We can validate this model by comparing the inference result with the true label y. Here any prediction value larger than 0.5 produced by the model is deemed as positive, otherwise it's negative.

```
let test_log x y =
  let p' = Owl.Regression.D.logistic ~i:true x y in
  let p = Mat.(p'.(0) @= p'.(1)) in
  let x = Mat.(concat_horizontal x (ones (row_num x) 1)) in
  let y' = Mat.(sigmoid (x *@ p)) in
  let y' = Mat.map (fun x -> if x > 0.5 then 1. else 0.) y' in
  let e = Mat.((mean' (abs (y - y')))) in
  Printf.printf "accuracy: %.4f\n" (1. -. e);;
```

```
# test_log x y;;
accuracy: 0.9910
```

This result shows that the trained model obtains a very high accuracy, albeit on the dataset used to train it.

As we said earlier, classification effectively seeks to draw a *decision boundary* in a hyperplane to separate categorically different groups of data points. For example, if we are using a linear model h within the sigmoid function, the linear model itself divides the points into two halves in the plane. Using Eq. 10.20 as an example, any x_0, x_1 that gives $h(x_0, x_1) > 0$ is taken as positive, otherwise it's negative. Therefore, the boundary line we need to draw is: $16x_0 + 12x_1 + 20 = 0$, or $x_1 = -(4x_0 + 5)/3$. We can visualise this decision boundary on a 2D-plane and how it divides the two groups of data.

```
open Owl

let data = Mat.concat_horizontal x y
let f x = -.(4. *. x +. 5.) /. 3.

let plot_logistic data =
  let neg_idx = Mat.filter_rows
    (fun m -> Mat.get m 0 2 = 0.) data in
  let neg_data = Mat.get_fancy
    [ L (Array.to_list neg_idx); R [] ] data in
  let pos_idx = Mat.filter_rows
    (fun m -> Mat.get m 0 2 = 1.) data in
  let pos_data = Mat.get_fancy
    [ L (Array.to_list pos_idx); R [] ] data in
  let h = Plot.create "reg_logistic.png" in
  Plot.(scatter ~h ~spec:[ Marker "#[0x2217]"; MarkerSize 5. ]
    (Mat.get_slice [[];[0]] neg_data)
    (Mat.get_slice [[];[1]] neg_data));
  Plot.(scatter ~h ~spec:[ Marker "#[0x2295]"; MarkerSize 5. ]
    (Mat.get_slice [[];[0]] pos_data)
    (Mat.get_slice [[];[1]] pos_data));
  Plot.plot_fun ~h ~spec:[ RGB (0,0,255); LineWidth 2. ]
    f (-5.) 5.;
  Plot.output h;;
```

The code above visualises the data, two types of points showing the negative and positive data, and the line shows the decision boundary we get from the logistic model. The result is shown in Fig. 10.9. Although there are some categorisations that are incorrect, the model appears to work well for most of the data.

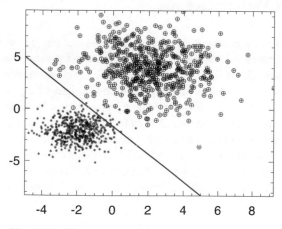

Fig. 10.9: Visualise the logistic regression dataset

Of course, we can use more than a simple linear model within the sigmoid function. For example, we can configure the model as $h_\theta(x) = \sigma(\theta_0 + \theta_1 x + \theta_2 x^2)$. If we use a non-linear polynomial model, the plane is divided by curves rather than straight lines.

Logistic regression uses the linear model. If you believe your data won't be linearly separable, or you need to be more robust to outliers, you might look to non-linear model, or the Support Vector Machines (SVMs, see below) [15].

10.5.4 Multi-class Classification

We have seen how to classify objects into one of two classes using logistic regression but what if we have more than two classes? One approach is called *one-vs-all*. Suppose we need to classify one object into three different classes. We might still use logistic regression but rather than just one binary classifier, we train three binary classifiers, each pitting one class against the other two. In the prediction phase, using an object as input, each classifier yields a probability as output. We then choose the largest probability as the classification result.

The multi-class classification problem is prevalent in the image recognition tasks, which often includes classifying one object in the image into one of tens, hundreds, or even thousands different classes. For example, a standard classification problem is to recognise hand-written numerals based on the MNIST dataset. It requires the model to determine which of ten digits a 28x28 grey scale image, represents. We will discuss this example in detail later in the beginning of the Neural Network chapter where you will see logistic regression is extended, and thus how the real-world multi-class classification problems are solved using neural networks.

10.6 Support Vector Machines

These are a group of supervised learning algorithms that can be used for classification and regression tasks. The SVM is one of the most powerful and widely used machine learning methods, used to solve real-world problems in many fields, including classification of text, image, satellite data, and protein data [16] [17] [18] [19]. As in logistic regression, the SVM creates a hyperplane (decision boundary) that separates data into classes. However, it tries to maximise the "margin" between the hyperplane and the boundary points of the classes in the data. By using "kernels", the SVM can also fit non-linear boundaries. SVM also works well with unstructured and semi-structured data such as text and images, with reduced risk of over-fitting. In this section, we will introduce the basic idea behind SVMs and how they work through examples.

Again, let's start with the objective cost function. It turns out that the cost function of the SVM is very similar to that of logistic regression in Eq. 10.18, with some modifications:

$$C_\theta(x, y) = \frac{1}{m} \sum_{i=1}^{m} (y^{(i)} g_0(\theta^T x^{(i)}) + (1 - y^{(i)}) g_1(\theta^T x^{(i)})) + \frac{1}{2} \sum_{j=1}^{m} \theta_j^2 \quad (10.21)$$

Functions g_0 and g_1 are simplifications of the logarithm function:

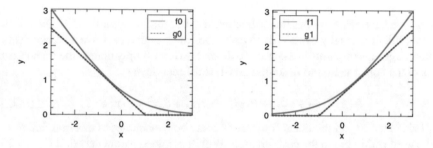

Fig. 10.10: Simplifying the cost function of logistic regression

Here $f_0(x) = -\log(\sigma(x))$ is what is used in the cost function of the logistic regression. This computation heavy logarithm is replaced with $g_0(x)$, a simple segmented function. Similarly, $f_1(x) = \log(1 - \sigma(x))$ is replaced by $g_1(x)$.

Another difference is that a regularisation term is added to the cost function in Eq. 10.21. Therefore, considering the properties of $g_0(x)$ and $g_1(x)$, by minimising this function, we are actually seeking parameters set θ to minimise $\sum_{j=1}^{m} \theta_j^2$, with the limitation that $\theta^T x > 1$ when $y = 1$, or $\theta^T x < -1$ when $y = 0$.

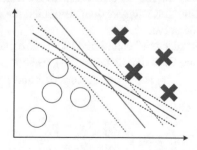

Fig. 10.11: Margins in the Supported Vector Machines

It turns out that by solving this optimisation problem SVM tends to obtain a large margin between different categories of data points. An example is shown in Fig. 10.11 of two possible decision boundaries that both effectively divide the two groups of training data. However, the blue boundary has a larger distance to the positive and negative training samples, denoted with dotted lines. These dotted lines indicate the *margin* of the SVM. As to the inference phase, any data x that makes $\theta^T x > 0$ is deemed positive, i.e. $y = 1$, or negative if $\theta^T x < 0$. Intuitively, a model with larger margin tends to better predict the test data.

10.6.1 Kernel and Non-Linear Boundary

So far we have talked about linear boundaries, but that is not a limit of SVMs. In fact we might normally use SVMs to train non-linear boundaries when categorising different groups of points in a space. To do that, we can simply update the linear part $\theta^T x$ in the cost function to make it a non-linear function, e.g.,

$$f_\theta(x) = \theta_0 + \theta_1 x_1 + \theta_2 x_2 + \theta_3 x_1 x_2 + \theta_4 x_1^2 + \theta_5 x_2^2 + \dots \qquad (10.22)$$

This function and the linear function $\theta^T x$ are both examples of a *kernel function*, to be used within $g()$ in the cost function. With the trained parameters θ, if Eq. 10.22 is larger than zero, then the inference result is positive, otherwise it's negative. However, this model is apparently not scalable with regard to the number of features of the input.

Currently, one common way to model this function is to choose k reference points in the space, and use the *distances* to these points as a feature. In other words, for data x that contains any number of features, the objective function is reduced to using a fixed number k features:

$$\theta_0 + \sum_{i=1}^{k} \theta_i d_i, \qquad (10.23)$$

where d_i is the "distance" of the current point to the reference point p_i. In the inference phase, if this function is larger than zero, then the point is predicted to be positive, otherwise it is negative.

So how exactly is this "distance" calculated? There are many ways to do that, and one of the most used is the Gaussian distance [20], as shown in Eq. 10.24. Here the total number of features is n. x_j is the j-th feature of x and $p_i^{(j)}$ is the j-th feature of reference point p_i.

$$d_i = \exp\left(-\frac{\sum_{j=1}^{n}(x_j - p_i^{(j)})}{2\sigma^2}\right). \qquad (10.24)$$

The intuition is that with suitable trained parameters θ this approach can represent different regions in the classification. For example, in Fig. 10.12 we choose only three reference points: p_0, p_1, and p_2. In this example, we set the parameters for p_1 and p_3 to be larger than that of p_2. The left-hand figure shows the summation of distances of a point to all three of them. The contour graph on the right then shows clearly how this model lead to a prediction that's obviously large around a region that's close to the point p_1 and p_3. That's the boundary we use to decide if a point is positive (close to p_1 or p_3) or negative. You can imagine how this non-linear boundary can be changed with new parameters.

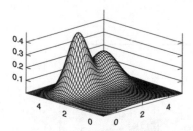 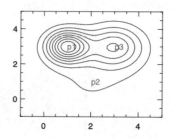

Fig. 10.12: Using the Gaussian kernel to locate non-linear boundary in categorisation

Only three reference points is not normally enough to support a complex non-linear boundary. Indeed, one common practice is to use every training set data point as the reference points.

In this example, the distance function d_i is the kernel function and the Gaussian kernel we have just used is a commonly used one, but other kernel functions such as the polynomial kernel, Laplacian kernel, etc., can also be used in Eq. 10.23. The previous linear model $\theta^T x$ is called the linear kernel, or "no kernel" as is sometimes called.

10.6.2 Example

The SVM is a very important and widely used machine learning method, and so it has highly efficient implementations in many libraries. For example, the Libsvm is an open source library that devotes solely to SVMs. It is widely interfaced to many languages such as Matlab and Python.

Using its optimisation engine, Owl provides basic support for SVM using linear kernel. Let's look at an example where we apply SVM to another randomly generated dataset as before but we now have labels 1 and -1. After applying `Regression.D.svm ~i:true x y` to the data we obtain:

```
val theta : Owl_algodiff_primal_ops.D.arr array =
[|          C0
R0 -0.438535
R1 -0.922763
;          C0
R0  5.7011 |]
```

That means the hypothesis we have for the function $\theta^T x$ is $f(x) = 5.7 - 0.43x_1 - 0.92x_2$. If $f(x) > 0$, the categorisation result is positive, otherwise it's negative. We can visualise this boundary line by setting $f(x) = 0$, as shown in Fig. 10.13. Here the y axis is x_2, and x axis is x_1.

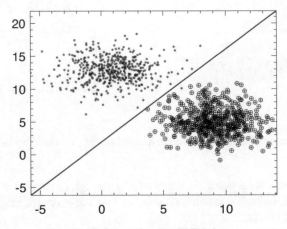

Fig. 10.13: Visualise the SVM dataset

10.7 Model Error and Selection

We have introduced least squares as a cost function for minimising the distance between model and data, but it is by no means the only way to assess how good is a model. We next discuss how to assess the quality of a model and compare different models. When testing a model, for each data point its real value is y while the predicted value is y'. The difference between these two is called the *residual*. In this section, when we say error, we actually mean residual, which should not be confused with the ϵ item in the linear model. The latter is the deviation of the observed value from the unobservable true value, while the residual is the difference between the observed value and the predicted value.

10.7.1 Error Metrics

First, let's look at two most commonly used metrics:

- **Mean absolute error** (MAE): average absolute value of residuals,

$$\text{MAE} = \frac{1}{n} \sum |y - y'|. \tag{10.25}$$

- **Mean square error** (MSE): average squared residuals,

$$\text{MSE} = \frac{1}{n} \sum (y - y')^2. \tag{10.26}$$

This is the method we used earlier in this chapter for linear regression. The part before applying average is called **Residual Sum of Squares** (RSS): RSS $= \sum (y - y')^2$.

The difference between using the MAE and MSE is different sensitivity to outliers: the MSE grows quadratically with error and so outliers are taken into consideration in the regression when minimising the MSE. In contrast, in the MAE each residual contributes proportionally to the metric, and so outliers do not have especially large impact on the model fitting. How to choose one of these metrics thus depends on how you want to treat the outliers in data.

Building on these two basic metrics, we can derive others:

- **Root mean squared error** (RMSE): the square root of the MSE. By applying square root, the unit of error reverts to the natural unit and is thus easier to interpret. This metric is analogous to the standard deviation and indicates how wide the residuals spread.
- **Mean absolute percentage error** (MAPE): MAPE changes the MAE into a percentage representation,

$$\text{MAPE} = \frac{1}{n} \sum |\frac{y - y'}{y}|. \tag{10.27}$$

It denotes the average distance between a model's predictions and the corresponding outputs in percentage format, making interpretation easier.

- **Mean percentage error** (MPE): similar to MAPE, but it does not use the absolute value,

$$\text{MPE} = \frac{1}{n} \sum \left(\frac{y - y'}{y} \right). \tag{10.28}$$

Without the absolute value, the metric can represent if the predicted value is larger or smaller than the observed value in data. Thus, unlike MAE and MSE, it's a relative measure of error.

10.7.2 Model Selection

It is common to see that in a multiple regression model, many variables are used in the data and modelling, but only a part of them are actually useful. For example, we might consider multiple factors such as amount of precipitation when choosing the location of McDonald's store, but many will make only marginal contribution to the overall profitability. By removing these redundant features, we can simplify a model and increase its interpretability. Regularisation is one way to downplay these features; in this section we briefly introduce another commonly used technique, *feature selection*. We have already briefly mentioned it in Sec. 10.2.2.

The basic idea of feature selection is simple: choose features from all the possible combinations, and evaluate the performance of each model using a metric such as RSS, before choosing the best. In more detail, suppose we have n features in a multi-variable regression, then for each $i = 1, 2, ...n$, test all the $\binom{n}{i}$ possible models with i variable(s), choose a best one according to its RSS, and we call this model M_i. Once this step is done, we can select the best one from the n models $M_1, M2, ...M_n$ using various methods.

You might have already spotted a major problem with this approach: computation complexity. To test all 2^n possibilities is a terribly large cost for even a medium number of features, and so computationally efficient approaches are proposed, one of which is *stepwise selection*.

Stepwise selection builds models based on existing best models. We start with one model with zero parameters (so it always predicts the same value regardless of the input) and assume it is the best model. From this starting point, we increase the number of features by one, choose among all the n possible models according to their RSS, and name the best one M_1. Based on M_1, we consider adding another feature, and so on. Once we have all the models $M_i, i = 1, 2, ...n$, we can select the best from among them using a suitable methods. This process is called "forward stepwise selection", and there is an analogous "backward stepwise selection" process where you build the model sequence from full features selection M_n down to M_1.

You might notice that we mention using various methods to select the best one from these n models – but what are they? An obvious answer is continue to use (e.g.,) RSS as the metric, but the problem is that the model with full features always has the smallest error and so gets selected every time. Instead, we need to estimate the test error. We can do that directly using a separate validation dataset. Otherwise we can adjust the training error such as RSS to include the bias caused by over-fitting. Methods include C_p, Akaike information criterion (AIC), Bayesian information criterion (BIC), adjusted R^2, etc. but these are beyond the scope of this book. We recommend specific textbooks such as James G. et al [6].

10.8 Summary

This chapter introduced different regression techniques, starting with the most basic, linear regression of one variable, before extending to multiple variables. We introduced the basic idea of linear regression, its theoretical support, and showed how it is supported in Owl with examples. Furthermore, we extended it to more complex regressions with non-linear models, such as the polynomial regression. Unlike linear regression, the logistic regression model categorises data into different groups by finding out a decision boundary. With a bit of change in its cost function, we ventured to introduce a more advanced machine learning technique, Support Vector Machines. They can be used for both linear and non-linear decision boundary by using different kernel functions. We have also talked about related issues, such as the regularisation, model error and selection.

References

1. Goldberger AS. Classical linear regression. Econometric theory. 1964:156-212.
2. Lemaréchal C. Cauchy and the gradient method. Doc Math Extra. 2012 Dec; 251(254):10.
3. Han J, Pei J, Kamber M. Data mining: concepts and techniques. Elsevier; 2011 Jun 9.
4. Bishop CM. Bishop-Pattern Recognition and Machine Learning-Springer 2006. Antimicrob. Agents Chemother. 2014:03728-14.
5. Ioffe S, Szegedy C. Batch normalization: Accelerating deep network training by reducing internal covariate shift. International conference on machine learning, 2015 Jun 1 (pp. 448-456). PMLR.
6. James G, Witten D, Hastie T, Tibshirani R. An introduction to statistical learning. New York: springer; 2013 Feb 11.
7. The Boston Housing Dataset [Internet]. University Toronto, 2021 [cited 2022, Jan 4]. Available from https://www.cs.toronto.edu/~delve/data/boston/bostonDetail.html.
8. Smith K. On the standard deviations of adjusted and interpolated values of an observed polynomial function and its constants and the guidance they give towards a proper choice of the distribution of observations. Biometrika. 1918 Nov 1;12(1/2):1-85.
9. Tetko IV, Livingstone DJ, Luik AI. Neural network studies. 1. Comparison of overfitting and overtraining. Journal of chemical information and computer sciences. 1995 Sep 1;35(5):826-33.
10. Alice SE, David CW. Penalty functions Handbook of Evolutionary Computation, Section C 5.2.
11. Hilt DE, Seegrist DW. Ridge, a computer program for calculating ridge regression estimates. Department of Agriculture, Forest Service, Northeastern Forest Experiment Station; 1977.
12. Friedman J, Hastie T, Tibshirani R. The elements of statistical learning. New York: Springer series in statistics; 2001.
13. Tibshirani R. Regression shrinkage and selection via the lasso. Journal of the Royal Statistical Society: Series B (Methodological). 1996 Jan;58(1):267-88.
14. Breiman L. Better subset regression using the nonnegative garrote. Technometrics. 1995 Nov 1;37(4):373-84.
15. Cortes C, Vapnik V. Support-vector networks. Machine learning. 1995 Sep;20(3):273-97.
16. Joachims T. Text categorization with support vector machines: Learning with many relevant features. European conference on machine learning. 1998 Apr 21 (pp. 137-142). Springer, Berlin, Heidelberg.

17. Barghout L. Spatial-taxon information granules as used in iterative fuzzy-decision-making for image segmentation. Granular computing and decision-making. 2015 (pp. 285-318). Springer, Cham.
18. Maity A. Supervised Classification of RADARSAT-2 Polarimetric Data for Different Land Features. arXiv preprint arXiv:1608.00501. 2016 Aug 1.
19. Leslie C, Eskin E, Noble WS. The spectrum kernel: A string kernel for SVM protein classification. Biocomputing. 2002, pp. 564-575.
20. Chang YW, Hsieh CJ, Chang KW, Ringgaard M, Lin CJ. Training and testing low-degree polynomial data mappings via linear SVM. Journal of Machine Learning Research. 2010 Apr 1;11(4).
21. Zou H, Hastie T. Regularization and variable selection via the elastic net. Journal of the royal statistical society: series B (statistical methodology). 2005 Apr;67(2):301-20.

Chapter 11
Neural Network

Abstract Originally conceived as far back as the 1940s, neural networks have received a lot of attention in recent years due to frequently astounding results. In this chapter, we will demonstrate step by step how to build up a practical neural network module with what we have learned so far, especially algorithmic differentiation, optimisation, and regression. After that, we will further present Owl's neural network module, including its design and usage. After studying the basic feedforward neural network, we will introduce some more advanced types of neural networks, including the Convolutional Neural Network, the Recurrent Neural Network, and Generative Adversarial Network.

The original idea of a "neural network" comes from attempts by computer scientists to model how biological neural systems work. The signal processing neurons carry out is modelled as computation, and the complex transmission and triggering of impulses are simplified to activations. In this chapter we start with the simplest neuron, the *perceptron*, and then go on to introduce how neural network can be built using regression as previously introduced in Chap. 10. We then proceed to introduce the neural network module from Owl, and three popular types of neural network it supports: convolutional neural network, recurrent neural network, and generative adversarial network. We will give each a quick introduction before conclude this chapter.

11.1 The Perceptron

Before diving into the complex neural network structures, let's briefly recount the simplest form of neural network: the *perceptron*. Its definition is actually very similar to that of logistic regression. Recall that logistic regression prediction function can be expressed as:

$$h_\theta(x) = g(\theta^T x),$$

where g is a sigmoid function, $g(x) = \frac{1}{1+e^{-x}}$. This function projects a number from $[-\infty, \infty]$ to $[0, 1]$. To get a perceptron all we need to do is to change the function to:

$$g(x) = \begin{cases} 1 & \text{if } \mathbf{w} \cdot \mathbf{x} + \mathbf{b} > \mathbf{0} \\ 0 & \text{otherwise} \end{cases}$$

Here w and b are the weight and bias parameters in the model that should be updated during the training. This function is called the *unit*, *Heaviside*, or *binary* step function. Rather than a range $[0, 1]$, the result is either 0 or 1, making it suitable for binary classification.

The perceptron is a basic unit of neural network. It was first proposed in the 1950s to perform binary image classification [1], and was designed to model how individual neuron in the brain works. Though initially promising, researchers quickly realised that perceptrons could not be trained to recognise many classes of patterns, which is almost always what is required for any image recognition task. Solving this problem requires introducing more layers of interconnected perceptrons, producing a *feedforward neural network*, which will be discussed in more detail in the two following sections.

11.2 Yet Another Regression

To some extent, any deep neural network is nothing but a regression problem in a very high-dimensional space. We seek to minimise its cost function by using higher-order derivatives. To better understand this, let's build a small neural network from scratch using only regression before looking into the `Neural` module provided by Owl. Following the logistic regression example in Chap. 10, we will build a simple neural network with one hidden layer, and train its parameters to perform hand-written digit recognition.

11.2.1 Model Representation

In logistic regression we use multiple parameters as one layer to decide if the input data belong to one type or the other, as shown in Fig. 11.1(a). The θ parameters have the same meaning as previously. We then extend it to multiple classes using a new hidden layer. In Fig. 11.1(b), the a_i represents the hidden layer. This model contains two sets of parameters. The θ_0 is a 3×3 matrix for the conversion $a = \theta_0^T x$, and θ_1 is a 3-element vector that implements $f(\theta_1^T a)$.

The data we will use come from MNIST dataset [2]. In the code below, we use the `Dataset` module in Owl to load MNIST images and labels data. The data are loaded as ndarrays, and we can verify their shape using the `shape` function.

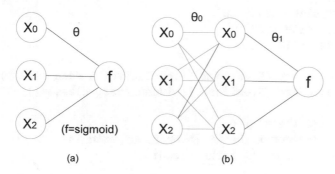

Fig. 11.1: Extend logistic regression to neural network with one hidden layer

```
# let x, _, y = Dataset.load_mnist_train_data_arr ();;
# let x_shape, y_shape =
  Dense.Ndarray.S.shape x, Dense.Ndarray.S.shape y;;

val x_shape : int array = [|60000; 28; 28; 1|]
val y_shape : int array = [|60000; 10|]
```

The labels are in *one-hot* format. For example, the y representation means that the first three labels are 5, 0, and 4.

```
val y : Owl_dense_matrix.S.mat =
        C0  C1  C2  C3  C4  C5  C6  C7  C8  C9
   R0    0   0   0   0   0   1   0   0   0   0
   R1    1   0   0   0   0   0   0   0   0   0
   R2    0   0   0   0   1   0   0   0   0   0
       ... ... ... ... ... ... ... ... ... ...
```

11.2.2 Forward Propagation

In our example we use a hidden layer of size 25, and require an output class of size 10, the number of digits we seek to recognise. As we will calculate derivatives when training parameters, we construct all the computation using the Algorithmic Differentiation module. The computation is a nested logistic regression:

$$h_\theta(\mathbf{x}) = f(\theta_1^T f(\theta_0^T \mathbf{x})).$$

Here θ denotes the collection of parameters θ_0 and θ_1. It can be implemented as:

```
open Algodiff.D
module N = Dense.Ndarray.D
```

```
let input_size = 28 * 28
let hidden_size = 25
let classes = 10

let theta0 = Arr (N.uniform [|input_size; hidden_size|])
let theta1 = Arr (N.uniform [|hidden_size; classes|])

let h theta0 theta1 x =
  let t = Arr.dot x theta0 |> Maths.sigmoid in
  Arr.dot t theta1 |> Maths.sigmoid;;
```

That's it. We can now classify an input 28×28 array into one of the ten classes ... except that we can't really because the parameters are set to random values. To actually be able to perform the classification accurately, we need to train the model to discover suitable values for θ_0 and θ_1.

11.2.3 Back Propagation

Training a network is essentially the minimisation of the cost function by adjusting the weights of each layer. The core of training is the back propagation which, as introduced in Chap. 8, involves propagating the error from the network output back to the input layer, i.e., evaluating the network in reverse. Back propagation is especially useful for functions where the number of input parameters is far larger than that of the output parameters – as is the case with neural networks.

Back propagation [3] is thus the core of all neural networks, and we can easily build up to it from scratch with the help of AD module. Recall from Chap. 10 that training is the process of iteratively seeking the parameters that minimise the cost function. In the case of this neural network, its cost function is similar to that of logistic regression (Eq. 10.18). Assuming we have m training data pairs, then cost function can be expressed as:

$$C_\theta(x, y) = \frac{1}{m} \sum_{i=1}^{m} (-y^{(i)} log(h_\theta(x^{(i)})) - (1 - y^{(i)}) log(1 - h_\theta(x^{(i)}))). \quad (11.1)$$

As we have explained in Chap. 10, the summation contains two parts $A + B$, where $A = -y^{(i)} log(h_\theta(x^{(i)}))$, and $B = (1 - y^{(i)}) log(1 - h_\theta(x^{(i)}))$. Depending on whether the value of y is 0 or 1, only one of them is used in a calculation. If $y = 1$, the A term is used. Here if the prediction results $y' = h_\theta(x^{(i)})$ is close to y, then A is close to 0 based on the property of logarithm operation; otherwise, if y' is close the opposite of y value, A tends to be a large number. Similar analysis applies to the B term if $y = 0$. In summary, the cost function penalises prediction results that are not close to y values. That's why it is suitable to be used in model training. Eq. 11.1 can be directly translated to OCaml code as:

```
let c t0 t1 x y =
  let z = h t0 t1 x in
  Maths.add
    (Maths.cross_entropy y z)
    (Maths.cross_entropy Arr.(sub (ones (shape y)) y)
       Arr.(sub (ones (shape z)) z));;
```

Here the operation `cross_entropy` y x means $-y \log(x)$. In Chap. 10, to find the parameters that minimise C, we iteratively apply:

$$\theta_i \leftarrow \theta_i - \alpha \frac{\partial C}{\partial \theta_i}$$

until it converges, and the same applies here. Although the partial derivative does not have an intuitive analytical solution, the partial derivatives of both parameters can be calculated using the AD module, and we have shown in Chap. 8 how it can be done in OCaml:

```
let x', y' = Dataset.draw_samples x y 1;;
let cost = c t0 t1 (Arr x') (Arr y');;
let _ = reverse_prop (F 1.) cost;;
let theta0' =  adjval t0 |> unpack_arr;;
let theta1' =  adjval t1 |> unpack_arr;;
```

That's it for one iteration. We get $\frac{\partial C}{\partial \theta_i}$, and can iteratively update the θ_0 and θ_1 parameters.

11.2.4 Feedforward Network

Although this example works well, it has several problems which we will now explain and revise the code to fix.

11.2.5 Layers

The first problem is that the previous example mixes all the primitive computation together. We should add the abstraction of a *layer* as the building block of neural network rather than manually encoding so many basic computations. The following code defines the *layer* and *network* types as OCaml records. Note that for each layer we add an extra *bias* parameter alongside the matrix multiplication. This bias vector influences the output without actually interacting with the data. Each linear layer performs the calculation in Eq. 11.2 where a is a non-linear activation function.

$$y = a(xw + b) \tag{11.2}$$

Each layer consists of three components: weight w, bias b, and activation function a.
A network is then just a collection of layers.

```
open Algodiff.S

type layer = {
  mutable w : t;
  mutable b : t;
  mutable a : t -> t;
}

type network = { layers : layer array };;
```

Despite the complicated internal structure, we can treat a neural network as a
function that takes input data and generates predictions. The question is how to
evaluate a network, by decomposing it into a sequence of evaluations of each layer.
The output of one layer is given as input to the next layer, propagating forwards until
it reaches the network's output layer. The following two lines show how to evaluate
a neural network in *forward mode*.

```
let run_layer x l = Maths.((x *@ l.w) + l.b) |> l.a
let run_network x nn = Array.fold_left run_layer x nn.layers;;
```

The run_network function generates what equals to the $h_\theta(x)$ function in the
previous section.

11.2.6 Activation Functions

Let's step back and check again the similarity between neural network and what
we think biological neuron works. So far we have connected every computation
node (or "neuron") to every node between two layers, but that's hardly how real
neurons work. Rather than activating all connecting neurons during information
transformation, different neurons are activated differently in an irregular fashion.
Biology aside, think about it mathematically: if we fully connect multiple layers,
we are in essence expressing multiple matrix multiplications, which is ultimately
equivalent to a single matrix multiplication once all the layers are collapsed.

Therefore, to make adding more layers actually mean something, we frequently
use *activation* functions to simulate how neurons work in biology, introducing *non-
linearity* into the network. Non-linearity is a useful property in our models as most
real world data demonstrate non-linear features. In fact, we have already seen two
types of activation function! The first is the Unit Step Function which works like a
simple on/off digital gate that allows part of neurons to be activated. The second is
the familiar sigmoid function which constrains its output to be in the interval [0, 1]
and so can be though of as a kind of probability of activation. There are many other

types of non-linear activation functions, shown in Fig. 11.2. The `tanh(x)` function computes $\frac{e^x - e^{-x}}{e^x + e^{-x}}$. Softsign computes $\frac{x}{1+|x|}$. The `relu(x)` computes:

$$\text{relu}(x) = \begin{cases} x & \text{if } \mathbf{x} > \mathbf{0} \\ 0 & \text{otherwise} \end{cases}$$

There is also the `softmax` function which takes a vector of K real numbers, and normalises it to a probability distribution consisting of K probabilities proportional to the exponential of the input numbers:

$$f_i(x) = \frac{e^{x_i}}{\sum_{k=1}^{K} e^{x_k}} \text{ for } i = 1, 2, \ldots, K.$$

We will keep using these activation functions in later network structures.

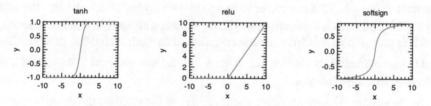

Fig. 11.2: Different activation functions used in neural networks

11.2.7 Initialisation

In this small example we use only two layers, `l0` and `l1`. `l0` uses a 784×40 matrix as weights, and `tanh` as its activation function. `l1` is the output layer and `softmax` is its activation function.

```
let l0 = {
  w = Maths.(Mat.uniform 784 40 * F 0.15 - F 0.075);
  b = Mat.zeros 1 40;
  a = Maths.tanh;
}
let l1 = {
  w = Maths.(Mat.uniform 40 10 * F 0.15 - F 0.075);
  b = Mat.zeros 1 10;
  a = Maths.softmax ~axis:1;
}
let nn = {layers = [|l0; l1|]};;
```

In Chap. 10 we saw that finding a good initial starting point for parameters values can be beneficial to the performance of gradient descent. You might think that using uniformly generated parameters would be fine, but that's not the case.

If you initialise with uniform or normal randomly distributed values, you will find that the elements in the outputs will get extremely large after a few layers. Even if by using `sigmoid` activation function we can force the output to within $[0, 1]$, it still means results are very near to 1, and the change in parameter values will be extremely small between each round of training, making training computationally very intensive. On the other hand, if we choose initial parameters that are close to 0, then the output result from the network itself will also be close to zero – the "vanishing gradient" problem. In either case the network cannot be trained efficiently.

There have been many attempts to solve this problem [4] [5] [6]. One common solution is to use ReLU as the activation function since it is more robust to this issue. As to initialisation itself, there are multiple heuristics that can be used. For example, the commonly used Xavier initialization approach [7] scales the uniformly generated parameters by $\sqrt{\frac{1}{n}}$. This parameter is shared by two layers, with n the size the first layer. This approach is especially suitable to be used with `tanh` activation function, and it is provided by the `Init.Standard` method in the initialisation module. The `Init.LecunNormal` is similar, but it uses $\sqrt{\frac{1}{n}}$ as the standard deviation of the Gaussian random generator.

Some authors [7] have proposed using $\sqrt{\frac{2}{n_0+n_1}}$ as the standard deviation in Gaussian random generation. Here n_0 and n_1 is the input and output size of the current layer. This can be used via `Init.GlorotNormal`. If we want to use the uniform generation approach, the parameters should be scaled by $\sqrt{\frac{6}{n_0+n_1}}$. For this example, we use `Init.GlorotUniform` or `Init.Tanh`. Owl does also still provide the vanilla uniform (`Init.Uniform`) and Gaussian (`Init.Gaussian`) randomisation, as well as allowing a custom method (`Init.Custom`) that accepts a custom function to manipulate the initial parameters.

11.2.8 Training

The loss function is constructed by executing the network in forward mode as the code below shows.

```
let loss_fun nn x y =
  let t = tag () in
  Array.iter (fun l ->
    l.w <- make_reverse l.w t;
    l.b <- make_reverse l.b t;
  ) nn.layers;
  Maths.(cross_entropy y (run_network x nn) /
    (F (Mat.row_num y |> float_of_int)));;
```

The `backprop` function uses the same procedure as the previous example. The partial derivative is calculated using `adjval`, and the parameter `w` and `b` of each layer are updated accordingly. It then uses the gradient descent method; for simplicity, the learning rate `eta` is fixed to a constant.

```
let backprop nn eta x y =
  let loss = loss_fun nn x y in
  reverse_prop (F 1.) loss;
  Array.iter (fun l ->
    l.w <- Maths.((primal l.w) -
      (eta * (adjval l.w))) |> primal;
    l.b <- Maths.((primal l.b) -
      (eta * (adjval l.b))) |> primal;
  ) nn.layers;
  loss |> unpack_flt;;
```

11.2.9 Test

Now we have all these parts in place, we can finally see how well our trained model works. The `test` function performs model inference and compares the predictions with labelled data. By doing so, we can evaluate the accuracy of a neural network.

```
let test nn x y =
  Dense.Matrix.S.iter2_rows (fun u v ->
    let p = run_network (Arr u) nn |> unpack_arr in
    Dense.Matrix.Generic.print p;
    Printf.printf "prediction: %i\n"
      (let _, i = Dense.Matrix.Generic.max_i p in i.(1))
  ) (unpack_arr x) (unpack_arr y);;
```

To put all these parts together, the following code starts the training for 999 iterations. We use the `Dataset` module in Owl to load training and test data, and draw a certain number of samples randomly from the loaded data.

```
let main () =
  let x, _, y = Dataset.load_mnist_train_data () in
  for i = 1 to 999 do
    let x', y' = Dataset.draw_samples x y 100 in
    backprop nn (F 0.01) (Arr x') (Arr y')
    |> Owl_log.info "#%03i : loss = %g" i
  done;
  let x, y, _ = Dataset.load_mnist_test_data () in
  let x, y = Dataset.draw_samples x y 10 in
  test nn (Arr x) (Arr y);;
```

When the training starts, our code keeps printing the value of loss function at the end of iteration. From the output, we can see the value of loss function keeps decreasing quickly after training starts.

```
01:04:14.632 INFO : #001 : loss = 2.54432
01:04:14.645 INFO : #002 : loss = 2.48446
01:04:14.684 INFO : #003 : loss = 2.33889
01:04:14.696 INFO : #004 : loss = 2.28728
01:04:14.709 INFO : #005 : loss = 2.23134
01:04:14.720 INFO : #006 : loss = 2.21974
01:04:14.730 INFO : #007 : loss = 2.0249
01:04:14.740 INFO : #008 : loss = 1.96638
...
```

After training is finished, we test the accuracy of this network. Here is one example where we input an image of a hand-written digit. The vector below shows the prediction. The model says with 90.14% chance it is a number 3, which is quite accurate.

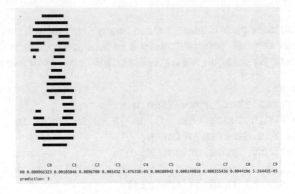

Fig. 11.3: Prediction from the model

11.3 Neural Network Module

The simple implementation we built in the previous section looks promising, but a neural network library cannot really leave it for the users to define so much for themselves. Owl provides the Neural Network module `Neural`, the use of which is actually very similar to the naive framework we just built, but with an improved API and more complete support for various types of neuron. In this section we introduce both how to use it and how it is built.

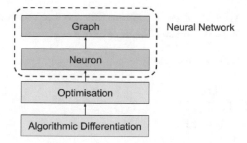

Fig. 11.4: Neural network module structure

Fig. 11.4 shows the basic architecture of the neural network module in the Owl library. The neural network module mainly consists of two submodules, Neuron and Graph, built on the Optimisation module, which in turn built on the Algorithmic Differentiation module. As we have seen in the previous section, for neural networks we need only describe the logic of the forward pass without worrying about the backward propagation at all thanks to Algodiff figuring it out automatically for us, reducing the potential for errors.

11.3.1 Neurons

Owl represents each neuron as a module that encapsulates the computation introduced above, then used in layers. In this section, we will explain the internal structure of a neuron module. We have already seen how the input and hidden layers can be connected; such computation can be abstracted into the FullyConnected neuron module, the type definition of which is shown below.

```
module FullyConnected = struct
  type neuron_typ = {
    mutable w : t;
    mutable b : t;
    mutable init_typ : Init.typ;
    mutable in_shape : int array;
    mutable out_shape : int array
  }
  ...
end
```

This module contains the two parameters we have seen: w and b, both of type t, the Algodiff array. To use this neuron, we need to specify its input and output shape. They are actually the length of input vector and the length of the hidden layer itself. The last part in the definition of this neuron init_typ describes the initialisation used. After the type definition, this module contains several standard functions that are shared by all the neuron modules.

```
let create ?inputs o init_typ =
  let in_shape =
    match inputs with
    | Some i -> [| i |]
    | None   -> [| 0 |]
  in
  { w = Mat.empty 0 o; b = Mat.empty 1 o;
    init_typ; in_shape; out_shape = [| o |] };;
```

After definition of its type, a neuron is created using the create function. Here we only need to specify the output shape, or the size of hidden layer o.

```
let connect out_shape l =
  assert (Array.length out_shape > 0);
  l.in_shape <- Array.copy out_shape;;
```

The input shape is actually inferred from the output of the previous layer in the connect function. We will see why we need the connect function later. Next we initialise the parameters:

```
let init l =
  let m = Array.fold_left (fun a b -> a * b) 1 l.in_shape in
  let n = l.out_shape.(0) in
  l.w <- Init.run l.init_typ [| m; n |] l.w;
  l.b <- Mat.zeros 1 n;;
```

The m is a flattened input size in case input ndarray is of multiple dimensions. The w parameter is initialised with predefined initialisation function, and we can just make b all zero, which means no bias at the beginning. Then we build the forward propagation procedure, in the run function:

```
let run x l =
  let m = Mat.row_num l.w in
  let n = Arr.numel x / m in
  let x = Maths.reshape x [| n; m |] in
  let y = Maths.((x *@ l.w) + l.b) in
  y;;
```

It is the familiar matrix multiplication and summation we have shown previously. The only thing we add is to reshape the possibly multi-dimensional input into a matrix.

Finally, we divide back propagation into several parts: tagging the parameters, obtaining derivatives, and updating parameters. They are also included in the neuron module:

```
let mktag t l =
  l.w <- make_reverse l.w t;
  l.b <- make_reverse l.b t

let mkpri l = [| primal l.w; primal l.b |]

let mkadj l = [| adjval l.w; adjval l.b |]

let update l u =
  l.w <- u.(0) |> primal';
  l.b <- u.(1) |> primal';;
```

That's about all for the structure of the `FullyConnected` neuron, and the other neurons share the same or very similar structure. Similar to other modules such as `Ndarray` and `Algodiff`, the `Neural.Neuron` module provides two sub-modules S and D for both single precision and double precision neural networks.

11.3.1.1 Neural Graph

The neuron is the core of the neural network module but we do not work directly with neurons but with neural networks, where individual neurons are instantiated into *nodes* and connected into a *graph* using the `Graph` module. A node in a neural network is composed of multiple parts:

```
type node = {
  mutable name : string;
  mutable prev : node array;
  mutable next : node array;
  mutable neuron : neuron;
  mutable output : t option;
  mutable network : network;
  mutable train : bool;
};;
```

Besides the neuron itself, a node also contains information such as its parents, children, output, the network it belongs to, a flag if this node is only for training, etc. In the `Graph` module, most of the time we need to deal with functions that build a node and connect it to existing network. For example:

```
let fully_connected ?name ?(init_typ = Init.Standard)
  outputs input_node =
  let neuron = FullyConnected
    (FullyConnected.create outputs init_typ) in
  let nn = get_network input_node in
  let n = make_node ?name [||] [||] neuron None nn in
  add_node nn [| input_node |] n;;
```

This function instantiates the FullyConnected neuron using its create function, and wraps it into a node n. The current network nn is built from its input node input_node. Then we add n as a child node to nn and connect it to its parents using the add_node function. This step uses the connect function of the neuron, and also update the child's input and output shape during connection.

Finally, after understanding the Graph module, we can now re-define the network in previous example with the neural network module:

```
open Neural.S
open Neural.S.Graph
open Neural.S.Algodiff

let make_network () =
  input [|28; 28; 1|]
  |> fully_connected 40 ~act_typ:Activation.Tanh
  |> linear 10 ~act_typ:Activation.(Softmax 1)
  |> get_network;;
```

The code above shows how convenient this module makes building neural networks. We can see how the input, the hidden layer, and the output from the previous example are concisely expressed. The linear layer in the code is similar to fully_connected, only that it accepts a one-dimensional input. The parameter act_typ specifies the activation function applied on the output of this node.

The network definition always starts with an input layer and ends with the get_network function which finalises and returns the constructed network. We can also see the input shape is reserved as a passed-in parameter so the shape of the data and the parameters will be inferred later whenever the input_shape is determined. You only need to provide the shape of the data in the input node, and the network can automatically infer the shape in the downstream nodes which saves a lot of efforts and significantly reduces the potential bugs.

11.3.2 Training Parameters

After a network is built, the last thing to do is to train the model by learning the weights. Again, we want to encapsulate all the manual back-propagation and parameter update into one simple function, primarily implemented in the minimise_network function in the Optimise module. This module provides the Params sub-module which maintains a set of training hyper-parameters. To avoid diving too deeply into the sea of implementation details, we focus on one update, the i-th iteration, and see how these hyper-parameters work. Let's start with the first step in this iteration: training data batching.

```
let xt, yt = batch_fun x y i;;
```

We use `Batch` module to specify how the training data are batched. In the definition of cost functions, we often assume that we must include all the data when updating parameters, described using the type constructor `Batch.Full`. However, for large scale training tasks, there might be millions of points of training data, which can not only cause memory issues but often also wastes time wait for all the data to be processed to get a single parameter update.

The most commonly used batching method is mini-batch, `Batch.Mini`, which uses only a small part of the training data [8]. As long as the data is completely used after certain number of iterations, this approach is mathematically equivalent to the full batch. This method is usually more efficient since training data are often correlated with each other and so iterate over all the training data is not necessary to obtain a good update.

If we continue the process to its ultimate end where only one data sample is used each time, we are using the *stochastic* batch method, `Batch.Stochastic`. This is not often a very good choice as the vectorised computation optimisation will not be efficiently used.

A final batching approach is `Batch.Sample` which is the same as mini-batch except that every mini-batch is randomly chosen from the training data. This is particularly useful if the training data are presented in some kind of order. For example, imagine that the training data for the MNIST task are ordered according to the digit value – you might end up with a model that only works for the lower digits like 0, 1, and 2 at the beginning.

```
let yt', ws = forward xt;;
```

Next, The `forward` function executes the computation layer by layer, accumulating the result in `yt'`, which is not simply an ndarray but an `Algodiff` data type that contains all the computation graph information. The `ws` is an array of all the parameters in the neural network.

```
let loss = loss_fun yt yt'
let loss = Maths.(loss / _f (Mat.row_num yt |> float_of_int));;
```

To compare how different the inference result `y'` is from the true label `y`, we need a loss function. Previously we have used the `cross_entropy`, but the `Loss` module provides other popular loss functions [9] [10]:

- `Loss.L1norm`: $\sum |y - y'|$
- `Loss.L2norm`: $\sum \|y - y'\|_2$
- `Loss.Quadratic`: $\sum \|y - y'\|_2^2$
- `Loss.Hinge`: $\sum \max(0, 1 - y^T y')$

In Chap. 10 we talked about regularisation and its benefits, introducing different types of regularisation methods. The `Regularisation` sub-module is provided in the optimisation module, and we can use `L1norm`, `L2norm`, or `Elastic_net` regularisation methods [11] in training. We can also choose not to apply regularisation method by using the `None` parameter. The code below adds a regularisation method to the existing loss function.

```
let reg =
  match params.regularisation <> Regularisation.None with
  | true  -> Owl_utils.aarr_fold
    (fun a w -> Maths.(a + regl_fun w)) (_f 0.) ws
  | false -> _f 0.
let loss = Maths.(loss + reg);;
```

The Graph module provides a train function that is a wrapper of this optimisation function. As a result, we can train the network by simply calling the train function.

```
let train () =
  let x, _, y = Dataset.load_mnist_train_data_arr () in
  let network = make_network () in
  let params = Params.config
    ~batch:(Batch.Mini 100)
    ~learning_rate:(Learning_Rate.Adagrad 0.005) 0.1
  in
  Graph.train ~params network x y |> ignore;
  network;;
```

The train function uses the Learning_rate sub-module. Until now we have been using constant learning rates, Learning_rate.Const, but this has problems: we want the gradient descent to be fast using large steps when training starts, decreasing to use small steps as it closes in on the minimum point so we can avoid overshooting. To achieve this we have the Decay and Exp_decay learning rate methods which both reduce the base learning rate according to the iteration number [12].

The first reduces the learning rate by a factor of $\frac{1}{1+ik}$, where i is the iteration number and k is the reduction rate; the second reduces the leaning rate by a factor of e^{-ik}.

The Learning_Rate.Adagrad [13] method we use here adapts the learning rate to the parameters as well as the iteration number: it uses smaller step for parameters associated with frequently occurring features, making it very suitable for sparse training data. It achieves this by storing all the past squared gradients and using RMSprop [14] to restrict the window of accumulated past gradients by keeping an exponentially decaying average of past squared gradients, managing the learning rate reduction strategy. Another method, Adam [15], keeps an exponentially decaying average of the gradients themselves.

The last thing to notice in the training parameters is the last number, 0.1, denoting the training epochs or how many times to repeat on the whole dataset. A value of 0.1 means that we process only a tenth of all the training data at a time.

After training is finished, finally we can test the trained parameter on test set, by comparing the accuracy of correct inference result. The code below shows how to do that. It is built around the Graph.model function that generate a functional model to perform inference, given a trained network and inputs.

```
let test network =
  let imgs, _, labels = Dataset.load_mnist_test_data () in
  let m = Dense.Matrix.S.row_num imgs in
  let imgs = Dense.Ndarray.S.reshape imgs [|m;28;28;1|] in

  let mat2num x = Dense.Matrix.S.of_array (
      x |> Dense.Matrix.Generic.max_rows
        |> Array.map (fun (_,_,num) -> float_of_int num)
    ) 1 m
  in

  let pred = mat2num (Graph.model network imgs) in
  let fact = mat2num labels in
  let accu = Dense.Matrix.S.(elt_equal pred fact |> sum') in
  Owl_log.info "Accuracy on test set: %f"
    (accu /. (float_of_int m));;
```

The final result shows that we can achieve an accuracy of 71.7% in only 0.1 epochs. Increasing the training epoch to 1 improves accuracy to 88.2%. Further increasing the epoch number to 2 improves accuracy again, to about 90%.

The Graph module also provides utility functions such as save, load, print, and to_string to help manipulating the neural network.

11.4 Convolutional Neural Network

So far we have seen how an example of fully connected feedforward neural network evolves step by step. However, there are many other kinds of neural network. In this section we will see how to improve the performance of the previous network dramatically by using a *convolutional neural network*, (CNN).

As its name suggests, this neural network is built on the convolution operation, a 2D version of the 1D convolution operation seen in Chap. 7. The 2D convolution is similar, the only difference being that the input and filter/kernel are now both matrices instead of vectors. The kernel matrix moves along the input matrix in both directions, and the sub-matrix on the input matrix is multiplied element-wise with the kernel. This operation is especially good at capturing the features in images, and is the key to image processing using neural networks [16].

To build a CNN we need more types of neurons, the most common of which are implemented by the neural network module, including convolution (both 2D and 3D), pooling [17], batch normalisation [18]. They are sufficient to support building many state-of-the-art neural network architectures.

As the MNIST handwritten recognition task is also a computer vision task, let's try it using a CNN. The code below creates a small convolutional neural network of six layers.

```
let make_network input_shape =
  input input_shape
  |> lambda (fun x -> Maths.(x / F 256.))
  |> conv2d [|5;5;1;32|] [|1;1|] ~act_typ:Activation.Relu
  |> max_pool2d [|2;2|] [|2;2|]
  |> dropout 0.1
  |> fully_connected 1024 ~act_typ:Activation.Relu
  |> linear 10 ~act_typ:Activation.(Softmax 1)
  |> get_network;;
```

The training method is exactly the same as before. But by using a CNN, the training accuracy achieved is about 93.3% within only one 0.1 epoch, compared to the 70% using the feedforward network previously.

As the CNN is such an important driving force in the computer vision field, Part III of this book includes three case studies using it: *image recognition*, *instance segmentation*, and *neural style transfer*. We will see CNN at play with more details in these application cases.

11.5 Recurrent Neural Network

In all the previous examples there is one obvious, perhaps trivial pattern that can be observed – given one input, the trained network generates an output – but not every real world task works like that. Often the input data is a sequence and the output is updated based not on the current input alone but also on earlier inputs from the sequence. For example, if we need to generate English based on French, or label each frame in a video, focusing on only the current word or frame is not enough.

That's where the Recurrent Neural Network (RNN) [3] comes in, accepting sequences as inputs and generating sequences as output. The basic structure of an RNN is quite simple: a neural network with loops allowing the output of one loop to be fed back in an earlier loop, enabling information from previous data in the sequence to be retained.

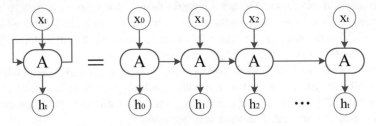

Fig. 11.5: Unroll the recurrent neural network

As shown in Fig. 11.5, an RNN can be unrolled into a chain of multiple connected neural networks. Here the x_i's are sequential input, and the h_i's are the *hidden states*, or outputs of the RNN. The functionality of a RNN therefore mainly relies on the processing logic in A.

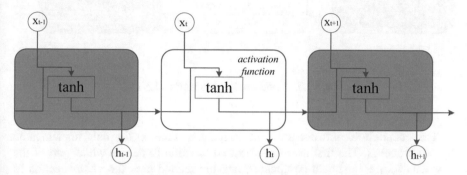

Fig. 11.6: Basic processing unit in classic recurrent neural network

In a vanilla recurrent neural network, the function can be quite simple:

$$h_i = \text{activation}(w(h_{i-1}x_i) + b). \tag{11.3}$$

This is exactly what we have seen in the feedforward networks. Here w and b are the parameters to be trained in this RNN. This process is shown in Fig. 11.6. The activation function in this calculation is usually the tanh function, so as to keep values within range of $[-1, 1]$.

However, this has problems: if you keep updating A based on a sequence of events, eventually information from past events will be flooded out by information about newer events. Similarly, in this RNN the old data in the sequence has a diminishing effect on the output, making the output less sensitive to the context. That's why current practical uses often use a special kind of RNN, the Long/Short Term Memory (LSTM).

Proposed by Hochreiter and Schmidhuber in 1997 [19] and widely used and refined by many subsequently, the basic idea of LSTM is simple. We still pass in the output from previous loop, but instead of taking it as-is, three choices are made: what to forget, what to remember, and what to output. In this way, useful information from old data can be kept longer, increasing the RNN's memory.

Let's see how it achieves this effect. The process unit of LSTM is shown in Fig. 11.7. It consists of three parts that correspond to the three choices listed above. Unlike standard RNN, each unit also takes in and produces a state C that flows along the whole loop process. This state is modified twice within the unit.

The first choice is made by the *forget gate layer*. This combines the output h_{t-1} from previous loop and the data x_i, and outputs a probability to decide how much of the existing information should be kept. This probability, as you may have guessed, is achieved using the sigmoid activation function, denoted by σ.

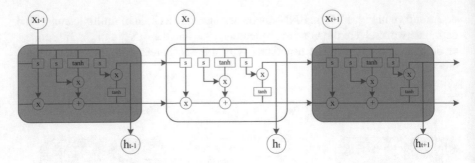

Fig. 11.7: Basic processing unit in LSTM

The second choice is to decide what to remember of the existing data, implemented by two branches. The first uses the `sigmoid` function to denote which part of the new data $h_{t-1} + x_t$ should be updated, and the second uses the `tanh` function to decide how much value to update for the vector. Both branches follow the procedure in Eq. 11.3, but with different w and b parameters.

By multiplying these two branches together, we know how much new information we should add to the information flow C. The flow C is therefore first multiplied with the output from the *forget gate* to remove unnecessary information, before adding the output from the second step to gain necessary knowledge.

The final choice is to decide what to output, by once more running a `sigmoid` function to decide which part of information flow C to keep. This is applied as a filter to a `tanh`-scaled information flow to finally obtain the output h_t.

LSTM is widely used in time-series related applications such as speech recognition, time-series prediction, human action recognition, robot control, etc. Using the neural network module, we can easily built a RNN that generates text by itself, following the style of the input text. The code below builds a LSTM network structure.

```
open Neural.S
open Neural.S.Graph

let make_network wndsz vocabsz =
  input [|wndsz|]
  |> embedding vocabsz 40
  |> lstm 128
  |> linear 512 ~act_typ:Activation.Relu
  |> linear vocabsz ~act_typ:Activation.(Softmax 1)
  |> get_network;;
```

That's it. The network is even simpler than that of the CNN. The only parameter we need to specify in building the LSTM is the length of vectors. However, the simpler architecture does not indicate the RNN is easy to compute. The RNN is notoriously known for the its high memory and computational complexity. It is also more difficult to scale up than other DNN architectures.

The LSTM has been refined in subsequent work proposal, and there are many variants of it. For example, the *Gated Recurrent Unit* (GRU) proposed in 2014 [20]. Compared to LSTM, the GRU structure consists of two parts: a "reset gate" that decides how much information to forget from the past, and an "update gate" behaving like a combination of LSTM's forget and input gates. It also merges the information flow C and output status h. With these changes, the GRU can achieve the same effect as LSTM with fewer operations, and therefore is a bit faster than LSTM in training. In the LSTM code above, we can just replace the lstm node with gru.

11.6 Generative Adversarial Network

We need to discuss one more type of neural network which isn't actually a particular type with new neurons like a DNN or RNN, but more of a (huge) family of networks exhibiting a particular pattern: the *Generative Adversarial Network* (GAN) [21]. The typical target for training a GAN is to automatically learn patterns in the input data so that the model can be used to generate "fake data" that look like they could have been drawn from the original data.

A GAN consists of two competing parts, *generator* and *discriminator*, that are constantly at war with each other. During training, the generator tries its best to synthesise data based on existing parameters, and the discriminator tries its best to separate the generated data from the true data. This contest is iterated until the discriminator can no longer tell the difference between the generated data and the true data – normally meaning that we human beings are also not very likely to be able to distinguish them.

Let's look at an example. Previously we have used the MNIST dataset extensively in image recognition task, but now let's try something different with it. Say we want to build a neural network that can produce a digit picture that looks like it's taken from the MNIST dataset. It doesn't matter which digit – the point is simply that it should be realistic but not actually in the original dataset. To generate such an image does not require a highly complex network structure. For example, we can use something like below:

```
open Neural.S
open Neural.S.Graph
open Neural.S.Algodiff

let make_generator input_shape =
  let out_size = Owl_utils_array.fold_left ( * ) 1
    input_shape in
  input input_shape
  |> fully_connected 256 ~act_typ:(Activation.LeakyRelu 0.2)
  |> normalisation ~decay:0.8
```

```
|> linear 512 ~act_typ:(Activation.LeakyRelu 0.2)
|> normalisation ~decay:0.8
|> linear 1024 ~act_typ:(Activation.LeakyRelu 0.2)
|> normalisation ~decay:0.8
|> linear out_size ~act_typ:Activation.Tanh
|> reshape input_shape
|> get_network;;
```

We connect multiple linear layers, activation layers, and normalisation layers, and don't even need to use any convolution layers. By now you should be familiar with this kind of network structure: this network accepts an ndarray of image shape 28×28, and outputs an ndarray of the same shape, i.e. a black and white image.

As well as this generator, the other half of the GAN is the discriminator. The structure is also quite simple:

```
let make_discriminator input_shape =
  input input_shape
  |> fully_connected 512 ~act_typ:(Activation.LeakyRelu 0.2)
  |> linear 256 ~act_typ:(Activation.LeakyRelu 0.2)
  |> linear 1 ~act_typ:Activation.Sigmoid
  |> get_network;;
```

The discriminator takes in an image as input and outputs only a single value through the sigmoid activation function – a probability that shows how good the discriminator thinks the outputs from generator are. An output of 1 means the discriminator think this output must be taken from MNIST while 0 means the input is obviously a fake.

The question is: how to train these two parts so that they mutually reinforce each other? We do so by carefully crafting the loss functions. The point of this step is to make the discriminator distinguish the output from generators from the true images as effectively as possible. Assume that each time we only take one picture from MNIST and training data. First, to train the discriminator we consider the *ground truth*: we know that the data taken from MNSIT must be true, so it is labelled 1, while we know that anything that comes from generator, no matter how good it is, must be a fake and so is labelled 0. By adding these two parts together, we can get a loss value for training the discriminator.

We also want to train the generator using the same batch of training data, so we reverse our strategy. We combine the generator network and discriminator network together, give it a random noise image as input, and label the true output as 1, even though we know that at the beginning the output from generator would be obviously not an MNIST image. The loss value is obtained by comparing the output of this combined network and the true label 1. During the training of this loss value, we need to make the discriminator un-trainable. The point of this step is to make generator produce images that can fool the discriminator as convincingly as possible.

That's all. In some ways it is like a small scale Darwinian simulation: by iteratively strengthening both parties, the generator can finally become so good that even a good discriminator cannot tell if an input image is faked by the generator or really taken from MNIST. At that stage, the generator is trained well and the job is done.

This approach has been successfully applied in many applications, such as Pix2Pix [22], face ageing [23], increase photo resolution [24], etc. For Pix2Pix, you give it a pencil-drawn picture of bag and it can render it into a real-looking bag. Or think about the popular application that creates an animation character that does not really exist previously [25]. In these applications, the generators are all required to generate images that just do not exist but somehow are real enough to fool the people to think that they do exist in the real world. A more thorough introduction to the recent development of GAN can be seen in Pan et al [26].

11.7 Summary

This chapter gave a detailed introduction to the theory behind neural networks, alongside the implementation of a simple neural network module. Starting with the most basic and earliest form of neural network, the perceptron, we manually built a feedforward network to perform multi-class handwritten digit recognition task by extending logistic regression. Next, we introduced Owl's Neural Network module, including how its core works and how it is built step by step. We use this module to solve the same example, including training and testing, and explained how to initialise training parameters in the regression module. We then introduced two important types of neural network: the convolutional neural network and the recurrent neural network in two of its variants, LSTM and GRU. We finished with a brief introduction to the basic idea behind Generative Adversarial Networks, another type of neural network that has gained a lot of momentum in research and application recently.

References

1. Rosenblatt F. The perceptron, a perceiving and recognizing automaton Project Para. Cornell Aeronautical Laboratory; 1957.
2. LeCun Y, Bottou L, Bengio Y, Haffner P. Gradient-based learning applied to document recognition. Proceedings of the IEEE. 1998 Nov; 86(11):2278-324.
3. Rumelhart DE, Hinton GE, Williams RJ. Learning representations by back-propagating errors. nature. 1986 Oct;323(6088):533-6.
4. Schmidhuber J. Learning complex, extended sequences using the principle of history compression. Neural Computation. 1992 Mar;4(2):234-42.
5. Hochreiter S, Schmidhuber J. Long short-term memory. Neural computation. 1997 Nov 15;9(8):1735-80.
6. He K, Zhang X, Ren S, Sun J. Deep residual learning for image recognition. Proceedings of the IEEE conference on computer vision and pattern recognition 2016 (pp. 770-778).
7. Glorot X, Bengio Y. Understanding the difficulty of training deep feedforward neural networks. Proceedings of the thirteenth international conference on artificial intelligence and statistics. 2010 Mar 31 (pp. 249-256). JMLR Workshop and Conference Proceedings.

8. Li M, Zhang T, Chen Y, Smola AJ. Efficient mini-batch training for stochastic optimization. Proceedings of the 20th ACM SIGKDD international conference on Knowledge discovery and data mining 2014 Aug 24 (pp. 661-670).
9. Tangian AS. Constructing a quasi-concave quadratic objective function from interviewing a decision maker. European Journal of Operational Research. 2002 Sep 16;141(3):608-40.
10. Rosasco L, De Vito E, Caponnetto A, Piana M, Verri A. Are loss functions all the same?. Neural computation. 2004 May 1;16(5):1063-76.
11. Zou H, Hastie T. Regularization and variable selection via the elastic net. Journal of the royal statistical society: series B (statistical methodology). 2005 Apr; 67(2):301-20.
12. You K, Long M, Wang J, Jordan MI. How does learning rate decay help modern neural networks?. arXiv preprint arXiv:1908.01878. 2019 Aug 5.
13. Duchi J, Hazan E, Singer Y. Adaptive subgradient methods for online learning and stochastic optimization. Journal of machine learning research. 2011 Jul 1; 12(7).
14. Tieleman, T. and Hinton, G. Lecture 6.5 - RMSProp, COURSERA: Neural Networks for Machine Learning. Technical report. 2012.
15. Kingma DP, Ba J. Adam: A method for stochastic optimization. arXiv preprint arXiv:1412.6980. 2014 Dec 22.
16. Zhang MZ, Ngo HT, Livingston AR, Asari VK. A high performance architecture for implementation of 2-D convolution with quadrant symmetric kernels. International Journal of Computers and Applications. 2008 Jan 1; 30(4):298-308.
17. Scherer D, Müller A, Behnke S. Evaluation of pooling operations in convolutional architectures for object recognition. International conference on artificial neural networks. 2010 Sep 15 (pp. 92-101). Springer, Berlin, Heidelberg.
18. Ioffe S, Szegedy C. Batch normalization: Accelerating deep network training by reducing internal covariate shift. International conference on machine learning. 2015 Jun 1 (pp. 448-456). PMLR.
19. Hochreiter S, Schmidhuber J. Long short-term memory. Neural computation. 1997 Nov 15; 9(8):1735-80.
20. Cho K, Van Merriënboer B, Gulcehre C, Bahdanau D, Bougares F, Schwenk H, Bengio Y. Learning phrase representations using RNN encoder-decoder for statistical machine translation. arXiv preprint arXiv:1406.1078. 2014 Jun 3.
21. Goodfellow I, Pouget-Abadie J, Mirza M, Xu B, Warde-Farley D, Ozair S, Courville A, Bengio Y. Generative adversarial nets. Advances in neural information processing systems. 2014;27.
22. Isola P, Zhu JY, Zhou T, Efros AA. Image-to-image translation with conditional adversarial networks. Proceedings of the IEEE conference on computer vision and pattern recognition 2017 (pp. 1125-1134).
23. Antipov G, Baccouche M, Dugelay JL. Face aging with conditional generative adversarial networks. In2017 IEEE international conference on image processing (ICIP). 2017 Sep 17 (pp. 2089-2093). IEEE.
24. Sajjadi MS, Scholkopf B, Hirsch M. Enhancenet: Single image super-resolution through automated texture synthesis. Proceedings of the IEEE International Conference on Computer Vision 2017 (pp. 4491-4500).
25. Jin Y, Zhang J, Li M, Tian Y, Zhu H, Fang Z. Towards the automatic anime characters creation with generative adversarial networks. arXiv preprint arXiv:1708.05509. 2017 Aug 18.
26. Pan Z, Yu W, Yi X, Khan A, Yuan F, Zheng Y. Recent progress on generative adversarial networks (GANs): A survey. IEEE Access. 2019 Mar 14; 7:36322-33.

Chapter 12
Vector Space Modelling

Abstract Dominant media types on the Internet include images, audio, video, and text. Many day-to-day tasks involve analysis of text and Natural Language Processing (NLP) is a powerful tool for extracting insights from text corpora. NLP is a very large topic with many interesting challenges. In this chapter we focus on information retrieval and specifically *topic modelling*.

12.1 Introduction

NLP is a field that aims to understand, interpret and manipulate human language. It draws on many disciplines, including linguistics, computer science, information engineering, and artificial intelligence. It is a complex topic as the rules that lie behind natural languages are not always easy to determine and express; abstract ideas such as sarcasm and humour remain difficult to convey to computers. Furthermore, much text generated in the real world lacks proper formal structure and does not easily fit into a traditional row and column structure. Unstructured data are difficult to manipulate. Broadly speaking, NLP includes these fields as listed below.

Information Retrieval [1], focused on recognising structured information such as key relations or event types from given unstructured information. Named Entity Recognition (NER) is a starting point for this category of application, extracting named objects such as people, places, and dates from a given text [2]. A higher level of information is about "relations", such as extracting the relation of "Washington D.C. *is the capital of* the USA". The two place entities "Washington D.C" and "USA" are connected by a predicate "is the capital of". Finally, an *Event* includes relationships between multiple entities. An event detection task requires obtaining information about the place, people, time, etc. involved.

Machine Translation [3] [25] is an important NLP field, involving automated translation of text or speech from one language to another. The basic idea uses a decoding algorithm to generate a possible translations, finding the one that maximises a utility function describing the translation quality between source language and the

L. Wang et al., *OCaml Scientific Computing*, Undergraduate Topics in Computer Science,
https://doi.org/10.1007/978-3-030-97645-3_12

generated translation. The malleability of human languages makes this task extremely challenging because words are often rephrased or reordered in practice.

Text Generation [4] [5] is another significant area, where text is generated to explain or describe certain inputs, combining information from multiple sources into a summary, or for interactive conversation with human participants. Among typical tasks we define Data-to-Text generation, aiming to describe data such as image or experiment results; and Text-to-Text generation focused on summarising or simplifying the given text, such as describing the main theme of an article.

Although linguistic approaches are widely used, NLP need not necessarily rely on analysing the text from a linguistic perspective. Deep learning approaches have achieved good performance on many tasks. Compared to the traditional input data such as image or number array, NLP tasks often need to deal with discrete and sequential data. The Recurrent Neural Network [6], as we briefly discussed in the neural network chapter, is thus widely used in this field.

We clearly cannot cover all of these subjects in just a single chapter. Here we focus on information retrieval, specifically *topic modelling*. This is a type of statistical modelling for discovering the representative keywords that occur in a collection of documents. Unlike text summarisation, it only chooses the keywords from a piece of given text.

12.2 Text Corpus Analysis

We term a large, structured collection of documents a *text corpus*. For example, widely used corpuses for English language include the Corpus of Contemporary American English [7], the Georgetown University Multilayer Corpus [8], etc. In this chapter, we will use a news dataset[1] crawled from the Internet, containing 130,000 pieces of news from various sources with each line in the corpus file representing a single document. For example the first line/document is:

```
a brazilian man who earns a living by selling space for tattoo
adverts on his body is now looking for a customer for his
forehead , it appears ... borim says when clients do n't pay
or cancel an ad , he crosses them out . " skinvertising "
caused a stir in the mid-2000s , when many dot.com companies
experimented with it...
```

To perform NLP tasks such as topic modelling, the first and perhaps most important challenge is to represent the text corpus in a format that models can process, rather than using natural language directly. For topic modelling, we first perform *tokenisation* on the input English text to represent each word as an integer index so that we can proceed by processing numbers instead of words.

---

[1] URL:https://github.com/ryanrhymes/owl_dataset/raw/master/news.txt.gz

12.2.1 Building a Text Corpus

The NLP module supports building a text corpus from an input text dataset. In this part we show how to build a corpus from a collection of documents, step by step.

The first step is to remove special characters (i.e., stopwords). Owl has pre-defined regular expressions `regexp_split` for special characters such as `,`, `?`, `\t`. Remove them, and then convert all the text to lowercase. The code below defines such a function, and the `Nlp.Corpus.preprocess` function applies it to all the input text.

```
let simple_process s =
  Str.split Owl_nlp_utils.regexp_split s
  |> List.filter (fun x -> String.length x > 1)
  |> String.concat " "
  |> String.lowercase_ascii
  |> Bytes.of_string

let preprocess input_file =
  let output_file = input_file ^ ".output" in
  Nlp.Corpus.preprocess simple_process input_file output_file;;
```

Based on the processed text corpus, we can build the *vocabulary* by assigning each word to a numeric index, and retaining the dictionary that maps between words and indices, using the `Nlp.Vocabulary.build` function.

```
let build_vocabulary input_file =
  let vocab = Nlp.Vocabulary.build input_file in
  let output_file = input_file ^ ".vocab" in
  Nlp.Vocabulary.save vocab output_file;;
```

The `build` function returns a vocabulary comprising three hash tables. The first maps a word to an index, and the second an index to a word. The third maps each index to its frequency, i.e. the number of occurrences in the input text. We can examine the words of highest frequency with:

```
let print_freq vocab =
  Nlp.Vocabulary.top vocab 10 |>
  Owl.Utils.Array.to_string ~sep:", " fst;;
```

Unsurprisingly, words like "the" and "a" are the most frequent:

```
- : string =
"the, to, of, a, and, in, \", s, that, on"
```

Changing `Nlp.Vocabulary.top` to `Nlp.Vocabulary.bottom` shows the lowest frequency words:

```
"eichorst, gcs, freeross, depoliticisation, humping, shopable,
appurify, intersperse, vyaecheslav, raphaelle"
```

However, when undertaking topic modelling we want neither the most frequent ones (which often convey little specific meaning) nor the least frequent ones (which will not typically be connected to the topic of the text). We thus remove the most and least frequently used words, in this case trimming the top and bottom 1% of words.

```
let trim_vocabulary vocab =
  Nlp.Vocabulary.trim_percent ~lo:0.01 ~hi:0.01 vocab;;
```

With a good vocabulary at hand, we can now tokenise a text.

```
let tokenise vocab text =
  String.split_on_char ' ' text |>
  List.map (Nlp.Vocabulary.word2index vocab);;
```

For example, tokenising "this is an owl book" gives the following:

```
tokenise vocab "this is an owl book";;
- : int list = [55756; 18322; 109456; 90661; 22362]
```

We can thus now tokenise the whole news collection.

```
let tokenise_all vocab input_file =
  let doc_s = Owl_utils.Stack.make () in
  Owl_io.iteri_lines_of_file
    (fun i s ->
      let t =
        Str.split Owl_nlp_utils.regexp_split s
        |> List.filter (Owl_nlp_vocabulary.exists_w vocab)
        |> List.map (Owl_nlp_vocabulary.word2index vocab)
        |> Array.of_list
      in
      Owl_utils.Stack.push doc_s i)
    input_file;
  doc_s;;
```

The process is simple: in the text corpus each line is a document and we iterate through the corpus line by line, removing special characters, selecting only those words that exist in the vocabulary, and mapping each to its integer index in the vocabulary. Even though this step is simple, it illustrates the typical starting point of text analysis before diving into any specific detail.

12.2.2 Using Corpus Module

In fact, we don't need to build a text corpus step by step – we can use the NLP.Corpus module. The Nlp.Corpus.build function performs both tasks above: building vocabulary and tokenising the input corpus, as well as allowing us to specify how to trim the high-frequency and low-frequency words. Here is an example:

```
let main () =
  let ids = Nlp.Corpus.unique "news.txt" "clean.txt" in
  Printf.printf "removed %i duplicates." (Array.length ids);
  let corpus = Nlp.Corpus.build
    ~lo:0.01 ~hi:0.01 "clean.txt" in
  Nlp.Corpus.print corpus;;
```

The Nlp.Corpus.unique function is another layer of pre-processing that removes possibly duplicated lines (documents), printing progress and ultimately, summary statistics about the corpus.

```
19:07:05.461 INFO : build up vocabulary ...
19:07:10.461 INFO : processed 13587, avg. 2717 docs/s
19:07:15.463 INFO : processed 26447, avg. 2644 docs/s
...
19:08:09.125 INFO : convert to binary and tokenise ...
19:08:34.130 INFO : processed 52628, avg. 2104 docs/s
19:08:39.132 INFO : processed 55727, avg. 1857 docs/s
...
corpus info
  file path   : news.txt
  # of docs   : 129968
  doc minlen  : 10
- : unit = ()
```

The corpus contains three parts: vocabulary, tokens, and text strings. By calling the build function, we save them for later use, as it creates several files in the current directory.

First, the vocabulary files news.txt.voc and news.txt.voc.txt contain several lines, each line containing a word and its corresponding index. Only difference is that the latter is a human readable form. We can obtain the vocabulary with Corpus.get_vocab.

The text corpus is tokenised into the news.txt.tok file, and the string format content saved as the binary file news.txt.bin, which reduces the file size compared to text format. To get the *i*-th document, we can use Corpus.get corpus i to get the text string, or Corpus.get_tok corpus i to get an integer array of the tokenised version of this document.

To access different documents efficiently by document index (line number), we track the accumulated length of the text corpus and token array after processing each document. These two types of indexes are saved in the news.txt.mdl file, which also contains the document id. We have seen the minlen value in the output of corpus information. Each document with less than 10 words will not be included in the corpus. The document id is an int array that shows the line number of each document in the original text corpus so that it can be traced back. The document id can be retrieved by Corpus.get_docid corpus.

The Corpus module provides three mechanisms to iterate through the text corpus: next, iteri, mapi. The next function is a generator that yields the next line of text document string in the text corpus until it hits the end of a file. The `iteri` and `mapi` functions work as in the normal `Array` module, respectively iterating and mapping all the documents in the corpus into another array. The `iteri_tok` and `mapi_tok` functions work similarly, except that they take integer arrays as input rather than string arrays.

The `Corpus` module is designed to support many text corpuses. With this tool in hand, we can proceed with the task of topic modelling.

12.3 Vector Space Models

Given the tokenised text corpus, we next need a mathematical model that is able to express abstract ideas such as "this sentence makes sense and that one does not", "these two documents are similar", or "the key word in that paragraph is such and such". To perform such NLP tasks we use a *Vector Space Model* (VSM) [9].

According to Wikipedia, a VSM is "an algebraic model for representing text documents (and any objects, in general) as vectors of identifiers". The basic idea is actually very simple. For example, let's assume we only care about just three topics in the news: *covid19*, *economics*, and *elections*. Then we can represent any news article as a three-element vector, each representing the weight of one topic in it. Thus the BBC news article "Coronavirus: Millions more to be eligible for testing"[2] might then be represented by the vector $(100, 3.5, 10)$. The specific values do not matter here; the point is only that, instead of a large chunk of text corpus, we need to only deal with this three-dimensional vector for further processing.

In general, a vector space model maps each document to a vector $d = (x_1, x_1, \ldots, x_N)$. This N-dimensional vector space is defined by N basic terms. Under this framework, we must make three important decisions. The first is to choose the meaning of each dimension, or the N basic concepts in the vector space – particular words, or topics, or something else? The second is to specify the weight of each dimension for a document. For example, in our simple example, why assign the first weight to 100 instead of 50? Finally, after learning the vector representation, we need to define their similarity, allowing us to cluster or search among documents. Common metrics of similarity include Euclidean distance and cosine similarity and will be discussed later.

12.3.1 Word Embedding and BERT

In this chapter, we focus on mapping documents to vector spaces but VSMs are not limited only to documents – we can also map a word into a vector representing a point in a certain vector space. This process is called *word embedding* [10]. With

---

[2] Coronavirus: Millions more to be eligible for testing, BBC. URL:https://www.bbc.co.uk/news/uk-52462928

a proper word embedding representation, similar words should cluster together, so that calculations on the words themselves such as the following make sense:

$$V_{\text{king}} - V_{\text{man}} + V_{\text{women}} \approx V_{\text{queen}}.$$

One of the most widely used methods for word embedding is word2vec [11]. For general purpose use, Google has already published a pre-trained word2vec-based word embedding vector set based on part of the GoogleNews dataset. This vector set contains 300-dimensional vectors for 3 million words and phrases.

In 2018, Google releases the BERT model [12], a breakthrough for the NLP research community as it performs extremely well on language-based tasks. The BERT team has already released a pre-trained model trained on massive datasets such as the Wikipedia text corpus.

The BERT model is a trained Transformer Encoder stack [25], the details of which are beyond this chapter, but we should note its use of the *Attention* mechanism. This helps to improves the performance of machine translation tasks by adding context awareness. For example, in the sentence:

```
...with a gesture of desperation, he tore the mask from his face
and hurled it upon the ground.
```

What does the "it" refer to? Gesture, mask, face, or ground? The Attention mechanism helps the translator connect "it" with "mask" in this case [13] [25]. In other words, Attention allows a model to look at other positions in the input sequence (that is, to use context) for information in order to give a better representation of a given word.

Based on this mechanism, we can understand the advantage of BERT compared to other embedding models such as Word2Vec: BERT is context-aware. So while Word2Vec always generates the same embeddings for a given word, BERT can give different results. For example, the word "change" may be similar to vectors of words such as "alter" and "vary", but also to "money" depending the context in the input sequence. Thus BERT takes sentences, not only words, as input and give vector representation of the whole input sentence. Also, we have already talked about the "vocabulary" of a model where, if the model vocabulary contains 100 words, any word outside this cannot be represented. In contrast, BERT learns representations at a finer granularity than single words, and so can represent words that are not in the vocabulary.

This chapter focuses on more basic concepts in topic modelling. Please read the original BERT paper [14] for more details.

12.3.2 Bag of Words (BOW)

Let us return to the theme of mapping documents into a vector space. We will now introduce a simple method to instantiate a VSM: the *Bag of Words*. This uses the vocabulary as the dimensions for, if there are N different words in the vocabulary, the vector space has N dimensions, and the mapping function simply counts how many times each word in the vocabulary appears in a document.

For example, suppose the vocabulary contains just five words "news", "about", "coronavirus", "test", and "cases". These become the five dimensions in the vector space. Then, the document "...we heard news a new coronavirus vaccine is being developed which is expected to be tested..." will be represented as [1, 1, 1, 1, 0] and the document "...number of positive coronavirus cases is 100 and cumulative cases..." will be projected to vector [0, 0, 1, 0, 2].

This Bag of Words method is easy to implement based on the text corpus. We first define a function to count term occurrences in a document, returning a hash table:

```
let term_count htbl doc =
  Array.iter
    (fun w ->
      match Hashtbl.mem htbl w with
      | true  ->
        let a = Hashtbl.find htbl w in
        Hashtbl.replace htbl w (a +. 1.)
      | false -> Hashtbl.add htbl w 1.)
    doc;;
```

The hash table contains the counts of all words in this document. Of course, we can also represent the returned results as an array of integers, though it would likely be sparse. Then we can apply this function to all the documents in the corpus using the map function:

```
let build_bow corpus =
  Nlp.Corpus.mapi_tok
    (fun i doc ->
      let htbl = Hashtbl.create 128 in
      term_count htbl doc;
      htbl)
    corpus;;
```

Based on this bag of words, the similarity between two vectors can be measured using different methods such as a simple dot product.

This method is easy and inexpensive to implement. For some tasks, especially those with no strict requirement for context or position of words, this method actually work quite well. For example, to cluster spam emails, we only need to specify proper keywords as dimensions, such as "income", "bonus", "extra", "cash", "free", "refund", "promise". We can then expect that spam emails will be tightly clustered and thus easy to recognise in this vector space using just the bag of words.

Actually, one method that is even simpler is the Boolean model. Instead of term frequency (count of word), record only whether or not a term was present. This is even simpler and can be useful in certain tasks, but it loses information about importance of the word indicated by its frequency.

On the other hand, this simple approach does have its own problems. In the previous example, if we want to know how close a document is to "news about coronavirus test cases", then the doc "...number of positive coronavirus cases is 100 and cumulative cases are 1000..." is scored the same as "I got some good news about your math test result.". It is surely not what we expected. Intuitively, words like "coronavirus" should matter more than common words like "test" and "about". For that we introduce an improved method in the next section.

12.4 Term Frequency–Inverse Document Frequency (TF-IDF)

In the previous section, we used the count of each term when converting a document to a vector, as this *term frequency* represents the frequency of certain terms in the document. We had the intuition that the importance of different words should carry different weight, which cannot be achieved by simply using term count. To remedy this, we introduce the idea of *Inverse Document Frequency* (IDF).

The basic idea is simple: the IDF represents how common a word is across all the documents in a corpus. The intuition is that, if a word is used in every document, it must be less important in determining a feature of a document. In contrast, if a word exists in only a couple of documents, it must be highly salient to the topic of those documents. In this way the IDF factor multiplied by the term frequency gives a more accurate metric for representing a document as vector called TF-IDF [15] [16].

Actually, the two parts TF and IDF simply provide frameworks for different computation methods. To compute the term frequency, we can use the count of words c, or the percentage of words in the current document $\frac{c}{N}$ where N is the total number of words in the document. Another method is *logarithm normalisation* which uses $\log(c + 1)$. We can even use the boolean count that sets the frequency of words that exist to 1 and ones that do not exist to 0. These methods are all defined in the Owl_nlp.Tfidf module.

```
type tf_typ =
  | Binary
  | Count
  | Frequency
  | Log_norm;;
```

The same goes for computing IDF. One way to measure how common a word w is across all documents is to compute $\log(\frac{N_D}{n_w})$, where N_D is the total number of documents and n_w is the number of documents containing term w, a metric falling in the range $[0, \infty)$ that increases with larger total numbers of documents or with smaller numbers of documents containing a specific word. An improved version is called Idf_Smooth, calculated as $\log(\frac{N_D}{n_w + 1})$, avoiding n_w being zero (causing divide errors), and also avoiding getting a 0 for a word just because it is used in every document. These methods are included in the type df_typ. Here the Unary method implies not using IDF, only term frequency.

```
type df_typ =
  | Idf
  | Idf_Smooth
  | Unary;;
```

The Owl_nlp.Tfidf module provides the TF-IDF methods, and the corpus we built in the previous section is used as input to it. Specifically, we use the Nlp.Tfidf.build function to build the TF-IDF model:

```
let build_tfidf corpus =
  let tf = Nlp.Tfidf.Count in
  let df = Nlp.Tfidf.Idf in
  let model = Nlp.Tfidf.build ~tf ~df corpus in
  Nlp.Tfidf.save model "news.tfidf";
  model;;
```

In this code snippet, we use a bag-of-words style word count method to calculate term frequency, and the normal logarithm method to compute inverse document frequency. The model can be saved for later use. After the model is built, we can search for similar documents given a sentence. As a random example, let's just use the first sentence in our first piece of news in the dataset as search target: "a brazilian man who earns a living by selling space for tattoo adverts on his body is now looking for a customer for his forehead".

```
let query model doc k =
  let typ = Owl_nlp_similarity.Cosine in
  let vec = Nlp.Tfidf.apply model doc in
  let knn = Nlp.Tfidf.nearest ~typ model vec k in
  knn;;
```

Recall the three ingredients for a vector space model: choosing the dimension topic words, mapping documents to vectors, and measurement of similarity. Here we use *cosine similarity* as a way to measure how aligned two vectors A and B are – we will discuss similarity measurement in detail later. Next, the vec returned by the apply function is an array of (int * float) tuples. For each, the integer is the tokenised index of a word in the input document doc, and the float is the corresponding TF-IDF value, based on the model obtained in the previous step. Finally, the nearest function searches all the documents and finds the vectors with the greatest similarity to the target document; setting $k = 10$ gives the top-10 results:

```
val knn : (int * float) array =
  [|(11473, -783.54606886327087); (87636, -669.765336035355);
    (121966, -633.9255557772090); (57239, -554.8385417996606);
    (15810, -550.9546813404825); (15817, -550.7752769121831);
```

```
(15815, -550.77527691218313); (83282, -547.3223855523124);
(44647, -526.07456742508884); (0, -496.924176137374445)|]
```

The returned result shows the id of the matched documents, which we can retrieve by e.g. `Owl_nlp.Corpus.get corpus 11473`. To save you the effort, we list here links to some of the original news stories that were considered similar to the target document:

1. *Every tatto tells a story*, doc id: 11473. [3]
2. *The Complete Guide to Using Social Media for Customer Service*, doc id: 87636. [4]
3. *Murder ink? Tattoos can be tricky as evidence*, doc id: 57239. [5]
4. *Scottish independence: Cinemas pull referendum adverts*, doc id: 15810. [6]
5. *The profusion of temporarily Brazilian-themed products*, doc id:44647. [7]

If you are interested in experimenting further, the input document comes from this BBC news, *Brazil: Man "earns a living" from tattoo ads*.[8] You can see that the search results are quite related to the input document, especially the first one which refers to exactly the same event reported in another piece of news, though the second result is somewhat unrelated. Perhaps the reason is that the word "customer" is heavily used in this document, and we might guess that this word is not frequently seen elsewhere in the text corpus. The fourth result is not about the tattoo man but features the topics "customer" and "adverts". The fifth result appears to be chosen because the infrequent word "brazilian" carries a lot of weight in TF-IDF. Note that we simply took a random sentence without any pre-processing or keyword design, and used the untrimmed version of text corpus. Even so, we managed to achieve a reasonably satisfactory matching.

12.5 Latent Dirichlet Allocation (LDA)

In the previous section we saw that by specifying a document and using it as a query, we could retrieve similar documents. The query document itself is actually treated as a collection of words, but real articles are rarely simply just collections of words.

---

[3] *Every tatto tells a story*, BBC News, URL: https://www.bbc.co.uk/news/magazine-27831231

[4] *The Complete Guide to Using Social Media for Customer Service*, Courtney Seiter, URL: https://buffer.com/resources/social-media-for-customer-service-guide

[5] *Murder ink? Tattoos can be tricky as evidence*, Dave Collins, URL: https://www.gazettenet.com/Archives/2014/06/tattoos-hg-060514

[6] *Scottish independence: Cinemas pull referendum adverts*, BBC News, URL: https://www.bbc.co.uk/news/uk-scotland-scotland-politics-27602601

[7] *The profusion of temporarily Brazilian-themed products*, BBC News, URL: https://www.bbc.co.uk/news/magazine-27879430

[8] Brazil: Man "earns a living" from tattoo ads, BBC News. URL: https://www.bbc.co.uk/news/blogs-news-from-elsewhere-27051009

More often than not, an article will cover one or more *topics*: the responsibility of government, the protection of the environment, and a recent protest in the city, etc. Each of these topics can hardly be completely covered by just one single word. This introduces the task of *topic modelling*: instead of proposing a search query to find similar content in text corpuses, we wish to cluster documents automatically according to several topics, each represented by several words.

One important method to perform topic modelling is called *Latent Dirichlet Allocation* (LDA). The trained model of LDA contains two matrices: "document-topic" containing the number of tokens assigned to each topic in each document, and the "word-topic table" holding the number of tokens assigned to each topic for each word. We will later see how they work in an example. But before that, let us introduce some background theory.

12.5.1 Models

Let's take a look at the model of LDA proposed by Blei et al [17]. It shows how the LDA "thinks" about the way a document is composed, and is expressed in Fig. 12.1 [9]. This model uses the plate notation for describing probabilistic graphical models to capture the dependencies among variables [18]. In Tbl. 12.1 we give the definition of the math notations used here and in this section.

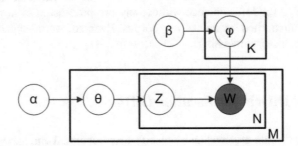

Fig. 12.1: Plate notation for LDA with Dirichlet-distributed topic-word distributions. This illustration from Wikipedia is licensed under CC BY-SA 3.0.

In this model, to infer the topics in a corpus, let's imagine a **generative process** to create a document. The core idea is that each document can be described by the distribution of topics, and each topic can be described by the distribution of words. This intuitively makes sense, since we don't need the text in order to find the topics of an article. The process is as follows.

---

[9] This illustration is used from Wikipedia unmodified, URL:https://en.wikipedia.org/wiki/File:Smoothed_LDA.png. It is licensed under CC BY-SA 3.0.

Table 12.1: Variable notations in the LDA model

Variable	Meaning
K	number of topics
D	number of documents in text corpus
V	number of words in the vocabulary
N	total number or words in all document
α	vector of length K, prior weight of the K topics in a document
β	vector of length V, prior weight of the V words in a topic
Θ	vector of length K, distribution of topics in a document
ϕ	vector of length V, distribution of words in a topic
Z	matrix of shape $D \times V$, topic assignment of all words in all documents
W	matrix of shape $D \times V$, token of words in all documents
$n_{d,k}$	how many times the document d uses topic k in the document-topic table
$m_{k,w}$	the number of times topic k uses word w in the topic-word table

1. Initialise the distribution of topics $\theta_d \sim \text{Dirichlet}(\alpha)$ in document d. Dirichlet(α) is a Dirichlet distribution parameterised by α. We will talk about it in detail later.
2. Initialise the distribution of words $\phi_k \sim \text{Dirichlet}(\beta)$ for topic k.
3. Iterate each document d and each word position w, and then perform the following steps:

   a. Pick a topic (one of the elements in Z) at random. This choice of topic is actually taken according to a categorical distribution, parameterised by θ. Formally, this step is represented as $Z_{d,w} \sim \text{Categorical}(\theta_d)$;
   b. From the words this topic contains, pick a word randomly according to ϕ. The picking process also follows categorical distribution, $W_{d,w} \sim \text{Categorical}(\phi_{Z_{d,w}})$.

After finishing this generative process, we now have a "fake" document. The total probability of the model is:

$$P(W, Z, \theta, \phi; \alpha, \beta) = \prod_{i=1}^{K} P(\phi_i; \beta) \prod_{j=1}^{D} P(\theta_j; \alpha) \prod_{t=1}^{N} P(Z_{j,t}|\theta_j)P(W_{j,t}|\phi_{Z_{j,t}}).$$

(12.1)

Eq. 12.1 corresponds to the above process and model in Fig. 12.1. It is a multiplication of three parts: the probability of θ across all the documents, of ϕ across all the topics, and of the generated words across all documents. The LDA hopes to make this generated document as close to a real document as possible. In other words, LDA tries to maximise the possibility Eq. 12.1 that this document can be generated from a set of topics when we are looking at real documents.

12.5.2 Dirichlet Distribution

We must add something to the generative process above – how are θ and ϕ generated? Randomly? No, that would make each document equally likely to contain any topic, something that rarely happens. An article cannot talk about all topics simultaneously. What we really hope is that a single document belongs to a single topic, a more likely real-world scenario; and the same for the word-topic table.

To that end, LDA uses the *Dirichlet Distribution*, a family of continuous multivariate probability distribution parameterised by a vector α [20]. For example, suppose we have just two topics, and so the tuple $(0, 1)$ means a document entirely concerns just the second topic while $(1, 0)$ means it entirely concerns just the first. We can run the `Stats.dirichlet_rvs` function to generate such a pair of floats. To demonstrate the impact of the α parameter, let's execute this several times and plot the first number in the tuple as the x-axis value, and the other as the y-axis value. The results of using two different α's are shown in Fig. 12.2. Both figures have the same number of dots but with a smaller α value the distribution is pushed to the corners, where it is concerns one topic or the other, while a larger α value makes concentrates around the middle, a mixture of both topics.

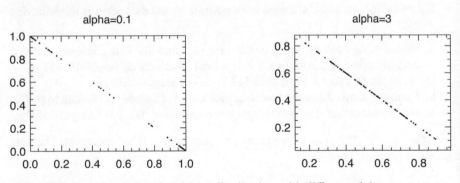

Fig. 12.2: Two dimensional dirichlet distribution with different alpha parameters

Therefore in the model in Fig. 12.1, we have two parameters α and β as prior weights to initialise Θ and ϕ respectively. We use reasonably small parameters to get skewed probability distributions, where only a small set of topics or words have high probability.

12.5.3 Gibbs Sampling

We now briefly introduce how the training algorithm works to obtain the topics using LDA. The basic idea is to iterate through the documents, initially assigning each word to a random topic from the Dirichlet distribution. We then repeatedly iterate over all the documents, looking at each word and trying to find a better topic for it. During this process, we assume that all the other topic assignments in the text corpus are correct except for the word we are currently looking at. We then move forward to the next word in this document, so on and so forth. In one iteration, we process all the words in all the documents in the same way. After enough iterations, we will have a relatively good topic assignment for each word, which is then enough to assign topics to each document.

In this general description the most important question is, how exactly do we update the topic assignment of a word when sampling a document? The answer is, we use *Gibbs Sampling* [21] to approximate the distribution of $P(Z|W; \alpha, \beta)$. For a current word, we expect to get a vector of length k where k is the number of topics. It represents a conditional probability distribution of a one word topic assignment conditioned on the rest of the model. Based on Eq. 12.1, the k-th element in this distribution vector can be derived, as shown in Eq. 12.2.

$$p(Z_{d,n} = k | Z_{-d,n}, W, \alpha, \beta) = \frac{n_{d,k} + \alpha_k}{\sum_{i=1}^{K}(n_{d,i} + \alpha_i)} \frac{m_{k,w_{d,n}} + \beta_{w_{d,n}}}{\sum_i m_{k,i} + \beta_i} \tag{12.2}$$

Here $w_{d,n}$ is the word we are currently looking at. To perform the sampling, we assume that only the current topic assignment to $w_{d,n}$ is wrong, so we remove it from the model before this round of iteration begins. Z is the topic assignment of all words in all documents, and W is the text corpus.

This computation is a product of two parts. As shown in Tbl. 12.1, the first part $n_{d,k}$ shows how many times the document d uses topic k while α_k is the prior weight of topic k in the document. It is effectively the percentage of words that are also assigned the same topic in the whole document, or a measure of how much this document "likes" topic k. The larger it is, the more likely we will assign the current word to topic k again. Similarly, the second part is the percentage of words that are also assigned the same topic in the whole document. This item thus indicates how each topic likes the word w. Larger numbers mean w will be assigned to this topic k again.

Finally, we multiply these two items to get the final distribution of probability of the word $w_{d,n}$, in the form of a vector of length K. Then we can uniformly draw a topic from this vector. We iterate this sampling process until the model is good enough.

12.5.4 Topic Modelling Example

Applications of topic modelling are widespread. For example, it can be used for summarising a large corpus of text data, text categorisation, spam filter, recommender system, or automatic tagging of articles, etc. [22] It can even be used to effectively discover useful structure in large collection of biological data [23].

The `Owl_nlp.Lda` module is used to perform the LDA method. We start with an example to demonstrate how LDA works.

```
let build_lda corpus topics =
  let model = Nlp.Lda.init ~iter:1000 topics corpus in
  Nlp.Lda.(train SimpleLDA model);
  model;;
```

The input to LDA is still the text corpus we have already built. We also need to specify how many topics we want the text corpus to be divided into. For the sake of argument, let's say we want 8 topics. We first initialise model using the `init` function, and then we train it. Let's take a look at the document-topic table in this model, as shown below.

```
val dk : Arr.arr =
     C0   C1   C2   C3   C4   C5   C6   C7
R0   13   13    4    7   11   12   14   16
R1   35  196   15   42   31   23  122    4
R2    7    9    3    1    3  163    2    4
R3   10   22   23  140   18   11   17  143
...
```

This matrix shows the distribution of topics in documents, with each row representing a document and each column a topic. For example, the fifth column of in the third document (R2) is considerably larger than most others suggesting that document predominantly talks only about topic 6. Similarly, in the fourth document, topics 4 and 8 are of equal coverage.

We can also check the topic-word table of this model:

```
val wk : Arr.arr =
     C0   C1   C2   C3   C4   C5   C6   C7
R0    1    0    0    0    0    0    0    0
R1    0    0    0    1    0    0    0    0
R2    0    0    0    0    3    0    0    0
R3    0    0    0    0    0    0    0    3
...
```

This is a sparse matrix where each row represents a word from the vocabulary, and each column a topic. Thus a topic can be represented as the words having the largest values in that column. For example, if we were to represent a topic by 10 words, the translation from the word-topic table to text representation would be:

```
let get_topics vocab wt =
  Mat.map_cols (fun col ->
    Mat.top col 10
    |> Array.map (fun ws ->
      Owl_nlp.Vocabulary.index2word vocab ws.(0))
  ) wt;;
```

As an example, we can take a look at the topics generated by the "A Million News Headlines" dataset[10] from Kaggle:

1. police child calls day court says abuse dead change market
2. council court coast murder gold government face says national police
3. man charged police nsw sydney home road hit crash guilty
4. says wa death sa abc australian report open sex final
5. new qld election ban country future trial end industry hour
6. interview australia world cup china south accused pm hill work
7. police health govt hospital plan boost car minister school house
8. new water killed high attack public farmers funding police urged

Here each topic is represented by its ten highest ranked words in the vocabulary, and so you might "feel" a common theme by connecting these dots together, even though some words may seem a bit far from this theme. Note that we cannot directly observe the topic, only documents and words, and so topics are *latent*. The word-topic matrix shows that each word has different weight in the topic and the words in a topic are ranked according to the weight. Now that we know what each topic talks about, we can cluster the documents by their most prominent topic, or just discover what topics are covered in a document.

We have introduced the basic mechanisms of LDA. There are many other works that extend it, such as SparseLDA [19] and LightLDA [24]. Although they may differ in details, they share the same basic theory.

12.6 Latent Semantic Analysis

Besides LDA, another common technique in topic modelling is Latent Semantic Analysis (LSA) [26]. Its purpose is the same as LDA: to obtain two matrices, the document-topic table and the word-topic table to show the probability distribution of topics in documents and words in topics. The difference is that, rather than using an iterative update approach, LSA explicitly builds the *document-word matrix* and then performs singular value decomposition (SVD) on it to get the two aforementioned matrices.

Assume the text corpus contains n documents and the vocabulary contains m words. The document-word matrix is then of size $n \times m$. We can use the simple word count as elements in this matrix, but as discussed above, the count of words does not reflect the significance of a word. A better way to fill in the document-word matrix is thus to use the TF-IDF approach for each word in a document.

Apparently, this matrix would be quite sparse and its row vectors have very high dimension so there is surely redundant information here. For example, if two documents talk about the same topic(s), then the words they contain will have

---

[10] A Million News Headlines, Kaggle, URL: `https://www.kaggle.com/therohk/million-headlines`

significant overlap. The SVD reduces the dimensionality and redundancy in this matrix, as we have talked about in Chap. 4.

SVD decomposes a matrix A into three: $A = USV^T$. In this specific context of semantic analysis, A is the document-word composition. We can think of the U as representing the relationship between document and topics, and V^T the relationship between topics and words. The columns of U and rows of V^T are both orthonormal bases, and the diagonal matrix S has eigenvalues along its diagonal, each representing the weight of a group of corresponding bases from U and V. We can thus throw away the bases of less weight, truncating to only K columns (rows) from each matrix. In that way we can preserve a large part of information from the original document-word table while choosing only a small number of topics. This process is shown in Fig. 12.3. Once we have the document-topic table U and the topic-word table V, using the model will be the same as in LDA example.

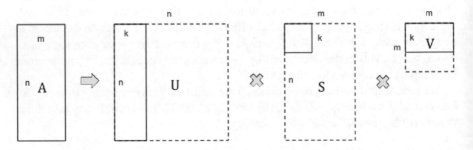

Fig. 12.3: Applying SVD and then truncating on document-word matrix to retrieve topic model

Compared to LDA, this process is easy to understand and implement, but SVD is also computationally intensive and hard to iterate with new data. The result may vary from corpus to corpus, some empirical studies [27] indicate it may not be as good as LDA in separating out the topic categories.

12.7 Search Relevant Documents

Topic models are effective tools for clustering documents based on their similarity or relevance. We can further use this tool to query relevant document given an input one. In this section, we will briefly introduce how to query models built using the previous topic modelling method.

12.7.1 Euclidean and Cosine Similarity

In the previous sections, we saw that topic modelling techniques map documents to a vector space of topics. We can use different metrics to compare the similarity between two vectors in these vector spaces. Two of the most commonly used are the *Euclidean* and *Cosine* distances. Suppose we have two vectors A and B, both of length of n. Then the Euclidean distance between these two are:

$$\sqrt{\sum_{i=1}^{n}(a_i - b_i)^2}. \tag{12.3}$$

The cosine similarity between two vectors A and B is defined as the dot product of two vectors divided by the product of the length of both vectors:

$$\cos(\theta) = \frac{A \cdot B}{\|A\|\|B\|}. \tag{12.4}$$

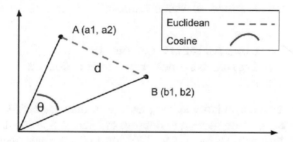

Fig. 12.4: Euclidean distance and cosine similarity in a two dimensional space

The `Nlp.Similarity` module implements both similarity metrics to be used in NLP. The relationship between the Euclidean distance and Cosine similarity is expressed in Fig. 12.4 where there are two points in this two dimensional space. The Euclidean distance measures the direct distance of these two points, while cosine similarity measures the angle between them and so is more suitable for cases where the magnitude of the vectors does not matter. For example, in topic modelling, we already have two vectors representing documents. If we multiply all the elements in one of them by a scalar 10, the Euclidean distance between these two would change greatly. However, since the probability distribution in the vector does not change, we don't actually expect the similarity between them to change so we would prefer in this case to use the cosine similarity as a measure.

12.7.2 Linear Searching

Suppose we have n documents, each represented by a vector of length m. They are then denoted with variable corpus, an array of arrays, each of which is a document. We provide a vector doc as the query, and search for the top-k most similar documents to it. First, we need a function to calculate pairwise distance for the whole model, and returns result in the form of array of (id, dist) where id is the original index of the document, and dist is the distance between a document in corpus and the query document.

```
let all_pairwise_distance typ corpus x =
  let dist_fun = Owl_nlp_similarity.distance typ in
  let l = Array.mapi (fun i y -> i, dist_fun x y) corpus in
  Array.sort (fun a b -> Stdlib.compare (snd a) (snd b)) l;
  l;;
```

The results are sorted according to whichever distance metric we use. Based on this routine we can find the k most relevant document:

```
let query corpus doc k =
  let typ = Owl_nlp_similarity.Cosine in
  let l = all_pairwise_distance typ corpus doc in
  Array.sub l 0 k;;
```

Here we use cosine similarity as the measure of distance between vectors, implemented by matrix multiplication to improve efficiency. Specifically, suppose we have the query document vector A, and the corpus of document vectors as before, and this array of arrays has already been converted to a dense matrix B, where each row vector represents a document. Then we can compute AB^T to get the cosine similarity efficiently. Of course, according to Eq. 12.4, we also need to make sure that A and each row r in B are normalised by its own L2-norm before computations, so that for any vector v we can have $\|v\| = 1$.

```
let query corpus doc k =
  let vec = Mat.transpose doc in
  let l = Mat.(corpus *@ vec) in
  Mat.bottom l k;;
```

Compared to the previous direct element-by-element multiplication, the matrix dot product is often implemented with highly optimised linear algebra library routines, such as those in OpenBLAS. These use various techniques such as multi-processing and multi-threading to give much better performance than naive pairwise computation according to the definition.

12.8 Summary

In this chapter we focused on topic modelling, an important natural language processing task, and introduced the basic ideas and how to implement them in OCaml. First, we introduced how to tokenise a text corpus to enable further mathematical processing, and then we introduced the basic ideas of the vector space model, and two ways to project documents into a vector space: Bag of words (BOW), and Term Frequency-Inverse Document Frequency (TF-IDF). The BOW is straightforward to understand and implement, while TF-IDF adds consideration of how special a word is across the whole text corpus and so usually gives more accurate representations.

We then presented two different methods to retrieve topics from the documents based on vector representation: Latent Dirichlet Allocation (LDA), and Latent Semantic Analysis (LSA). LDA relies on a generative model to iteratively tuning with the training data, while LSA applies singular value decomposition to a document-word matrix. Once we have modelled the topics, we can compare similarity between documents and search for similar documents in the text corpus using different measures of vector distances, of which cosine similarity is commonly encountered in text analysis.

References

1. Jansen BJ, Rieh SY. The seventeen theoretical constructs of information searching and information retrieval. Journal of the American Society for Information Science and Technology. 2010 Aug;61(8):1517-34.
2. Nadeau D, Sekine S. A survey of named entity recognition and classification. Lingvisticae Investigationes. 2007 Jan 1;30(1):3-26.
3. Bahdanau D, Cho K, Bengio Y. Neural machine translation by jointly learning to align and translate. arXiv preprint arXiv:1409.0473. 2014 Sep 1.
4. Lewis M, Liu Y, Goyal N, Ghazvininejad M, Mohamed A, Levy O, Stoyanov V, Zettlemoyer L. Bart: Denoising sequence-to-sequence pre-training for natural language generation, translation, and comprehension. arXiv preprint arXiv:1910.13461. 2019 Oct 29.
5. Radford A, Narasimhan K, Salimans T, Sutskever I. Improving language understanding by generative pre-training.
6. Rumelhart DE, Hinton GE, Williams RJ. Learning representations by back-propagating errors. nature. 1986 Oct;323(6088):533-6.
7. Corpus of Contemporary American English [Internet]. Mark D, 2008 [cited 2022, Jan 4]. Available from: https://www.english-corpora.org/coca
8. Amir Z. The GUM corpus: creating multilayer resources in the classroom. Language Resources and Evaluation. 2017, 51(3), 581-612.
9. Salton G, Wong A, Yang CS. A vector space model for automatic indexing. Communications of the ACM. 1975 Nov 1;18(11):613-20.
10. Teller V. Speech and language processing: an introduction to natural language processing, computational linguistics, and speech recognition.
11. Mikolov T, Le QV, Sutskever I. Exploiting similarities among languages for machine translation. arXiv preprint arXiv:1309.4168. 2013 Sep 17.
12. Open Sourcing BERT: State-of-the-Art Pre-training for Natural Language ProcessingThe Boston Housing Dataset [Internet]. Google AI Blog [cited 2022, Jan 4]. Available from https://ai.googleblog.com/2018/11/open-sourcing-bert-state-of-art-pre.html.

13. Ramachandran P, Parmar N, Vaswani A, Bello I, Levskaya A, Shlens J. Stand-alone self-attention in vision models. arXiv preprint arXiv:1906.05909. 2019 Jun 13.

14. Turc I, Chang MW, Lee K, Toutanova K. Well-read students learn better: On the importance of pre-training compact models. arXiv preprint arXiv:1908.08962. 2019 Aug 23.

15. Luhn HP. A statistical approach to mechanized encoding and searching of literary information. IBM Journal of research and development. 1957 Oct;1(4):309-17.

16. Jones KS. A statistical interpretation of term specificity and its application in retrieval. Journal of documentation. 1972.

17. Blei DM, Ng AY, Jordan MI. Latent dirichlet allocation. the Journal of machine Learning research. 2003 Mar 1;3:993-1022.

18. Buntine WL. Operations for learning with graphical models. Journal of artificial intelligence research. 1994 Dec 1;2:159-225.

19. Yao L, Mimno D, McCallum A. Efficient methods for topic model inference on streaming document collections. InProceedings of the 15th ACM SIGKDD international conference on Knowledge discovery and data mining 2009 Jun 28 (pp. 937-946).

20. Kotz S, Balakrishnan N, Johnson NL. Continuous multivariate distributions, Volume 1: Models and applications. John Wiley & Sons; 2004 Apr 5.

21. Geman S, Geman D. Stochastic relaxation, Gibbs distributions, and the Bayesian restoration of images. IEEE Transactions on pattern analysis and machine intelligence. 1984 Nov(6):721-41.

22. Blei DM. Probabilistic topic models. Communications of the ACM. 2012 Apr 1; 55(4):77-84.

23. Zheng B, McLean DC, Lu X. Identifying biological concepts from a protein-related corpus with a probabilistic topic model. BMC bioinformatics. 2006 Dec; 7(1):1-0.

24. Yuan J, Gao F, Ho Q, Dai W, Wei J, Zheng X, Xing EP, Liu TY, Ma WY. Lightlda: Big topic models on modest computer clusters. Proceedings of the 24th International Conference on World Wide Web 2015 May 18 (pp. 1351-1361).

25. Vaswani A, Shazeer N, Parmar N, Uszkoreit J, Jones L, Gomez AN, Kaiser L, Polosukhin I. Attention is all you need. Advances in neural information processing systems 2017 (pp. 5998-6008).

26. Dumais ST. Latent semantic analysis. Annual review of information science and technology. 2004;38(1):188-230.

27. Stevens K, Kegelmeyer P, Andrzejewski D, Buttler D. Exploring topic coherence over many models and many topics. Proceedings of the 2012 joint conference on empirical methods in natural language processing and computational natural language learning. 2012 Jul (pp. 952-961).

Part III
Use Cases

In the previous two parts, we studied basic numerical techniques and advanced techniques of data analysis using Owl. In this final part, we bring everything learnt together and demonstrate four case studies built in OCaml. Three of them are the exemplar applications of deep neural networks, namely image recognition, instance segmentation, and neural style transfer. The last case demystifies a real recommender system that searches for similar articles in a text corpus using a vector space model. The purpose of these cases are two-fold. First, they bring what we have learnt previously into action; second, these fascinating applications can encourage our readers to further explore the potential of the functional programming by applying various techniques in data analysis. We hope you enjoy building these applications as much as we do.

Chapter 13
Case Study: Image Recognition

Abstract How can a computer take an image and answer questions like "is there a cat or a dog in this picture?" In recent years the machine learning community has made tremendous progress in tackling this problem. In particular, Deep Neural Networks (DNN) are able to achieve extraordinary performance on visual recognition tasks, matching or even exceeding human performance in some domains. Having already introduced neural networks in the previous chapter, we will now discuss one specific use case built on that module: using the InceptionV3 architecture to perform image classification.

13.1 Types of Networks

InceptionV3 is a widely-used DNN architecture for image classification that can attain significant accuracy with a relatively small number of parameters. It did not arise in isolation however; using DNNs to perform image recognition is a stream of work dating back more than 20 years. In this chapter, we first introduce how image classification architectures evolved until Inception. Surveying these related works will help us to understand how InceptionV3 architectures are built.

13.1.1 LeNet

In Chap. 10 on regression, we introduced a simple neural network of three layers, and used it to recognise the simple handwritten digits from the MNIST dataset. Each pixel was an input feature, with each image represented as an ndarray. For black and white images such as those in the MNIST dataset, each bitmap is interpreted as a matrix where each cell holds a value between 0 and 255. For colour images, each cell can instead be interpreted as a 3-dimensional array with each dimension corresponding to three channels: blue, green, and red.

However, we cannot rely on adding more fully connected layers to perform high precision image classification. One important improvement that the Convolution Neural Network (CNN) makes is that it uses filters in the convolution operation. As a result, instead of using the whole image as an array of features, the image is divided into a number of tiles. These tiles then serve as the basic features of the network's prediction.

The next building block is a *pooling layer*. Recall from the neural networks chapter that both average pooling and max pooling can aggregate information from multiple pixels into one and "blur" the input image or feature. But why is this so important? By reducing the size of input, pooling helps to reduce the number of parameters and thus the amount of computation required accordingly. Blurring the features also helps to avoid over-fitting to training data. In the end, we only connect high-level features with full connection layers. This is the structure proposed by LeCun et al. [1]. This whole process is depicted in Fig. 13.1.

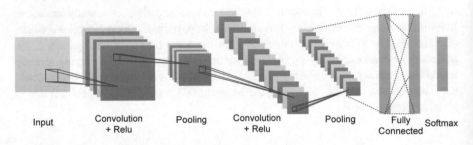

Input   Convolution   Pooling   Convolution   Pooling   Fully   Softmax
          + Relu   + Relu   Connected

Fig. 13.1: The network structure of LeNet

In general, layers of convolution retrieve information gradually moving from details at finer granularity to more abstract level. In a DNN, The lower layers of neurons retrieve information about simple shapes such as edges and points. Moving higher into the network, the neurons capture more complex structures, such as the tire of a car, or the face of a cat. In the top layers, the neurons retrieve and abstract complex ideas, such as "car", "cat", etc., finally generating the classification result.

13.1.2 AlexNet

The next breakthrough comes from AlexNet [2]. The authors introduce better *non-linearity* in the network with the ReLU activation. Recall that from Chap. 10 that although linear regression can be effective, for many applications we need more complex methods such as polynomial regression. The same principle holds here: to process complex input data such as images, we need non-linear activation functions in addition to linear operations such as matrix multiplication and summation.

There are many possible activation functions. Compared to the activation functions previously seen, such as `tanh` ($f(x) = \tanh(x)$), or `sigmoid` ($f(x) = (1 + e^{-x})^{-1}$), the computation of ReLU is simple: $f(x) = max(0, x)$ [3] [4]. In terms of training time with gradient descent, ReLu is much faster and so is more frequently used. The accuracy loss in gradient computation is small, while it makes the training faster. For example, Krizhevsky et al. [2] show that a four-layer CNN with ReLUs reaches a 25% training error rate on CIFAR-10 and six times faster than an equivalent network with `tanh` neurons.

Another improvement that AlexNet proposes is the `dropout` layer, used to address over-fitting issue. This operation is only used during training phase and is removed from the model during inference. It randomly selects some elements in the input ndarray and set their values to zero, deactivating the information contained in those points. By doing so, the network intentionally and randomly drops parts of training examples giving better generalisation. It is similar to the regularisation method we use in linear regression.

The one more thing that we need to note from AlexNet is that by going deeper, making the network "longer", we increase accuracy. So instead of just convolution and pooling, we now build convolution followed by convolution and then pooling, and repeat this process again. A deeper network captures finer features, and this was a trend followed by successive architectures.

13.1.3 VGG

The VGG network [5] is the next step after AlexNet. The most notable change introduced by VGG is that it uses small kernel sizes such as 3×3 rather than 11×11 with a large stride of 4 in AlexNet. Using multiple small kernels is much more flexible than using only one large one. For example, for an input image, by applying two 3×3 kernels with slide size of 1, that equals to using a 5×5 kernel. If stacking three 3×3, it equals using one 7×7 convolution.

By replacing large kernels with multiple small kernels, the number of parameters is significantly reduced. In the previous two examples, replacing one 5×5 with two 3×3, we reduce the parameters by $(1 - 2 * 3 * 3/(5 * 5)) = 28\%$. By replacing the 7×7 kernel, we save parameters by $(1 - 3 * 3 * 3/(7 * 7)) = 45\%$.

With this reduction of parameter size we can build a network with more layers, which tends to improve accuracy. The VGG networks come with two variants, VGG16 and VGG19, which essentially have the same network structure, but the VGG19 is deeper. An extra benefit is that using small kernels increases non-linearity. Imagine an extreme case where the kernel is as large as the input image, then the whole convolution is just one big matrix multiplication, a complete linear operation. As we have just explained in the previous section, we hope to reduce the linearity in training CNN to accommodate more real-world problems.

13.1.4 ResNet

We keep saying that building deeper networks is the trend. However, you may encounter many challenges as a network becomes deeper: the deeper you go, the more you will experience the "vanishing gradient" problem [6] [7]. This problem is that during the back-propagation phase in a very deep network, the repeated multiplication operations will make gradients very small, reducing training performance.

The ResNet [8] proposes an "identity shortcut connection" that skips one or more layers and combines with predecessor layers. It is called a residual block, as shown in Fig. 13.2.

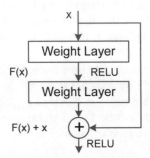

Fig. 13.2: Residual block in the ResNet

We can see that there is an element-wise addition that combines the information of the current output with its predecessors two layers before. It solves the vanishing gradient problem by stacking layers, as the error can now be backpropagated through multiple paths. The authors show that during training the deeper layers do not produce an error higher than its predecessor in lower layer.

Furthermore, note that a residual block aggregates features from different levels of layers, rather than simply stacking them. This pattern proves to be useful and is also used in the Inception architectures. The ResNet can also be constructed by stacking for different layers, so that we have ResNet50, ResNet101, ResNet152, etc.

13.1.5 SqueezeNet

All these architectures, including Inception, seek to push classification accuracy forward. At some point we are confronted with the trade-off between accuracy and model size. We have seen that sometimes reducing the parameter size can help the network to go deeper, and thus increase accuracy. However, with the growing trend of edge computing there are situations where extremely small deep neural networks are more practical, so that they can be easily distributed and deployed on less powerful devices. In such cases, a small reduction in accuracy may be acceptable.

There are more and more efforts devoted to that direction [9] [10], and the SqueezeNet [11] is just one of them. It claims to achieve the AlexNet level of accuracy, but with 50 times fewer parameters. There are three design principles embedded in SqueezeNet. The first is one we are already familiar with: using a lot of 1x1 convolutions. The second also sounds familiar: using the 1x1 convolutions to reduce the output channels before feeding the result to the more complex 3×3 convolutions, allowing them to have smaller input channels therefore fewer parameters. These two principles are incorporated into a basic building block called the "fire module". The third design principle is to delay the down-sampling to the later layers in a network. Recall that the down-sampling layers such as maxpooling "blur" the information intentionally. The intuition is that, if we only use them occasionally and in the deeper layer of the network, we can have large activation maps that preserve more information and make the detection result more accurate.

13.1.6 Capsule Network

Research on image detection network structures is on-going. Besides the number of parameters and detection accuracy, other requirements arise. For example, how to recognise objects, e.g. a car, from different perspectives, or the "Picasso problem" where we need to recognise an object the features of which are intentionally distorted or misplaced. These problems show one deficiency in the existing image classification approach: the lack of connection between features. It may recognise a "nose" feature, and an "eye" feature, and then the object is recognised as a human face, even though the nose may be above the eyes. The *Capsule Network* [12] is proposed to address this: rather than using scalars to represent features, it uses a vector that includes information such as orientation and object size. This "capsule" thus captures the relative relationship between features.

There are many more networks that we cannot cover here, but hopefully you can see that there are some common themes in the development of image recognition architectures. We now turn to the main topic of this chapter: how InceptionV3 is designed.

13.2 Building the InceptionV3 Network

Proposed by Christian Szegedy et. al. in [13],InceptionV3[1] is one of Google's more recent efforts to perform image recognition. It was trained for the ImageNet Large Visual Recognition Challenge [14], a standard task in computer vision, where models try to classify entire image dataset into 1000 classes, like "zebra", "dalmatian", and

---

[1] Rethinking the Inception Architecture for Computer Vision, Authors:Christian Szegedy et al. URL:https://arxiv.org/abs/1512.00567

"dishwasher". Compared with previous DNN models, InceptionV3 is one of the most complex neural network architectures in computer vision.

The design of image recognition networks is about the trade-off between computation cost, memory usage, and accuracy. Increasing model size and computation cost tend to improve accuracy, but the benefits of a given size quickly diminish. To solve this problem the Inception architecture aims to perform well with strict constraints on memory and computational budget. The design follows several principles, such as balancing the width and depth of the network, and performing spatial aggregation over lower dimensional embeddings which may lead to small losses in the representational power of networks. The resulting InceptionV3 network architectures[2] have high performance and a relatively modest computation cost compared to simpler, more monolithic architectures. We can see that the whole network can be divided into several parts, and the Inception module A, B, and C are both repeated based on one structure.[3]

13.2.1 InceptionV1 and InceptionV2

InceptionV3 was developed from two previous similar architectures. To understand it, we first need to know the characteristics of its predecessors.

The first version of Inception, GoogLeNet [15], combined convolutions with different filter sizes on the same input, then concatenated the resulting features. Let's use an image of birds as an example. If you stick with a normal square filter, then perhaps the features such as "feather" is more difficult to capture than if you use a "thin" filter e.g., 1×7. By aggregating information from different features, we can extract features from multiple levels at each step.

Of course, adding extra filters increases computational complexity. To mitigate this effect, the Inception network proposes using 1×1 convolutions to reduce the dimensions of feature maps. For example, we want to apply a 3×3 filter to an input ndarray of size [|1; 300; 300; 768|] and the output channel size is set to 320. Instead of applying a convolution layer of [|3; 3; 768; 320|] directly, we first reduce the dimension to, say, 192 by using a small convolution layer [|1; 1; 768; 192|], and then apply a [|3; 3; 192; 320|] convolution layer to get the final result. By reducing input dimension before the more complex computation with large kernels, the computation complexity is greatly reduced.

For those who are confused about the meaning of this array: the format of an image as ndarray is [|batch; column; row; channel|], and that the format of a convolution operation is mainly represented as [|kernel_column; kernel_row; input_channel; output_channel|]. Here we ignore the other parameters such as slides and padding since we focus only on the change of channels of feature map.

---

[2] Advanced Guide to InceptionV3, Google Cloud TPU Documentation, URL: https://cloud.google.com/tpu/docs/inception-v3-advanced.

[3] Actually, that's where the name "Inception" comes from: as in the film, like dreams, you can have stacked these basic units layer by layer.

In an updated version of GoogLeNet, the InceptionV2 (or BN-inception), utilises the "Batch Normalisation" layer [16]. We have seen how normalising input data plays a vital role in improving the efficiency of gradient descent in Chap. 9. Batch normalisation follows a similar path, only it now works between each layer instead of just at the input data. This layer rescales each mini-batch with the mean and variance of this mini-batch.

Imagine that we train a network to recognise "horse", but most of the training data are actually black or blown horse. The network's performance on white horses might be quite poor as a result – a form of over-fitting. The batch normalisation layer adds noise to input by scaling. As a result, the content at deeper layer becomes less sensitive to content in earlier layers. Overall, the batch normalisation layer greatly improves the efficiency of training.

13.2.2 Factorisation

Now that we understand how the image recognition architectures are developed, it's about time to see how these factors are used in InceptionV3. We use Owl in OCaml to construct this network.

```
open Owl
open Owl_types
open Neural.S
open Neural.S.Graph

let conv2d_bn ?(padding=SAME) kernel stride nn =
  conv2d ~padding kernel stride nn
  |> normalisation ~training:false ~axis:3
  |> activation Activation.Relu;;
```

The conv2d_bn is a basic building block used in this network, consisting of a convolution layer, a normalisation layer, and a relu activation layer. We have already introduced how these different types of layer work. You can think of it as an enhanced convolution layer.

Based on this basic block, the aim in building the InceptionV3 network is to go deeper, but to do so the authors introduce three types of *Inception Modules* as units for stacking layers. Each factorises large kernels into smaller ones. Let's look at each in turn.

```
let mix_typ1 in_shape bp_size nn =
  let branch1x1 = conv2d_bn [|1;1;in_shape;64|] [|1;1|] nn in
  let branch5x5 = nn
    |> conv2d_bn [|1;1;in_shape;48|] [|1;1|]
    |> conv2d_bn [|5;5;48;64|] [|1;1|]
  in
```

```
  let branch3x3dbl = nn
    |> conv2d_bn [|1;1;in_shape;64|] [|1;1|]
    |> conv2d_bn [|3;3;64;96|]    [|1;1|]
    |> conv2d_bn [|3;3;96;96|]    [|1;1|]
  in
  let branch_pool = nn
    |> avg_pool2d [|3;3|] [|1;1|]
    |> conv2d_bn [|1;1;in_shape; bp_size |] [|1;1|]
  in
  concatenate 3 [|branch1x1; branch5x5;
    branch3x3dbl; branch_pool|];;
```

The mix_typ1 structure implements the first type of Inception module. In branch3x3dbl branch, it replaces a 5×5 kernel convolution layer with two 3×3 convolution layers, following the design of the VGG network. Of course, as we have explained, both branches first use a 1×1 convolution to reduce dimensionality before the complex convolution computations.

In some implementations the 3×3 convolutions can be further factorised into 3×1 and then 1×3 convolutions. There is more than one way to do the factorisation. You might be thinking it is also a good idea to replace the 3×3 convolution with two 2×2 convolutions. While we could do that, it would save only about 11% of the parameters, compared to the 33% saved by current practice.

```
let mix_typ4 size nn =
  let branch1x1 = conv2d_bn [|1;1;768;192|] [|1;1|] nn in
  let branch7x7 = nn
    |> conv2d_bn [|1;1;768;size|] [|1;1|]
    |> conv2d_bn [|1;7;size;size|] [|1;1|]
    |> conv2d_bn [|7;1;size;192|] [|1;1|]
  in
  let branch7x7dbl = nn
    |> conv2d_bn [|1;1;768;size|]  [|1;1|]
    |> conv2d_bn [|7;1;size;size|]  [|1;1|]
    |> conv2d_bn [|1;7;size;size|]  [|1;1|]
    |> conv2d_bn [|7;1;size;size|]  [|1;1|]
    |> conv2d_bn [|1;7;size;192|]  [|1;1|]
  in
  let branch_pool = nn
    |> avg_pool2d [|3;3|] [|1;1|]
    |> conv2d_bn [|1;1; 768; 192|] [|1;1|]
  in
  concatenate 3 [|branch1x1; branch7x7;
    branch7x7dbl; branch_pool|];;
```

As shown in the code above, mix_typ4 shows the Type B inception module, another basic unit. It still separates into three branches and then concatenates them together. The feature of this type of branch is that it factorises a 7×7 convolution into a sequence of a 7×1 then a 1×7 convolution. Again, this change saves (49 - 14) / 49 = 71.4% parameters. If you have doubts about this replacement, you can do a simple experiment as blow.

```
open Neural.S
open Neural.S.Graph

let network_01 =
  input [|28;28;1|]
  |> conv2d ~padding:VALID [|7;7;1;1|] [|1;1|]
  |> get_network

let network_02 =
  input [|28;28;1|]
  |> conv2d ~padding:VALID [|7;1;1;1|] [|1;1|]
  |> conv2d ~padding:VALID [|1;7;1;1|] [|1;1|]
  |> get_network;;
```

Checking the output log, you can see that both networks give the same output shape. This factorisation is intuitive: convolution of size does not change the output shape at that dimension. The two convolutions actually perform feature extraction along each dimension (height and width) respectively.

```
let mix_typ9 input nn =
  let branch1x1 = conv2d_bn [|1;1;input;320|] [|1;1|] nn in
  let branch3x3 = conv2d_bn [|1;1;input;384|] [|1;1|] nn in
  let branch3x3_1 = branch3x3
    |> conv2d_bn [|1;3;384;384|] [|1;1|] in
  let branch3x3_2 = branch3x3
    |> conv2d_bn [|3;1;384;384|] [|1;1|] in
  let branch3x3 = concatenate 3
    [| branch3x3_1; branch3x3_2 |] in
  let branch3x3dbl = nn
    |> conv2d_bn [|1;1;input;448|] [|1;1|]
    |> conv2d_bn [|3;3;448;384|] [|1;1|] in
  let branch3x3dbl_1 = branch3x3dbl
    |> conv2d_bn [|1;3;384;384|] [|1;1|] in
  let branch3x3dbl_2 = branch3x3dbl
    |> conv2d_bn [|3;1;384;384|] [|1;1|] in
  let branch3x3dbl = concatenate 3
    [|branch3x3dbl_1; branch3x3dbl_2|] in
  let branch_pool = nn
    |> avg_pool2d [|3;3|] [|1;1|]
    |> conv2d_bn [|1;1;input;192|] [|1;1|] in
  concatenate 3
    [|branch1x1; branch3x3; branch3x3dbl; branch_pool|];;
```

The final inception module is slightly more complex, but by now you should be able to understand its construction in the code above. It aggregates information from four branches. The 1×1 convolutions are used to reduce dimensionality and computational complexity. Note that in both the branch3x3 and branch3x3dbl branches, both 3×1 and 1×3 convolutions are used, not as a replacement of 3×3, but as separate branches. This module promotes high dimensional representations.

Together, these three modules make up the core of the InceptionV3 architecture. By applying different techniques and designs, the modules take less memory and are less prone to over-fitting problem. Thus they can be stacked together to enable the whole network to go deeper.

13.2.3 Grid Size Reduction

From the very beginning of the design of CNNs, we need to reduce the size of feature maps while increasing the number of channels of the feature map. We have explained how it is done by combining the pooling layer and convolution layer, but the reduction solution constructed this way is often either too greedy or too computationally expensive.

The Inception architecture proposes a grid size reduction module. It feeds the same input feature map into two sets of pipelines, one using the pooling operation, and the other using only convolution layers. These two types of layers are then not stacked together but concatenated vertically, as shown in the following code.

```
let mix_typ3 nn =
  let branch3x3 = conv2d_bn [|3;3;288;384|]
    [|2;2|] ~padding:VALID nn in
  let branch3x3dbl = nn
    |> conv2d_bn [|1;1;288;64|] [|1;1|]
    |> conv2d_bn [|3;3;64;96|] [|1;1|]
    |> conv2d_bn [|3;3;96;96|] [|2;2|] ~padding:VALID
  in
  let branch_pool = max_pool2d [|3;3|] [|2;2|]
    ~padding:VALID nn in
  concatenate 3 [|branch3x3; branch3x3dbl; branch_pool|];;
```

mix_typ3 builds the first grid size reduction module. This replacement strategy can perform efficient reduction without losing too much information. Similarly, an extended version of this grid size reduction module is also included.

```
let mix_typ8 nn =
  let branch3x3 = nn
    |> conv2d_bn [|1;1;768;192|] [|1;1|]
    |> conv2d_bn [|3;3;192;320|] [|2;2|] ~padding:VALID
  in
```

```
let branch7x7x3 = nn
  |> conv2d_bn [|1;1;768;192|] [|1;1|]
  |> conv2d_bn [|1;7;192;192|] [|1;1|]
  |> conv2d_bn [|7;1;192;192|] [|1;1|]
  |> conv2d_bn [|3;3;192;192|] [|2;2|] ~padding:VALID
in
let branch_pool = max_pool2d [|3;3|] [|2;2|]
  ~padding:VALID nn in
concatenate 3 [|branch3x3; branch7x7x3; branch_pool|];;
```

mix_typ8 is the second grid size reduction module in the deeper part of the network. It uses three branches instead of two, and each convolution branch is more complex. The 1×1 convolutions are again used. But in general it still follows the principle of performing reductions in parallel and then concatenating them together, performing an efficient feature map reduction.

13.2.4 InceptionV3 Architecture

Having introduced the separate units, we can finally use them to build the entire network in the following code:

```
let make_network img_size =
  input [|img_size;img_size;3|]
  |> conv2d_bn [|3;3;3;32|] [|2;2|] ~padding:VALID
  |> conv2d_bn [|3;3;32;32|] [|1;1|] ~padding:VALID
  |> conv2d_bn [|3;3;32;64|] [|1;1|]
  |> max_pool2d [|3;3|] [|2;2|] ~padding:VALID
  |> conv2d_bn [|1;1;64;80|] [|1;1|] ~padding:VALID
  |> conv2d_bn [|3;3;80;192|] [|1;1|] ~padding:VALID
  |> max_pool2d [|3;3|] [|2;2|] ~padding:VALID
  |> mix_typ1 192 32 |> mix_typ1 256 64 |> mix_typ1 288 64
  |> mix_typ3
  |> mix_typ4 128 |> mix_typ4 160
  |> mix_typ4 160 |> mix_typ4 192
  |> mix_typ8
  |> mix_typ9 1280 |> mix_typ9 2048
  |> global_avg_pool2d
  |> linear 1000 ~act_typ:Activation.(Softmax 1)
  |> get_network;;
```

We need to mention just one last thing: *global pooling*. Global Average/Max Pooling calculates the average/max output of each feature map in the previous layer. For example, if you have an [|1;10;10;64|] feature map, then this operation can make it to be [|1;1;1;64|]. This operation may seem very simple, but it works at the very end of a network, replacing the fully connected layer. The parameter size of the fully connected layer has always been a problem; by replacing it with a non-trainable operation, the parameter size is greatly reduced without significantly affecting performance. The global pooling layer is also more robust to spatial translations in the data and is claimed to mitigate the over-fitting problem in a fully connected layer [18].

13.3 Preparing Weights

Just building a network structure is not enough – the weights must still be trained. This can be done by training on a large amount of image data for days or weeks or more on powerful machine clusters [17].

Training is usually done via supervised learning using a large set of labelled images. Although InceptionV3 can be trained from many different labelled image sets, ImageNet [14] is a commonly used dataset for this purpose. In this dataset, the basic unit is a "synset", a concept described by multiple words or word phrases. ImageNet aims to provide about 1,000 images for each of the 100,000 synsets in WordNet. Training the InceptionV3 model takes more than 1 million images from ImageNet. The training of this model can take hundreds of hours on multiple high-performance GPUs.

However, not everyone has access to such large resource. Another option is often more viable: importing weights from pre-trained TensorFlow models, currently widely available in model collections. In essence, weights are lists of Owl ndarrays, implemented using `Bigarray` in OCaml. So the basic idea is to find a intermediate representation so that we can translate the ndarray in NumPy to OCaml's `Bigarray`. In our implementation, we choose to use the HDF5[4] as this intermediate data exchange format. In Python, we use the h5py[5] library, whereas in OCaml we can use hdf5-ocaml[6].

The method to save or load HDF5 data files are fixed, but the methods to retrieve data from model files depend on the type of original files. For example, if we choose to import weights from a TensorFlow model, we can utilise the following code snippet to obtain the weight data of each layer:

```
...
reader = tf.train.NewCheckpointReader(checkpoint_file)
for key in layer_names:
    data=reader.get_tensor(key).tolist()
...
```

In a Keras, it is even more straightforward:

---

[4] HDF5, URL:https://portal.hdfgroup.org/display/HDF5/HDF5
[5] HDF5 for Python, URL:https://www.h5py.org/
[6] HDF5-OCaml, URL:https://github.com/vbrankov/hdf5-ocaml

```
...
for layer in model.layers:
    weights = layer.get_weights()
...
```

In OCaml, we first create a `Hashtable` and read all the HDF5 key-value pairs into it. Each value is saved as a double precision Owl ndarray.

```
...
let h = Hashtbl.create 50 in
let f = H5.open_rdonly h5file  in
for i = 0 to (Array.length layers - 1) do
  let w = H5.read_float_genarray f layers.(i) C_layout in
  Hashtbl.add h layers.(i) (Dense.Ndarray.Generic.cast_d2s w)
done;
...
```

We can then use the `mkpar` and `update` methods in the Neural Network module to load these values from the hashtable to networks:

```
...
let wb = Neuron.mkpar n.neuron in
Printf.printf "%s\n" n.name;
wb.(0) <- Neuron.Arr (Hashtbl.find layers n.name);
Neuron.update n.neuron wb
...
```

When you are importing weights from other platforms, make sure the name of corresponding layers match. Other differences must also be handled such as the `beta` and `gamma` weights in the batch normalisation layer being represented as two different values in TensorFlow model while belonging to the same layer in Owl. Also, channel order may have to be swapped in an ndarray during this weight conversion, either the "channel first" or "channel last" order.

But this is one-off work. Once you have successfully updated the network with weights, the weights can be saved using `Graph.save_weights` without having to repeat all these steps again. We have already prepared the weights for the InceptionV3 model and other similar models, and the users don't have to worry about these trivial model exchanging details.

13.4 Processing Image

Image processing in Owl is challenging as OCaml does not provide powerful functions to manipulate images. Though there are image processing libraries such as CamlImages[7], we don't want to add extra dependencies to Owl itself.

---

[7] CamlImages, URL:http://gallium.inria.fr/camlimages

To this end we choose the non-compressed image format PPM [19]. A PPM file is a 24-bit color image represented in text. It stores each pixel with a number from 0 to 65535, which specifies the color of the pixel. Therefore, we can just use ndarray in Owl and convert that directly to PPM image without using any external libraries. We only need to take care of header information during this process.

For example, here is the code for converting a 3-dimensional array in Owl img into a PPM file. We first need to get the content from each of the three colour channels using slicing, such as the blue channel:

```
let b = N.get_slice [[];[];[0]] img in
let b = N.reshape b [|h; w|] in
...
```

Here h and w are the height and width of the image. We then need to merge all three matrices into a large matrix of size [|h; 3*w|].

```
let img_mat = Dense.Matrix.S.zeros h (3 * w) in
Dense.Matrix.S.set_slice [[];[0;-1;3]] img_mat r;
...
```

After rotating this matrix by 90 degrees, we must rewrite it to a large byte variable.

```
let img_str = Bytes.make (w * h * 3) ' ' in
let ww = 3 * w in
for i = 0 to ww - 1 do
  for j = 0 to h - 1 do
    let ch = img_arr.(i).(j) |> int_of_float |> char_of_int in
    Bytes.set img_str ((h - 1 -j) * ww + i) ch;
  done
done;
```

Finally we build another byte string containing the metadata such as height and width, according to the specification of PPM format. Concatenating the metadata and data together, we then write the image into a file.

Reading a PPM image file into an ndarray in Owl is similar: we process the PPM file line by line with input_line function, ignoring metadata such as version and comments. We obtain important metadata such as width and height from the header.

```
...
let w_h_line = input_line fp in
let w, h = Scanf.sscanf w_h_line "%d %d" (fun w h -> w, h) in
...
```

According to this information, we read the rest of data into large bytes with Pervasive.really_input. Note that in 32bit OCaml, the maximum size will be about 16 MB.

```
...
let img = Bytes.make (w * h * 3) ' ' in
really_input fp img 0 (w * h * 3);
close_in fp;
...
```

We must then re-arrange the data in the bytes into a matrix.

```
let imf_o = Array.make_matrix (w * 3) h 0.0 in

let ww = 3 * w  in
for i = 0 to ww - 1 do
  for j = 0 to h - 1 do
    imf_o.(i).(j) <- float_of_int
        (int_of_char (Bytes.get img ((h - 1 - j ) * ww + i)));
  done
done;
```

This can then be further processed into an ndarray with proper slicing.

```
...
let m = Dense.Matrix.S.of_arrays img in
let m = Dense.Matrix.S.rotate m 270 in
let r = N.get_slice [[];[0;-1;3]] m in
let g = N.get_slice [[];[1;-1;3]] m in
let b = N.get_slice [[];[2;-1;3]] m in
...
```

There are several other functions such as reading in an ndarray from the PPM file; the full image processing code can be viewed in the code repository of this chapter.

Of course, most of the time we have to deal with images of more common formats such as PNG and JPEG. To convert from these formats to PPM or vice versa, we use the tool ImageMagick[8]. ImageMagick is a free and open-source tool suite for image related tasks, such as converting, modifying, and editing images. It can read and write over 200 image file formats, including the PPM format. Therefore, we preprocess the images by converting its format to PPM with the command `convert`. Computation time is related to input size, and you may wish to limit the size of images which can be done with the command `convert -resize`.

Another important preprocessing step is to normalise the input. Instead of processing the input ndarray whose elements ranges from 0 to 255, we must preprocess it so that all elements fall into the range $[-1, 1]$, as shown in the code below.

```
let normalise img =
  let img = Arr.div_scalar img 255. in
  let img = Arr.sub_scalar img 0.5 in
  let img = Arr.mul_scalar img 2. in
  img;;
```

---

[8] ImageMagick, URL:https://www.imagemagick.org

13.5 Running Inference

Having built the network and loaded the proper weights, it's time to actually use the model to perform inferencing. Actually the most important part is just one function: `Graph.model`. In this section we show how to build a service around this function so that a user can perform all the inference steps, from input image to output classification result as simply as possible. First let's look at the code:

```
#!/usr/bin/env owl
open Owl

#zoo "9428a62a31dbea75511882ab8218076f"

let _ =
  let img = "panda.png" in
  let labels = InceptionV3.infer img in
  let top = 5 in
  let labels_json  = InceptionV3.to_json ~top labels in
  let labels_tuples = InceptionV3.to_tuples labels in

  Printf.printf "\nTop %d Predictions:\n" top;
  Array.iteri (fun i x ->
    let cls, prop = x in
    Printf.printf "Prediction #%d (%.2f%%) : %s\n" i
      (prop *. 100.) cls;
  ) labels_tuples;;
```

The code first load the whole Inception network structure and weights from a gist[9] using Owl's Zoo module. This is a useful utility to import functions and modules in a OCaml script. This script should be first uploaded to Github Gist, which will then be assigned a Gist id. And then we can use the `zoo` primitive and this Gist id at the beginning of the current OCaml script or the `utop` environment to import its content. In this example, the InceptionV3 module provides three APIs.

- `infer` performs image recognition tasks over provided images. It takes a string specifying the location of a local image and returns a 1×1000 N-dimensional array, with each element a float number between 0 and 1, indicating the probability that the image belongs to one of the 1000 classes from ImageNet.
- `to_json` converts the inferred result to a JSON encoded string. Parameter `top` provides an integer value specifying how many of the top-N most probable labels to return. Default value is 5.
- `to_tuples` converts the inferred result to an array of tuples, containing a label name (string) and the probability of the target image being in that class (a number of float type between 0 and 1).

---

[9] An implementation of InceptionV3 network with Owl, Author:jzstark, URL:https://gist.github.com/jzstark/9428a62a31dbea75511882ab8218076f

After loading the InceptionV3 model using the#zoo primitive, the user must provide an absolute path to the local input image. Here we use the extend_zoo_path utility function to automatically find the "panda.png" image contained in the Gist itself. With this preliminary work done, we can finally run inference on the image using the neural network model by calling infer, a wrapper around the Graph.model function. The result is the top-N ($N = 5$ by default) most likely classifications as an array of tuples. Each tuple consists of a string indicating the classification, and a float from 0 to 100 representing the probability of the input image being this type as a percentage.

Having completed the whole script, we can wrap the code into a zoo gist so all a user must do to invoke it is the single line:

```
owl -run 6dfed11c521fb2cd286f2519fb88d3bf
```

That's it. This one-liner will produce an image classification for the embedded "panda.png" image, and here is the output:

```
Top 5 Predictions:
Prediction #0 (96.20%) : giant panda, panda, panda bear,
  coon bear, Ailuropoda melanoleuca
Prediction #1 (0.12%) : lesser panda, red panda, panda,
  bear cat, cat bear, Ailurus fulgens
Prediction #2 (0.06%) : space shuttle
Prediction #3 (0.04%) : soccer ball
Prediction #4 (0.03%) : indri, indris, Indri indri,
  Indri brevicaudatus
```

13.6 Applications

Building an image classification application is not the end itself – it is used in many other applications [20] [21] [22]:

- Face recognition. If you have a Facebook account, it must be very familiar to you how your friends are tagged in your uploaded photo.
- Photo organisation. Take Eden Photo[10], a Mac application, as an example. It automatically places tags and keywords on your photos, providing powerful photo categorisations.
- Self-driving cars. Image recognition is no doubt a core piece of the technology involved in producing a self-driving car, allowing objects on the road to be identified.
- Medical imaging. This technology can be used to enhance processes such as cancer detection and MRI processing by automatically detecting signs diseases in medical images.

---

[10] Heavenly simple photo organization, URL:https://imagga.com/success-stories/eden-case-study

That's just a small number of examples using image recognition. It's also a fundamental step in computer vision tasks including instance segmentation [23] and image style transfer [24] that we will discuss in the following chapters.

13.7 Summary

This case study demonstrated the full cycle of performing image classification using the InceptionV3 architecture. We first reviewed how state-of-the-art DNN architectures developed, and then the practical details including how the network is constructed, how model weights are imported, image processing in OCaml, and how the inference is performed. We also showed how this process can be greatly simplified using the Zoo system.

References

1. LeCun Y, Bottou L, Bengio Y, Haffner P. Gradient-based learning applied to document recognition. Proceedings of the IEEE. 1998 Nov; 86(11):2278-324.
2. Krizhevsky A, Sutskever I, Hinton GE. Imagenet classification with deep convolutional neural networks. Advances in neural information processing systems. 2012; 25:1097-105.
3. Brownlee J. A gentle introduction to the rectified linear unit (ReLU). Machine learning mastery. 2019 Jan 9;6.
4. Liu D. A practical guide to relu. Medilum, November. 2017 Nov.
5. Simonyan K, Zisserman A. Very deep convolutional networks for large-scale image recognition. arXiv preprint arXiv:1409.1556. 2014 Sep 4.
6. Hochreiter S. Untersuchungen zu dynamischen neuronalen Netzen [in German] Diploma thesis. TU Münich. 1991.
7. Hochreiter S, Bengio Y, Frasconi P, Schmidhuber J. Gradient flow in recurrent nets: the difficulty of learning long-term dependencies.
8. He K, Zhang X, Ren S, Sun J. Deep residual learning for image recognition. Proceedings of the IEEE conference on computer vision and pattern recognition 2016 (pp. 770-778).
9. Zhang X, Zhou X, Lin M, Sun J. Shufflenet: An extremely efficient convolutional neural network for mobile devices. Proceedings of the IEEE conference on computer vision and pattern recognition 2018 (pp. 6848-6856).
10. Howard AG, Zhu M, Chen B, Kalenichenko D, Wang W, Weyand T, Andreetto M, Adam H. Mobilenets: Efficient convolutional neural networks for mobile vision applications. arXiv preprint arXiv:1704.04861. 2017 Apr 17.
11. Iandola FN, Han S, Moskewicz MW, Ashraf K, Dally WJ, Keutzer K. SqueezeNet: AlexNet-level accuracy with 50x fewer parameters and< 0.5 MB model size. arXiv preprint arXiv:1602.07360. 2016 Feb 24.
12. Sabour S, Frosst N, Hinton GE. Dynamic routing between capsules. arXiv preprint arXiv:1710.09829. 2017 Oct 26.
13. Szegedy C, Vanhoucke V, Ioffe S, Shlens J, Wojna Z. Rethinking the inception architecture for computer vision. Proceedings of the IEEE conference on computer vision and pattern recognition 2016 (pp. 2818-2826).
14. Olga R*, Jia D*, Hao S, Jonathan K, Sanjeev S, Sean M, Zhiheng H, Andrej K, Aditya K, Michael B, Alexander B and Li F. (* = equal contribution) ImageNet Large Scale Visual Recognition Challenge. International Journal of Computer Vision, 2015.

15. Szegedy C, Liu W, Jia Y, Sermanet P, Reed S, Anguelov D, Erhan D, Vanhoucke V, Rabinovich A. Going deeper with convolutions. Proceedings of the IEEE conference on computer vision and pattern recognition 2015 (pp. 1-9).
16. Ioffe S, Szegedy C. Batch normalization: Accelerating deep network training by reducing internal covariate shift. International conference on machine learning 2015 Jun 1 (pp. 448-456). PMLR.
17. Raina R, Madhavan A, Ng AY. Large-scale deep unsupervised learning using graphics processors. Proceedings of the 26th annual international conference on machine learning 2009 Jun 14 (pp. 873-880).
18. Lin M, Chen Q, Yan S. Network in network. arXiv preprint arXiv:1312.4400. 2013 Dec 16.
19. Murray JD, VanRyper W. Encyclopedia of graphics file formats. Sebastopol: O'Reilly. 1996.
20. Libby C, Ehrenfeld J. Facial Recognition Technology in 2021: Masks, Bias, and the Future of Healthcare. Journal of Medical Systems. 2021 Apr; 45(4):1-3.
21. Taeihagh A, Lim HS. Governing autonomous vehicles: emerging responses for safety, liability, privacy, cybersecurity, and industry risks. Transport reviews. 2019 Jan 2; 39(1):103-28.
22. Shen D, Wu G, Suk HI. Deep learning in medical image analysis. Annual review of biomedical engineering. 2017 Jun 21;19:221-48.
23. Yi J, Wu P, Jiang M, Huang Q, Hoeppner DJ, Metaxas DN. Attentive neural cell instance segmentation. Medical image analysis. 2019 Jul 1;55:228-40.
24. Gatys LA, Ecker AS, Bethge M. A neural algorithm of artistic style. arXiv preprint arXiv:1508.06576. 2015 Aug 26.

Chapter 14
Case Study: Instance Segmentation

Abstract Computer vision is a field that automates tasks such as ascribing high-level descriptions to images and videos. It has been applied to a wide variety of domains ranging from the highly technical such as automatic tagging of satellite images or analysis of medical imaging [1] [2], to the more mundane such as categorising pictures in your phone or creating an emoji from a picture of your face. This field has seen tremendous progress since 2012, when A. Krizhevsky et al. first used deep learning in computer vision and crushed their opponents in the ImageNet challenge [3]. We have already discussed the image recognition task in the previous chapter. Here we introduce another classical computer vision task, *Instance Segmentation*, which labels objects within an image. We will discuss its connection with other similar applications, how the deep neural networks are constructed in OCaml, and how such a network when loaded with pre-trained weights, can be used to process users' input images.

14.1 Introduction

In the previous chapter, we introduced how various DNNs can be applied to classify a single object in an image. However, these neural networks can easily get confused if they are applied on an image with multiple objects, which happens all too often in real world images. So we need other methods to handle this case.

For example, *Object Detection* is another classical computer vision task [4]. Given an image that contains multiple objects, we seek to classify individual objects and locate each one using a bounding box. Similarly, the *Semantic Segmentation* task requires classifying the pixels in an image into different categories [5]. Each segment is recognised by a "mask" that covers the whole object. All possible objects are shown using different masks, but it does not categorise what those objects are.

In 2017, the *Mask R-CNN* (Mask Region-based Convolutional Neural Network) architecture was published [6]. With sufficient training, it can solve these problems at once, detecting objects in an image, labelling each of them, and providing a binary mask for the image to determine which pixels belong to which objects. This task is called *Instance Segmentation* [7]. As a preliminary example and visual motivation, the example in Fig. 14.1[1] show the output image this network generates.

Fig. 14.1: Example of using Mask R-CNN: *Regent Street, London* by Mark Pegrum, licensed under CC BY 2.0.

In this example, the pre-trained Mask R-CNN (MRCNN) network processes a normal picture, segments objects (people, sheep, bag, car, bus, umbrella, etc.) from the picture, and recognised them with confidence represented by a number between 0 and 1.

Image segmentation has many important applications including in medical imaging (locating tumours, detecting cancer cells . . .), traffic control systems, and identifying objects in satellite images [8] [9] [10] [11] [12]. We will next explain how this network can be built in OCaml mainly using the `Neural` module in Owl.

14.2 Object Detection Architectures

Before explaining the details of Mask R-CNN, we briefly introduce how deep network architectures for object detection are developed.

---

[1] Original image source: *Regent Street, London* by Mark Pegrum, URL:https://www.flickr.com/photos/ozmark17/3864732604. This photo is licensed under CC BY 2.0. Here we take part of the original photo, and have applied instance segmentation on it.

14.2.1 R-CNN

The idea of using a CNN to enhance object detection task was first proposed in 2014 [13]. It proposes the "Regions with CNN features" (R-CNN) object detection system. The processing pipeline is divided into several phases. The first phase of any object detection task is to locate possible objects of interest in an input image. One way to do this is to use a sliding window to detect the objects. R-CNN uses a different approach called "regions": for each input image it first generates a number of (e.g., 2000) *region proposals* that are independent of the object categories used – just rectangle regions of the image, of different aspects and sizes. The content in each region is then checked to see if it contains any object of interest.

Region proposal methods had been well studied but R-CNN chose to use the *Selective Search* method [14]. The basic idea of this method is to first generate a lot of small segmentations of the input image using the method of Felzenszwalb *et al.* [18]. Then smaller regions are iteratively combined into larger ones, based on their "similarity", which can be measured in colour, size, etc. Finally, the large regions are used to generate the candidate rectangle region proposals.

Each region proposal is then processed by a CNN to get a 4096-dimension feature vector. This CNN takes input of fixed size 227×227 and thus each region, regardless of its shape, is warped into this required size before being processed by the CNN. As to the output feature vector, it is processed by a trained SVM model [15] to be classified into the accepted results.

14.2.2 Fast R-CNN

R-CNN was a big success. Compared to the earlier state-of-the-art, it improves the mean average precision by more than 30%. However, it has several drawbacks. For one, R-CNN pipeline consists of several parts; therefore training takes multiple stages, including the training of the CNN, the SVM models, etc. Moreover, training R-CNN network is expensive in both space and time. Besides, since the CNN inference pass needs to be performed on all region proposals, the whole object detection process is slow.

Given these problems, Girshick et al. [16] proposed the *Fast R-CNN* approach. Similar to R-CNN, the region proposals are first generated based on the input images. But to reduce training costs, the Fast R-CNN consists of one CNN network that can be trained in a single stage. Furthermore, it does not need to perform a forward pass on each region proposal. Instead, it first computes a convolutional feature map for the whole input image, and then projects the region of interest (RoI) from the input image to this feature map, deciding which part of it should be extracted. Such a region on the feature map is pooled by a "RoI Pooling" layer into smaller feature map of fixed size, which is then turned into a feature vector by several fully-connected layers.

The feature vectors are fed into a branch. One output of this branch contains the classification, and the confidence of that classification of the object in that region. The other specifies the rectangle location of the object, encoded by 4 real-valued numbers. The output on this branch contains such 4-number tuple for all of the object categories in this task. Compared to R-CNN, this method does not require a lot of space for feature caching, and it proves to be about 9 times faster in training and 213 times faster in inference [16].

14.2.3 Faster R-CNN

In Fast R-CNN, region proposals are first generated based on the input image, and object detection is performed using both the convolutional feature map and the region proposal. However, note that the feature map, which abstractly represents features in the input image, may already contain sufficient information not only to perform objection detection but also to find regions where the objects may be. That important observations leads to the development of *Faster R-CNN* method [17].

Based on the Fast R-CNN design, this work introduces a new network structure called the *Region Proposal Network* (RPN). This extra network operates on the feature maps that are generated by the CNN in Fast R-CNN, generating region proposals that are passed to the RoI pooling step for successive detections. The RPN uses a sliding window over the feature map. At each sliding-window location, 9 different proposals that are centred around this location are checked, to see their coordinates (represented by 4 real-value numbers) and the probability that an object exists in this given region. Each such proposal is called an *anchor*.

The RPN and the basic Fast R-CNN architecture are trained separately and iteratively. In this work, the authors trained the RPN and used the generated proposals to train Fast R-CNN; the trained network is then used to initialise RPN, so on and so forth. Thus there is no need to use external methods to produce region proposals. Everything is performed in a unified network for the object detection task.

14.2.4 Mask R-CNN

To perform the task of not just objection detection but also semantic segmentation, Mask R-CNN keeps the architecture of Faster R-CNN, adding only one extra branch in the final stage of the RoI feature layer. Where previously outputs included object classification and location, now a third branch contains information about the mask of the detected object in the RoI. Therefore, the Mask R-CNN can retrieve the rectangle bound, classification results, classification possibility, and the mask of that object, for any RoI, in a single pass. In the next section, we will introduce the Mask R-CNN architecture in detail.

14.3 Mask R-CNN Network

This section outlines the main parts of the architecture of Mask R-CNN, explaining how it differs from its predecessors. For more detailed explanation, consult the original paper [6]. The OCaml implementation of the inference model is available in the code repository.[2]

14.3.1 Building Mask R-CNN

After a quick introduction to MRCNN architecture and its development, let's look at the code to understand how it is constructed. Feel free to skip this part if you are eager to simply run the code to process an image.

```
open Owl

module N = Dense.Ndarray.S

open CGraph
open Graph
open AD

module RPN = RegionProposalNetwork
module PL = ProposalLayer
module FPN = FeaturePyramidNetwork
module DL = DetectionLayer
module C = Configuration

let image_shape = C.get_image_shape () in
let inps = inputs
  ~names:[|"input_image"; "input_image_meta"; "input_anchors"|]
  [|image_shape; [|C.image_meta_size|]; [|num_anchors; 4|]|] in
let input_image = inps.(0)
and input_image_meta = inps.(1)
and input_anchors = inps.(2) i;;
```

The network accepts three inputs, each representing images, meta data, and number of anchors (the rectangular regions). The Configuration module contains a list of constants that will be used in building the network.

---

[2] Owl Mask R-CNN, URL:https://github.com/pvdhove/owl-mask-rcnn. Work in this chapter was conducted by Pierre Vandenhove during his internship in the OCamlLab at the Computer Laboratory, University of Cambridge. The code was ported from the Keras/TensorFlow implementation.

14.3.2 Feature Extractor

The picture is first fed to a convolutional neural network to extract features of the image. The first few layers detect low-level features of an image, such as edges and basic shapes. As we go deeper into the network, these simply features are assembled into higher level features such as "people", "cars" Five of these layers (called "feature maps") of various sizes, both high- and low-level, are then passed on to the next parts. This implementation uses Microsoft's ResNet101 network [19] as a feature extractor.

```
let tdps = C.top_down_pyramid_size in
let str = [|1; 1|] in
let p5 = conv2d [|1; 1; 2048; tdps|] str
  ~padding:VALID ~name:"fpn_c5p5" c5 in

let p4 =
add ~name:"fpn_p4add"
  [|upsampling2d [|2; 2|] ~name:"fpn_p5upsampled" p5;
    conv2d [|1; 1; 1024; tdps|]
      str ~padding:VALID ~name:"fpn_c4p4" c4|] in
let p3 =
add ~name:"fpn_p3add"
  [|upsampling2d [|2; 2|] ~name:"fpn_p4upsampled" p4;
    conv2d [|1; 1; 512; tdps|]
      str ~padding:VALID ~name:"fpn_c3p3" c3|] in
let p2 =
add ~name:"fpn_p2add"
  [|upsampling2d [|2; 2|] ~name:"fpn_p3upsampled" p3;
    conv2d [|1; 1; 256; tdps|]
      str ~padding:VALID ~name:"fpn_c2p2" c2|] in

let conv_args = [|3; 3; tdps; tdps|] in
let p2= conv2d conv_args str ~padding:SAME ~name:"fpn_p2" p2 in
let p3= conv2d conv_args str ~padding:SAME ~name:"fpn_p3" p3 in
let p4= conv2d conv_args str ~padding:SAME ~name:"fpn_p4" p4 in
let p5= conv2d conv_args str ~padding:SAME ~name:"fpn_p5" p5 in
let p6= max_pool2d [|1; 1|] [|2; 2|]
  ~padding:VALID ~name:"fpn_p6" p5 in

let rpn_feature_maps = [|p2; p3; p4; p5; p6|] in
let mrcnn_feature_maps = [|p2; p3; p4; p5|];;
```

The features are extracted by combining both ResNet101 and the Feature Pyramid Network [20]. ResNet extracts features of the image (early layers extract low-level features, later layers extract high-level features). The Feature Pyramid Network creates a second pyramid of feature maps from top to bottom so that every map has access to high and low level features. This combination achieves excellent gains in both accuracy and speed.

14.3.3 Proposal Generation

To locate objects in an image, about 250,000 overlapping rectangular regions or anchors are generated.

```
let nb_ratios = Array.length C.rpn_anchor_ratios in
let rpns = Array.init 5 (fun i ->
  RPN.rpn_graph rpn_feature_maps.(i)
  nb_ratios C.rpn_anchor_stride
  ("_p" ^ string_of_int (i + 2))) in
let rpn_class = concatenate 1 ~name:"rpn_class"
                (Array.init 5 (fun i -> rpns.(i).(0))) in
let rpn_bbox = concatenate 1 ~name:"rpn_bbox"
               (Array.init 5 (fun i -> rpns.(i).(1)));;
```

Single RPN graphs are applied on different features in rpn_features_maps, and the results from these networks are concatenated. For each bounding box on the image, the RPN returns the likelihood that it contains an object, called its *objectness*, and a refinement for the anchor, both are represented by three dimensional ndarrays.

Next, in the proposal layer, the 1000 best anchors are selected according to their objectness. Anchors that overlap too much with each other are eliminated, to avoid detecting the same object multiple times. Each selected anchor is also refined in case it was not perfectly centred around the object.

```
...
let rpn_rois =
  let prop_f = PL.proposal_layer
    C.post_nms_rois C.rpn_nms_threshold in
  MrcnnUtil.delay_lambda_array [|C.post_nms_rois; 4|]
    prop_f ~name:"ROI"
    [|rpn_class; rpn_bbox; input_anchors|] in
...
```

In rpn_rois, the proposal layer picks the top anchors from the RPN output, based on non-maximum suppression and anchor scores.

14.3.4 Classification

All anchor proposals from the previous layer are resized to a given fixed size and fed into a 10-layer neural network. The network assigns each of them the probability that it belongs to each class. The network is pre-trained on fixed classes. Changing the set of classes requires re-training the whole network. Note that this step does not take as much time for each anchor as a full-fledged image classifier such as Inception, since it reuses the pre-computed feature maps from the Feature Pyramid Network. The class with the highest probability is chosen for each proposal, and thanks to the class predictions, the anchor proposals are even more refined. Proposals classified in the background class are deleted. Eventually, only the proposals with an objectness over some threshold are kept, and we have our final detections, each with a bounding box and a label. This process can be described by code as below.

```
...
let mrcnn_class, mrcnn_bbox =
FPN.fpn_classifier_graph rpn_rois
  mrcnn_feature_maps input_image_meta
  C.pool_size C.num_classes C.fpn_classif_fc_layers_size in

let detections = MrcnnUtil.delay_lambda_array
  [|C.detection_max_instances; 6|]
  (DL.detection_layer ())
  ~name:"mrcnn_detection"
  [|rpn_rois; mrcnn_class; mrcnn_bbox; input_image_meta|] in
let detection_boxes = lambda_array
  [|C.detection_max_instances; 4|]
  (fun t -> Maths.get_slice [[]; []; [0;3]] t.(0))
  [|detections|]
...
```

A Feature Pyramid Network classifier associates a class to each proposal and further refines the bounding box for that class. The only thing left to do then is to generate a binary mask for each object. This is handled by a small convolutional neural network which produces a small square of values between 0 and 1 for each detected bounding box. This square is resized to the original size of the bounding box with bilinear interpolation, and pixels with a value over 0.5 are tagged as being part of the object.

```
let mrcnn_mask = FPN.build_fpn_mask_graph
  detection_boxes mrcnn_feature_maps
  input_image_meta C.mask_pool_size C.num_classes;;
```

Finally, the output contains detection results and masks from the previous steps.

```
outputs ~name:C.name [|detections; mrcnn_mask|];;
```

14.4 Run the Code

After getting to know the internals of the MRCNN architecture, we can now run the code to see it work. An example of using the MRCNN code is in evalImage.ml, which is included in the code repository of this book. The core part of this script is listed below.

```
open Mrcnn

let src = "image.png" in
let fun_detect = Model.detect () in
let Model.({rois; class_ids; scores; masks}) = fun_detect src in
let img_arr = Image.img_to_ndarray src in
let filename = Filename.basename src in
let format = Images.guess_format src in
let out_loc = out ^ filename in
Visualise.display_masks img_arr rois masks class_ids;

let camlimg = Image.img_of_ndarray img_arr in
Visualise.display_labels camlimg rois class_ids scores;
Image.save out_loc format camlimg;
Visualise.print_results class_ids rois scores;;
```

The most important step is to apply the Model.detect function on the input images, returning the regions of interest, the classification result ID of the object in each region, the classification certainty scores, and a mask that shows the outline of that object in the region. With this information, the Visualise module runs for three passes on the original image: the first for adding bounding boxes and object masks, the second for adding the numbers close to the bounding box, and the final for printing out the resulting images from the previous two steps.

In this example, the pre-trained weights on 80 classes of common objects are provided, which have been converted from the TensorFlow implementation mentioned above. As to the execution speed, processing one image with a size of 1024 × 1024 pixels takes between 10 and 15 seconds on a moderate laptop.

14.5 Summary

This chapter introduced Instance Segmentation, another powerful computer vision application that is used in many fields. It encompasses several different tasks, including image recognition, semantic segmentation, and object detection. This task can be performed with the Mask R-CNN network architecture. We explained in detail how this can be constructed in OCaml, and we provided example code and a pre-trained model so that you can play with this powerful application yourself.

References

1. Bapu JJ, Florinabel DJ. Automatic annotation of satellite images with multi class support vector machine. Earth Science Informatics. 2020 Sep 1; 13(3):811-9.
2. McAuliffe MJ, Lalonde FM, McGarry D, Gandler W, Csaky K, Trus BL. Medical image processing, analysis and visualization in clinical research. InProceedings 14th IEEE Symposium on Computer-Based Medical Systems. CBMS 2001 2001 Jul 26 (pp. 381-386). IEEE.
3. Krizhevsky A, Sutskever I, Hinton GE. Imagenet classification with deep convolutional neural networks. Advances in neural information processing systems. 2012. 25:1097-105.
4. Dasiopoulou S, Mezaris V, Kompatsiaris I, Papastathis VK, Strintzis MG. Knowledge-assisted semantic video object detection. IEEE Transactions on Circuits and Systems for Video Technology. 2005 Sep 26;15(10):1210-24.
5. Guo D, Pei Y, Zheng K, Yu H, Lu Y, Wang S. Degraded image semantic segmentation with dense-gram networks. IEEE Transactions on Image Processing. 2019 Aug 26;29:782-95.
6. He K, Gkioxari G, Dollár P, Girshick R. Mask r-cnn. Proceedings of the IEEE international conference on computer vision. 2017 (pp. 2961-2969).
7. Minaee S, Boykov YY, Porikli F, Plaza AJ, Kehtarnavaz N, Terzopoulos D. Image segmentation using deep learning: A survey. IEEE Transactions on Pattern Analysis and Machine Intelligence. 2021 Feb 17.
8. Pham DL, Xu C, Prince JL. Current methods in medical image segmentation. Annual review of biomedical engineering. 2000 Aug;2(1):315-37.
9. Wu W, Chen AY, Zhao L, Corso JJ. Brain tumor detection and segmentation in a CRF (conditional random fields) framework with pixel-pairwise affinity and superpixel-level features. International journal of computer assisted radiology and surgery. 2014 Mar;9(2):241-53.
10. Patil BG, Jain SN. Cancer cells detection using digital image processing methods. International Journal of Latest Trends in Engineering and Technology. 2014 Mar;3(4):45-9.
11. Grover S, Saxena VS, Vatwani T. Design of intelligent traffic control system using image segmentation. International Journal of Advances in Engineering & Technology. 2014 Nov 1;7(5):1462.
12. Barbieri AL, De Arruda GF, Rodrigues FA, Bruno OM, da Fontoura Costa L. An entropy-based approach to automatic image segmentation of satellite images. Physica A: Statistical Mechanics and its Applications. 2011 Feb 1;390(3):512-8.
13. Girshick R, Donahue J, Darrell T, Malik J. Rich feature hierarchies for accurate object detection and semantic segmentation. InProceedings of the IEEE conference on computer vision and pattern recognition 2014 (pp. 580-587).
14. Uijlings JR, Van De Sande KE, Gevers T, Smeulders AW. Selective search for object recognition. International journal of computer vision. 2013 Sep;104(2):154-71.
15. Cortes C, Vapnik V. Support-vector networks. Machine learning. 1995 Sep; 20(3):273-97.
16. Girshick R. Fast r-cnn. InProceedings of the IEEE international conference on computer vision 2015 (pp. 1440-1448).
17. Ren S, He K, Girshick R, Sun J. Faster r-cnn: Towards real-time object detection with region proposal networks. Advances in neural information processing systems. 2015; 28:91-9.
18. Felzenszwalb PF, Huttenlocher DP. Efficient graph-based image segmentation. International journal of computer vision. 2004 Sep; 59(2):167-81.
19. He K, Zhang X, Ren S, Sun J. Deep residual learning for image recognition. InProceedings of the IEEE conference on computer vision and pattern recognition 2016 (pp. 770-778).
20. Lin TY, Dollár P, Girshick R, He K, Hariharan B, Belongie S. Feature pyramid networks for object detection. InProceedings of the IEEE conference on computer vision and pattern recognition. 2017, pp. 2117-2125.

Chapter 15
Case Study: Neural Style Transfer

Abstract Neural Style Transfer (NST) is an exciting DNN-based application that creates arts. In this chapter we introduce this application in detail: its theory, the implementation, and examples of its use. NST is has been extended in many ways, one of which is the fast style transfer, and we also introduce how this application works with examples.

15.1 Content and Style

What is Neural Style Transfer (NST)? It is a process of using DNN to merge the semantic content of one image to different artistic styles, and a very neat application of deep neural networks. It is simple and yet beautiful. Imagine that we take two images A and B as input where, in this case, A is a daily street view photo while B is "The Starry Night" by Vincent van Gogh. We call A the *content image* and B the *style image*. What an NST can produce is a new street view image in the style of that particular Van Gogh. To create one with a different style, the user can simply replace image B and re-run the NST. Impressionism, abstractionism, classical art, you name it. Similarly, you can apply the same "Starry Sky" style to any other image. We will show such an example in Sec. 15.1.3.

Gatys et. al. first proposed using DNNs to let programmes to create artistic images of high perceptual quality. For more detailed and formal exposition, please refer to the original paper [14]. Here we only introduce the intuition behind how this Neural Style Transfer algorithm works.

In NST, we seek an image whose content is similar to one image but whose artistic style is similar to another. As a first step, we must express this aim in a mathematical form so it is amenable to the techniques we have been discussing. Let's assume for a moment we already know that style transfer can be formalised as an optimisation problem. Given a content image c and a style image s, our target is then to create an output image x that minimises:

L. Wang et al., *OCaml Scientific Computing*, Undergraduate Topics in Computer Science,
https://doi.org/10.1007/978-3-030-97645-3_15

$$g(x) = \texttt{content\_distance}(x, c) + \texttt{style\_distance}(x, s)$$

You may remember from earlier chapters that training is in fact an optimisation process. However, do not mistake the optimisation in NST for either regression or DNN training. For example, in a regression problem we consider the function f_w that contains parameter w and the training process optimises w to minimise $f_w(x)$. The optimisation in NST, on the other hand, is more like a traditional optimisation problem where we have a function f and, starting from an initial input x_0, we iteratively update it until we obtain a satisfying input x that minimises the target function f.

Now return to the problem at hand. While we human beings can feel the style of paint and visually recognise the content of a picture, how do we mathematically express "content" and "style", and measure how close two pictures are regarding content or style? The answer is: we use convolution networks. DNNs, especially the ones that are used for computer vision tasks, are a convenient tool for capturing the characteristics of an image. We demonstrated in the previous chapter how CNNs are good at spotting the "features" in an image layer by layer, and we have introduced several CNN architectures to perform image detection tasks. In NST we choose to use VGG19 to capture the features in an image, since it follows a simple linear stack structure and is proved to have good performance. Next, we divide our task into two parts: content reconstruction, and style recreation.

15.1.1 Content Reconstruction

From the previous chapter, we know that CNN extracts features layer by layer until the features are so abstract that it can produce a label such as "car" or "apple". Therefore, we can use the feature map to reconstruct content of an image. Which layer's output should we use as a suitable indication of the image's content? Let's perform a simplified version of NST where we care only about re-constructing the content of the input image, so our target is to minimise:

$$f(x) = \texttt{content\_distance}(x, c)$$

As an example, we use Fig. 15.1 as the target content[1]. Suppose we choose the output of the idx-th layer as the chosen feature map to represent the content. First, we need to compute the target feature map using the code below.

```
let fill_content_targets x net =
  let selected_topo = Array.sub nn.topo 0 (idx + 1) in
  run' selected_topo x;;
```

---

[1] *Tourists in Nanzen-Ji Hojo* by Benh Lieu Song, URL: https://ccsearch.creativecommons. org/photos/f4024dc8-ce39-4e86-acfd-47532fef824d. It is licensed under CC BY-SA 2.0. No changes were made to this image here.

Fig. 15.1: Example content image: *Tourists in Nanzen-Ji Hojo* by Benh Lieu Song, licensed under CC BY-SA 2.0.

The function `fill_content_targets` takes the content image and the VGG network as input, and returns the target feature map as output. We only need to compute the feature map of the target content image once.

Here the `run'` function is implemented by accumulating the inference result along the selected part of the network, from the network input until the chosen layer, instead of processing the whole network:

```
let run' topo x =
  let last_node_output = ref (F 0.) in
  Array.iteri (fun i n ->
    let input  - if i = 0 then x else !last_node_output in
    let output = run [|input|] n.neuron in
    last_node_output := output;
  ) topo;
  !last_node_output;;
```

Then we can start optimising the input image x. Let's set the initial x to be a "white noise" image that only contains random pixels. This image has the same shape as the content image.

```
let input_shape = Dense.Ndarray.S.shape content_img in
Dense.Ndarray.S.(gaussian input_shape |> scalar_mul 0.256);;
```

The feature map of the input image x is still calculated using the same process shown in the function `fill_content_targets`. We call the resulting feature map response. The loss value can be calculated with the L2Norm of the difference between two feature maps, and then normalised with the feature map size.

```
let c_loss response target =
  let loss = Maths.((pow (response - target) (F 2.))
    |> sum') in
```

```
let _, h, w, feature = get_shape target in
let c = float_of_int ( feature * h * w ) in
Maths.(loss / (F c));;
```

Once the loss value is calculated, we can apply optimisers. Here we use the `minimise_fun` from the `Optimise` module. The target function can be described as:

```
let g x =
  fill_losses x;
  c_loss response target
```

All it does is what we just described: first calculate the feature map `response` of the input image at a certain layer, and then compute the distance between itself and the target content feature map as the loss value.

Finally, we can perform the optimisation iterations:

```
let state, img = Optimise.minimise_fun
  params g (Arr input_img) in
let x' = ref img in

while Checkpoint.(state.current_batch < state.batches) do
  Checkpoint.(state.stop <- false);
  let a, img = Optimise.minimise_fun ~state params g !x' in
  x' := img
done
```

We keep updating the image x for a fixed number of iterations. In particular we use the Adam adaptive learning rate method [2] as it proves to be quite effective in style transfer optimisation:

```
let params = Params.config
  ~learning_rate:(Learning_Rate.Adam
    (learning_rate, 0.9, 0.999))
  ~checkpoint:(Checkpoint.Custom chkpt)
  iter
```

Using the process above, we return to the problem of choosing a suitable layer as the indication of the image content. In this 38-layer VGG network, some frequently used practical choices are these layers: 2, 7, 12, 21, 30. We can then compare the optimisation results to see the effect of image reconstruction. Each one is generated after 100 iterations.

Fig. 15.2 shows that the content information is accurately maintained in the lower layers. Along the processing hierarchy of the network, the feature map produced by the lower layer cares more about small features at the pixel level, while the higher layer gives more abstract higher-level information but fewer details to help with content reconstruction.

Layer 2   Layer 7   Layer 12   Layer 23   Layer 30

Fig. 15.2: Contents reconstruction from different layers

15.1.2 Style Recreation

We now explore the other end of this problem similarly, caring only about recreating an image with the style of another input image. That is to say, we optimise the input image with this target to minimise:

$$h(x) = \texttt{style\_distance}(x, s)$$

As an example, we will use the famous "The Great Wave off Kanagawa"[2] by by the Japanese artist Katsushika Hokusai as our target style image. The basic approach is the same as before. First, we compute the style representation of the target image using the output from one or more layers, before computing the style representation of the input image following the same method. The normalised distance between these two ndarrays are used as the optimisation target.

The difference is that, unlike the content representation, we cannot directly take one filter map from a particular layer as the style representation. Instead, we need to compute the correlations between different filter maps. This correlation can be represented by the *Gram matrix* [3], which intuitively captures the "distribution of features" of feature maps from a certain layer. The (i, j)-th element of a Gram matrix is computed by element-wise multiplication of the the i-th and j-th channels in the feature maps, before summing across both width and height. This process can be simplified as a matrix multiplication. The result is normalised with the size of the feature map. The code is shown below.

```
let gram x =
  let _, h, w, feature = get_shape x in
  let new_shape = [|h * w; feature|] in
  let ff = Maths.(reshape x new_shape) in
  let size = F (float_of_int (feature * h * w)) in
  Maths.((transpose ff) *@ ff / size);;
```

Now that we have a method to represent the "style" of an image, we can calculate the loss value during optimisation. It is very similar to the case of content recreation,

---

[2] *The Great Wave off Kanagawa* by Katsushika Hokusai. Wikipedia page: https://en.wikipedia.org/wiki/The_Great_Wave_off_Kanagawa

the only difference is that we use the distance between Gram matrices instead of the feature maps from a certain layer as loss value.

```
let s_loss response_gram target_gram =
  let loss = Maths.(
    ((response_gram - target_gram) (F 2.))
    |> pow |> sum'
  ) in
  let s = Algodiff.shape target_gram in
  let c = float_of_int (s.(0) * s.(1)) in
  Maths.(loss / (F c));;
```

However, note that for optimisation we usually use the loss values from multiple layers, instead of only one layer, as the optimisation target for style reconstruction:

```
let h x =
  fill_losses x;
  Array.fold_left Maths.(+) (F 0.) style_losses;;
```

Here the `fill_losses` function computes style losses at different layers, storing them into the `style_losses` array. Their summation is used as optimisation target. The rest of the process is the same as that in the content reconstruction.

In this example, we choose the same five layers from the VGG19 network: layers 2, 7, 12, 21, and 30. We then compute the aggregated loss value of the first layer, the first two layers, the first three layers, the first four layers, and all layers, as five different optimisation targets. Optimising these five different targets, the resulting images are shown in Fig. 15.3.

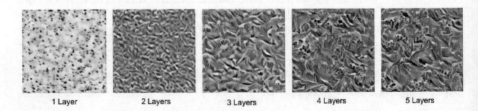

 1 Layer   2 Layers   3 Layers   4 Layers   5 Layers

Fig. 15.3: Style reconstruction from different layers

As the result shows, lower level features tend to contain low level information focused on pixels, so reconstructing styles according to them results in a fragmented white-noise-like representation without any recognisable style. By adding more deep layer features however, the style can be gradually reconstructed. The fifth generated image shows a quite obvious wave-like style.

(a) (b)

Fig. 15.4: Combining content and style reconstruction to perform NST

15.1.3 Combining Content and Style

Having seen these two extremes, recreating only content and only style, it's straight-forward to understand the theory of style transfer: to synthesise an image that has content similar to one image and style close to another, we simply add the loss value of content and styles as the final optimisation target.

One thing we should note when combining contents and style is the proportion of each part, and the choice of layers as representation. This problem is actually more artistic than scientific, so here we simply follow current practice for parameter configuration. Please refer to the original paper about the effect of parameter tuning. As suggested by previous results, we use the feature maps from the 23rd layer for content recreation, and combine the output of layers 2, 7, 12, 21, and 30 in VGG19 to represent the style feature of an image. When combining the loss values, we multiply the style loss with a weight number, and then add it to the content loss. Practice shows that a weight of 20 shows good performance.

You might also be wondering: why not choose the 2nd layer if it shows the best content reconstruction result? The intuition is that we don't want the synthesised image to be too close to the content image in content, because that would mean less style. Therefore we use a layer from the middle of CNN which shows to keep most of the information for content reconstruction.

Combining all these factors together, Fig. 15.4 shows the result of running our code and creating an artistic view based on the original image. All the code (about 180 lines) is included in the code repository, along with the pre-trained weight file for VGG19. As with the image detection applications, it also relies on the tool `ImageMagick` for image format conversion, resizing, and general manipulation. We list only part of it above, omitting many implementation details such as garbage collection to focus on the core idea of the application itself. We therefore suggest you to play with the code itself with images or parameters of your choice.

15.1.4 Running NST

To make the code above easier to use, this NST application provides a simple interfaces. Here is an example showing how to use it with two lines of code:

```
#zoo "6f28d54e69d1a19c1819f52c5b16c1a1";;

Neural_transfer.run
  ~ckpt:50
  ~src:"path/to/content_img.jpg"
  ~style:"path/to/style_img.jpg"
  ~dst:"path/to/output_img.png" 250.;;
```

Similar to the image detection application, the command can be simplified using the Zoo system. The first line downloads a gist and imports the code as an OCaml module, and the rest use the run function to produce an output image at the specified path. The syntax is quite straightforward: N is specified by the ckpt parameter, and its default value is 50 iterations, the src and style parameters specify the two input images, the dst parameter says where the output is produced, and the final parameter specifies how many iterations the optimisation algorithm runs. Typically $100 \sim 500$ iterations are enough.

This module also supports saving intermediate images every N iterations (e.g. path/to/output_img_N.png). If users are already happy with the intermediate results, they can terminate the program early without waiting for the final output image.

That's all. Now you can play with the code conveniently. If you don't have suitable input images at hand, the gist also contains several example content and style images for you to get started.

15.2 Extending NST

Neural Style Transfer has attracted a lot of attention since its publication. It is the core technology of many successful industrial applications, most notably for photo rendering [4] [5] [6] [7] [8]. For example, the Prisma Photo Editor[3] features transforming your photos into paintings of hundreds of styles.

There are many research work that aim to extend NST [9] [10] [11] [12] [13]. One such example is *Deep Photo Style Transfer* [1]. The idea is that, instead of using an art image, we can use another normal image as style reference. For example, we have a normal daylight street view in New York as a content image, and we want to use the night view of London as reference to synthesise an image of the night view of New York.

---

[3] Prisma Labs, Inc., URL: https://prisma-ai.com/

The authors identify two key challenges. The first is that, unlike in NST, we hope to only change to colours of the style image while keeping the content un-distorted, so as to create a "real" image as much as possible. For this challenge, the authors propose to add a regularisation item to the existing optimisation target "content distance + style distance". This item, depending on only input and output images, penalises image distortion and seeks an image transform that is locally affine in colour space. The second challenge is that we don't want the styles to be applied globally. For example, we only want to apply the style of an sunset sky to a blue sky, not a building. For this problem, the authors propose to coarsely segment input images into several parts before applying style transfer on them separately. If you are interested to check the original paper, the resulting photos are indeed beautifully and realistically rendered.

Another similar application is the "image-to-image translation" which broadly involves translating an input image into a certain output image. Style transfer and image colourisation can be seen as examples of this, with others including applications that change the lighting/weather in a photo. Isola et. al. [15] propose using Generative Adversarial Networks (GAN) to provide a general framework for this task. As we have introduced in the neural network chapter, GANs have two important components: the generator and the discriminator. During training, the generator synthesises images based on existing parameters, and the discriminator tries its best to separate the generated data from true data. This process is iterated until the discriminator can no longer tell the difference between the two. This approach has been successfully applied in many applications, such as Pix2Pix, face ageing [16], increasing photo resolution [17], etc.

15.3 Fast Style Transfer

One disadvantage of NST is that it could take a very long time to render an image, and if you want to change to another content or style image, you have to wait a long time for the training again. If you want to render some of your best (or worst) selfies fast and show them off to your friends, NST is perhaps not the best choice.

This problem leads to another application: Fast Neural Style Transfer (FST). FST sacrifices flexibility in that you cannot choose style images at will. But as a result, you only need to feed your content images to a DNN, finish an inference pass, and the output will be the rendered styled image as you expected. The best part is that, one inference pass is much faster than repeatedly running a training phase.

15.3.1 Building FST Network

The Fast Style Transfer paper [18] proposes building and training an *image transformation network*. Image transformation is not a new idea. It accepts some input image and transforms it into a certain output image. One way to do that is to train a feedforward CNN. This method is applied in different applications such as colourising greyscale photos or image segmentation [19] [20]. In this work the authors use a similar approach to solve the style transfer problem.

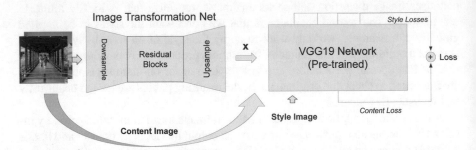

Fig. 15.5: System overview of the image transformation network and its training.

Fig. 15.5 shows a system overview of the image transformation network and its training. It can be divided into two parts. The first includes the image transformation network architecture. To synthesise an image of the same size as input image, it first uses down-sampling layers, and then the up-sampling layers. One benefit of first down-sampling images is to reduce the computation, which enables building a deeper network. We have already seen this design principle in Chap. 13.

Instead of using the normal pooling or upsampling layer in CNN, here the convolution layers are used for down/up-sampling. The main reason is to keep the image information as much as possible during the whole transformation process. Specifically, we use the *transpose convolution* operation for upsampling. Opposite to a normal convolution, in transpose convolution, feature maps size increases, and it still maintains the connectivity pattern in convolution.

```
open Owl
open Neural.S
open Neural.S.Graph
open Neural.S.Algodiff
module N = Dense.Ndarray.S

let conv2d_layer ?(relu=true) kernel stride nn  =
  let result =
    conv2d ~padding:SAME kernel stride nn
    |> normalisation ~decay:0. ~training:true ~axis:3
  in
  match relu with
  | true -> (result |> activation Activation.Relu)
  | _    -> result
```

```
let conv2d_trans_layer kernel stride nn =
  transpose_conv2d ~padding:SAME kernel stride nn
  |> normalisation ~decay:0. ~training:true ~axis:3
  |> activation Activation.Relu;;
```

In the code above, combined with batch normalisation and relu activation layers, we construct two building blocks: the conv2d_layer and conv2d_trans_layer. Think of them as enhanced convolution and transpose convolution layers. The benefit of adding these two types of layers is discussed in Chap. 13.

What connects these two parts are multiple residual blocks as proposed in the ResNet architecture. The authors claim that using residual connections makes it easier to keep the structure between output and input. It is an especially attractive property for style transfer neural networks. Specifically, the authors use the residual structure proposed in the ResNet. All the convolution layers use the common 3×3 kernel size. This residual block can be implemented with the conv2d_layer unit we have built.

```
let residual_block wh nn =
  let tmp = conv2d_layer [|wh; wh; 128; 128|] [|1;1|] nn
    |> conv2d_layer ~relu:false [|wh; wh; 128; 128|] [|1;1|]
  in
  add [|nn; tmp|];;
```

The wh parameter normally takes a value of 3. The residual block, as in ResNet, is repeatedly stacked. With these three different parts ready, we can finally combine them together. Note how the output channel of each convolution increases, stays the same, and then decreases symmetrically. Before the final output, we use the tanh activation layer to ensure all the values are between [0, 255] for the output image.

```
let make_network h w =
  input [|h;w;3|]
  |> conv2d_layer [|9;9;3;32|] [|1;1|]
  |> conv2d_layer [|3;3;32;64|] [|2;2|]
  |> conv2d_layer [|3;3;64;128|] [|2;2|]
  |> residual_block 3
  |> residual_block 3
  |> residual_block 3
  |> residual_block 3
  |> residual_block 3
  |> conv2d_trans_layer [|3;3;128;64|] [|2;2|]
  |> conv2d_trans_layer [|3;3;64;32|] [|2;2|]
  |> conv2d_layer ~relu:false [|9;9;32;3|] [|1;1|]
  |> lambda (fun x -> Maths.((tanh x) * (F 150.) + (F 127.5)))
  |> get_network;;
```

After constructing the image transformation network, let's look at the training process. In previous work, when training a image transformation network, the output will normally be compared pixel-wise with the ground-truth image as the loss value. That is not an ideal approach here since we cannot know what is a "correct" style-transferred image in advance. Instead, the authors are inspired by the NST work and use the same training process with a pre-trained VGG19 network to compute the loss. They call it the *perceptual loss* in contrast to the per-pixel loss, since high level perceptual information is contained in this loss.

The output image x from image transformation network is the one to be optimised. The input image itself is the content image, and we provide another fixed style image. We can then proceed to calculate the final loss by computing the distance between image x and the input with regard to content and styles. All of these are the same as in NST except that where we trained an image before, now we train the weights of the image transformation network during back-propagation. Note that this process means that we can only train one set of weights for one style. Considering that the artistic styles are relatively fixed compared to the unlimited number of content images, and speed improvement by orders of magnitude, fixing the styles is an acceptable trade-off.

Even better, this training phase is one-off. We can train the network once and the reuse it in the inference phase again and again. We refer you to the original paper if you want to know more details about the training phase. In our implementation, we directly convert and import weights from a TensorFlow implementation.[4] We finish by showing how to use it to perform fast style transfer.

15.3.2 Running FST

Like NST and image classification, we have wrapped all the above into a gist,[5] to provide a simple user interface. Here is an example:

```
#zoo "f937ce439c8adcaea23d42753f487299";;

FST.list_styles ();; (* all supported styles *)
FST.run ~style:1 "path/to/content_img.png"
  "path/to/output_img.jpg";;
```

The `run` function takes one content image and outputs a new image file, name specified by the user. The image can be of any popular format, JPEG, PNG, etc. This gist also contains exemplar content images for you to use. A set of trained weights for the FST DNN represents a unique artistic style. We have already included six different weight files for demonstration, and you just need to pick one and load it into the DNN without worrying about how to train these weights.

---

[4] *Fast Style Transfer in TensorFlow*, Logan Engstrom, URL: https://github.com/lengstrom/fast-style-transfer

[5] *Fast Style Transfer OCaml implementation*, Jianxin Zhao, URL: https://gist.github.com/jzstark/f937ce439c8adcaea23d42753f487299

Currently we support six art styles: "Udnie" by Francis Picabia, "The Great Wave off Kanagawa" by Hokusai, "Rain Princess" by Leonid Afremov, "La Muse" by Picasso, "The Scream" by Edvard Munch, and "The Shipwreck of the Minotaur" by J. M. W. Turner.

(a)   (b)   (c)

Fig. 15.6: Fast style transfer examples.

You can now render any image into a nice art style fast, in maybe half a minute (even faster if you are using GPU or other accelerator). As an example, we use a photo taken by one of the authors as input image (Fig. 15.6a). We then apply FST on this input image with the styles shown above. The rendered the input photo with two different styles ("Udnie" and "Wave off Kanagawa") are shown in Fig. 15.6b and Fig. 15.6c. Each of them rendered in only seconds on a moderate CPU desktop.

15.4 Summary

In this chapter we focus on two interesting DNN applications that centre around the idea of using DNN to generate images with artistic styles. We cover the basic theoretical support behind these applications, and introduce implementation details with Owl codes. By following the step-by-step introduction, you can see that even such complex and even magical applications can be constructed with no more knowledge than what we have introduced so far in this book. We encourage you to play with these codes yourselves. Partly because it is so beautiful, there are a range of similar work on DNN-based artistic image generation. Hopefully this chapter can give you a firm foundation to explore more in this field.

References

1. Luan F, Paris S, Shechtman E, Bala K. Deep photo style transfer. Proceedings of the IEEE conference on computer vision and pattern recognition. 2017 (pp. 4990-4998).
2. Kingma DP, Ba J. Adam: A method for stochastic optimization. arXiv preprint arXiv:1412.6980. 2014 Dec 22.
3. Horn RA, Johnson CR. Matrix analysis. Cambridge University Press. 2012 Oct 22.
4. Ruder M, Dosovitskiy A, Brox T. Artistic style transfer for videos. German conference on pattern recognition. 2016 Sep 12 (pp. 26-36). Springer, Cham.
5. Champandard AJ. Deep forger: Paint photos in the style of famous artists. 2015.
6. Mao H, Cheung M, She J. Deepart: Learning joint representations of visual arts. Proceedings of the 25th ACM international conference on Multimedia 2017 Oct 19 (pp. 1183-1191).
7. Semmo A, Trapp M, Döllner J, Klingbeil M. Pictory: combining neural style transfer and image filtering. ACM SIGGRAPH 2017 Appy Hour. 2017 Jul 30 (pp. 1-2).
8. Dürschmid T, Söchting M, Semmo A, Trapp M, Döllner J. Prosumerfx: Mobile design of image stylization components. InSIGGRAPH Asia 2017 Mobile Graphics & Interactive Applications 2017 Nov 27 (pp. 1-8).
9. Dumoulin V, Shlens J, Kudlur M. A learned representation for artistic style. arXiv preprint arXiv:1610.07629. 2016 Oct 24.
10. Chen D, Liao J, Yuan L, Yu N, Hua G. Coherent online video style transfer. Proceedings of the IEEE International Conference on Computer Vision 2017 (pp. 1105-1114).
11. Li Y, Fang C, Yang J, Wang Z, Lu X, Yang MH. Universal style transfer via feature transforms. arXiv preprint arXiv:1705.08086. 2017 May 23.
12. Verma P, Smith JO. Neural style transfer for audio spectrograms. arXiv preprint arXiv:1801.01589. 2018 Jan 4.
13. Atarsaikhan G, Iwana BK, Narusawa A, Yanai K, Uchida S. Neural font style transfer. 14th IAPR International Conference on Document Analysis and Recognition (ICDAR). 2017 Nov 9 (Vol. 5, pp. 51-56). IEEE.
14. Gatys LA, Ecker AS, Bethge M. A neural algorithm of artistic style. arXiv preprint arXiv:1508.06576. 2015 Aug 26.
15. Isola P, Zhu JY, Zhou T, Efros AA. Image-to-image translation with conditional adversarial networks. Proceedings of the IEEE conference on computer vision and pattern recognition 2017 (pp. 1125-1134).
16. Antipov G, Baccouche M, Dugelay JL. Face aging with conditional generative adversarial networks. In2017 IEEE international conference on image processing (ICIP). 2017 Sep 17 (pp. 2089-2093). IEEE.
17. Ledig C, Theis L, Huszár F, Caballero J, Cunningham A, Acosta A, Aitken A, Tejani A, Totz J, Wang Z, Shi W. Photo-realistic single image super-resolution using a generative adversarial network. Proceedings of the IEEE conference on computer vision and pattern recognition. 2017 (pp. 4681-4690).
18. Johnson J, Alahi A, Fei-Fei L. Perceptual losses for real-time style transfer and super-resolution. European conference on computer vision 2016 Oct 8 (pp. 694-711). Springer, Cham.
19. An J, Kpeyiton KG, Shi Q. Grayscale images colorization with convolutional neural networks. Soft Computing. 2020 Apr; 24(7):4751-8.
20. Wang J, Wang Z, Tao D, See S, Wang G. Learning common and specific features for RGB-D semantic segmentation with deconvolutional networks. European Conference on Computer Vision 2016 Oct 8 (pp. 664-679). Springer, Cham.

Chapter 16
Case Study: Recommender System

Abstract In this chapter, we will build a search engine **Sofia** using latent semantic analysis. Sofia is a content-based filtering system that captures the semantics contained in the unstructured web text. Sofia looks for similar articles in a text corpus when a query document arrives. You will find that much of the basic theory has been covered in Chap. 12. In this chapter, we focus on introducing the full workflow of this system, and how to address a series of technical challenges in building Sofia.

16.1 Introduction

In our daily life, we rely heavily on recommendations, where intelligent content provision attempts to match a user's profile of interests to the best candidates from a large repository of options. There are several parallel efforts in integrating intelligent content provision and recommendation in applications. They differentiate between each other by the main technique used to achieve the goal.

Initial efforts relied on the semantic web stack, [1] which requires adding explicit ontology information to all web pages so that ontology-based applications can use ontological reasoning to interconnect content semantically. Though the semantic web has a well-defined architecture, it suffers from the fact that most web pages are unstructured or at best semi-structured HTML files, and content providers lack of motivation to adopt this technology into their websites. Therefore, even though the relevant research remains active in academia, the actual progress of adopting ontology-based methods in real-life applications has stalled.

Collaborative Filtering (CF), [2] first coined in 1992, is a thriving research area and a second possible solution. Recommenders built on top of CF exploit the similarities in users' rankings to predict a user's preference for specific content. CF attracts considerable interest recently due to the popularity of online shopping (e.g., Amazon, eBay, Taobao, etc.) and content services (e.g., YouTube, Vimeo, Dailymotion, Twitter, etc.). [3] [4] However, recommender systems rely on user behaviours rather than the content itself as input to bootstrap the service, and are usually constrained

© The Author(s), under exclusive license to Springer Nature Switzerland AG 2022
L. Wang et al., *OCaml Scientific Computing*, Undergraduate Topics in Computer Science,
https://doi.org/10.1007/978-3-030-97645-3_16

within a single domain. Cross-domain recommenders [5] have made progress lately, but the complexity and scalability need further investigation.

Search engines can be considered a third alternative, though a user must explicitly extract the keywords from the page before launching a search, typically iteratively. The ranking of the search results is based on multiple ranking signals such as link analysis on the underlying graph structure of interconnected pages such as PageRank. [6] Such graph-based link analysis is based on the assumption that those web pages of related topics tend to link to each other, and the importance of a page often positively correlates to its degree. The indexing process is modelled as a random walk of the graph derived from the linked pages and must be pre-compiled offline.

The fourth alternative is to use information retrieval (IR) techniques. [7] In general, a text corpus is transformed to the suitable representation depending on the specific mathematical models (e.g., set-theoretic, algebraic, or probabilistic models), [8] based on which a numeric score is calculated for ranking. Different from the previous CF and link analysis, the underlying assumption of IR is that the text (or information in a broader sense) contained in a document indicates its (latent) topics. The relatedness of any two given documents can be calculated with a well-defined metric function over these topics. [9] [10] Since topics can have a direct connection to context, context awareness therefore becomes the most significant advantage in IR, which has been integrated into Hummingbird, Google's new search algorithm [1].

In the rest of this chapter, we introduce **Sofia**, a system built on top of latent semantic analysis that automatically searches for similar articles when a user queries the system by providing a query article. This chapter is based on published papers by Liang Wang [11, 12].

16.2 Key Components And Pipeline

Sofia consists of a backend and a frontend. In the backend, Sofia implements an LSA-based index and search service. The frontend is a Chrome plugin interacting with users during web browsing. Fig. 16.1 illustrates the general workflow and internal design of the system. The frontend is currently implemented as a lightweight extension in Chrome browser. The browser extension only sends the page URL into the service whenever a new webpage is visited. The backend service retrieves the content of the given URL and responds with the most relevant documents from its database. Finally, the results are then rendered on to the page being browsed. The user only needs to interact with the frontend when checking the list of recommendations in case any are of interest.

The backend exposes a simple RESTful API, given below, allowing great flexibility to different frontends. By loosely coupling with the backend, it becomes easy to mash-up new services on top of Sofia. In the code below, lines 1 and 2 give

---

[1] SEO Learning Center, "*Google Hummingbird*", URL: https://moz.com/learn/seo/ google-hummingbird

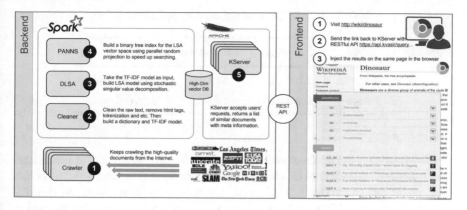

Fig. 16.1: Sofia architecture with components numbered based on their order in the workflow

an example request to Sofia service. type=0 indicates that info contains a URL, otherwise info contains a piece of text if type=1. Lines 4–9 present an example response from the server, which contains the meta-info of a list of similar articles. Note that the frontend can refine or rearrange the results based on the metadata, e.g., similarity or timestamp.

```
POST https://api.Sofia/query?type=0&info=url
```

```
{"results": [
  {"title": document title,
   "similarity": similarity metric,
   "page_url": link to the document,
   "timestamp": document create date}
]}
```

The backend system implements indexing and searching functionality using five components: *Crawler, Cleaner, DLSA, PANNS* and *KServer*. Three of these (Cleaner, DLSA and PANNS) are wrapped into one library since all are implemented on top of Apache Spark, and this library covers text cleaning, database building, and indexing. We briefly present the main tasks in each component as below.

Crawler collects raw documents from the web and then compiles them into different datasets. We use English Wikipedia dump and public news feeds as data sources. Tbl. 16.1 summarises the basic statistics of the data sets. Multiple instances of the Crawler run in parallel on different machines. Simple fault-tolerant mechanisms like periodic backup have been implemented to improve the robustness of crawling process. In addition to the text body, the Crawler also records the timestamp, URL and title of the retrieved news as meta information, which can be further utilised to refine the search results.

Table 16.1: Two datasets are used in Sofia evaluation

Dataset	# of entries	Raw text size	Article length
Wikipedia	3.9×10^6	47.0 GB	Avg. 782 words
News	4.6×10^5	1.62 GB	Avg. 648 words

Cleaner cleans the unstructured text corpus and converts the corpus into a term frequency-inverse document frequency (TF-IDF) model. In the preprocessing phase, we clean the text by removing HTML tags and stop words, de-accenting, tokenising, etc. The dictionary refers to the vocabulary of a language model and its quality directly impacts the model performance. To build the dictionary, we exclude both extremely rare and extremely common terms, and keep only the 10^5 most popular ones as *features*. More precisely, a term is considered rare if it appears in less than 20 documents, while it is considered as common if it appears in more than 40% of documents.

DLSA builds up an LSA-based model from the previously constructed TF-IDF model. [13] Technically, the TF-IDF itself is already a vector space language model but it is not used directly because it contains too much noise and the dimensionality is too high to process efficiently even on a modern computer. To convert a TF-IDF to a LSA model, DLSA's algebraic operations involve large matrix multiplications and time-consuming Singular Value Decomposition (SVD). We initially tried to use MLlib to implement DLSA. [14] However, MLlib is unable to perform SVD on a data set of 10^5 features with limited RAM, and so we have to implement our own stochastic SVD on Apache Spark using rank-revealing technique. The DLSA will be discussed in detail in later section.

PANNS builds the search index to enable fast k-NN search in high dimensional LSA vector spaces. Though dimensionality has been significantly reduced from TF-IDF (10^5 features) to LSA (10^3 features), k-NN search in a 10^3-dimension space is still a great challenge especially when we try to provide responsive services. A naive linear search using one CPU takes over 6 seconds to finish in a database of 4 million entries, which is unacceptably long for a realistic service. PANNS[2] implements a parallel RP-tree algorithm which makes a reasonable trade-off between accuracy and efficiency. PANNS is the core component in the backend system.

KServer runs within a web server, processes user requests, and replies with a list of similar documents. KServer uses the index built by PANNS to perform fast search in the database. The ranking of the search results is based on the cosine similarity metric. A key performance metric for KServer is the service time. We wrap KServer into a Docker image and deploy multiple KServer instances on different machines to achieve better performance. We also implement a simple round-robin mechanism to balance the request loads among the multiple KServers.

---

[2] Source code: https://github.com/ryanrhymes/panns.

16.3 Reducing Dimensionality

After the crawler and cleaner finish processing data and building the TF-IDF model, DLSA and PANNS continue to build language models and indexing the high dimensional data sets. In the remaining sections, we focus on introducing the various techniques used in Sophia, mostly in DLSA and PANNS components. First, we take a look at how to reduce the model dimensionality in DLSA, after a brief recap of LSA from Chap. 12.

The vector space model is an algebraic language model where each document is represented with a row vector. Each element in the vector represents the weight of a term in the dictionary calculated in a specific way such as the frequency of a term in a document or the slightly more complicated TF-IDF. The length of the vector is determined by the size of the dictionary, i.e., the number of features. A text corpus containing m documents and a dictionary of n terms will be converted to an $A = m \times n$ row-based matrix. Informally, we say that A grows taller if the number of documents (i.e., m) increases, and grows fatter if we add more terms (i.e., n) in the dictionary.

The core operation in LSA is to perform SVD. For that we need to calculate the covariance matrix $C = A^T \times A$, which is a $n \times n$ matrix, usually much smaller than A. This operation is a computational bottleneck: m can be very large (if we have a lot of documents) or n can be very large (if we have many features for each document). For the first problem, we can easily parallelise the calculation of C by dividing A into k smaller chunks of size $[\frac{m}{k}] \times n$, so that the final result can be obtained by aggregating the partial results as $C = \sum_{i-1}^{k} A_i^T \times A_i$.

However, the second problem is more serious. For example, the popular MLlib is Apache Spark's machine learning library, with APIs in multiple programming languages. However, the SVD function in MLlib is only able to handle tall thin matrices up to some hundreds of features. In most language models there may be hundreds of thousands features – e.g., 10^5 in our case. The covariance matrix C becomes too big to fit into the physical memory, so the native SVD operation in MLlib of Spark fails as the first subfigure of Fig. 16.2 shows.

In linear algebra, a matrix can be approximated by another of lower rank while still retaining approximately the important properties of the matrix for the problem at hand. In other words, we can use a thinner matrix B to approximate the original A. The corresponding technique is referred to as rank-revealing QR estimation [15] which we will not present in detail. The basic observation is that the columns are sparse and quite likely linearly dependent. If we can find the rank r of a matrix A and find suitable r columns to replace the original matrix, we can then approximate it. A TF-IDF model having 10^5 features often contains a lot of redundant information. Therefore, we can effectively thin the matrix A then fit C into the memory. Figure Fig. 16.2 illustrates the algorithmic logic in DLSA, which is essentially a distributed stochastic SVD implementation.

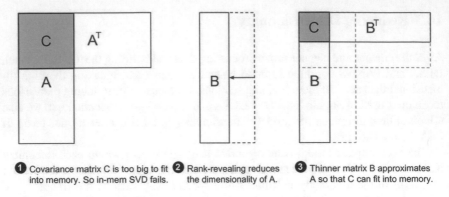

❶ Covariance matrix C is too big to fit   ❷ Rank-revealing reduces   ❸ Thinner matrix B approximates
into memory. So in-mem SVD fails.   the dimensionality of A.   A so that C can fit into memory.

Fig. 16.2: Rank-revealing reduces dimensionality to perform in-memory SVD

To sum up, we apply rank-revealing QR to reduce the size of TF-IDF model matrix, so that we can calculate a LSA model more efficiently, in which we know the document-topic and topic-word probability distribution.

16.4 Random Projection

With a LSA model at hand, finding the most relevant document is equivalent to finding the nearest neighbours for a given point in the derived vector space, which is often referred to as k-NN (k nearest neighbours) problem. The distance is usually measured with the cosine similarity of two vectors. In Chap. 12 we have seen how to use linear search in the LSA model. However, neither naive linear search nor conventional k-d tree is capable of performing efficient search in such high dimensional space even though the dimensionality has been significantly reduced from 10^5 to 10^3 by LSA.

However, we need not locate the exact nearest neighbours in practice. In most cases, slight numerical error (reflected in the language context) is not noticeable at all from the user's perspective. By sacrificing some accuracy, we can achieve a significant gain in searching speed.

To optimise the search, instead of searching in all the existing vectors, we can pre-cluster the vectors according to their distances, each cluster with only a small number of vectors. For an incoming query, as long as we can put this vector into a suitable cluster, we can then search for close vectors only in that cluster.

Fig. 16.3 gives a simple example on a 2-dimensional vector space. First, a random vector x is drawn and all the points are projected onto x. Then we divide the whole space into two halves at the mean value of all projections (i.e., the blue circle on x) to reduce the problem size. For each new subspace, we draw another random vector for projection, and this process continues recursively until the number of points in the space reaches the predefined threshold on cluster size.

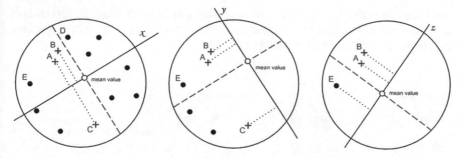

Fig. 16.3: Projection on different random lines

In the implementation, we can construct a binary tree to facilitate the search. Technically, this can be achieved by any tree-based algorithms such as that used in a database. Given a tree built from a database, we answer a nearest neighbour query q in an efficient way, by moving q down the tree to its appropriate leaf cell, and then return the nearest neighbour in that cell. In Sofia, we use the Randomised Partition tree (RP-tree) introduced by Dasgupta et al. [16] to do it. The general idea of the RP-tree algorithm used here is clustering the points by partitioning the space into smaller subspaces recursively.

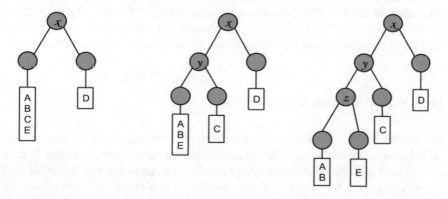

Fig. 16.4: Construct a binary search tree from the random projection

Fig. 16.4 illustrates how binary search can be built according to the dividing steps shown above. You can see the five nodes in the vector space are put into five clusters/leaves step by step. The information of the random vectors such as x, y, and z are also saved. Once we have this tree, given another query vector, we can put it into one of the clusters along the tree to find the cluster of vectors that are close to it.

This efficiency is traded-off against search accuracy. One common type of mis-classification is that we might separate close vectors into different clusters. As we can see in the first subfigure of Fig. 16.3, though the projections of *A*, *B*, and *C* seem close to each other on x, C is actually quite distant from A and B. The reverse can also be true: two nearby points are unluckily divided into different subspaces, e.g., points B and D in the left panel of Fig.16.3.

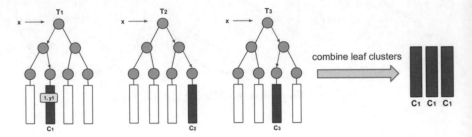

Fig. 16.5: Aggregate clustering result from multiple RP-trees

Such mis-classifications become arbitrarily rare as the iterative procedure continues by drawing more random vectors and performing corresponding splits. In the implementation, we build multiple RP-trees and expect that the randomness in tree construction will introduce extra variability in the neighbours that are returned by several RP-trees for a given query point. This can be taken as an advantage in order to mitigate the second kind of misclassification while searching for the nearest neighbours of a query point in the combined search set. As shown in Fig. 16.5, given an input query vector x, we find its neighbours in three different RP-trees, and the final set of neighbour candidates comes from the union of these three different sets.

16.5 Vector Storage Optimisation

You may have noticed that we must store all the random vectors that are generated in the non-leaf nodes of the tree. That means storing a large number of random vectors at every node of the tree, each with a large number of features, introducing significant storage overhead. For a corpus of 4 million documents, if we use 10^5 random vectors (i.e., a cluster size of $\frac{4 \times 10^6}{2 \times 10^5} = 20$ on average), and each vector is a 10^3-dimension real vector (32-bit float number), the induced storage overhead is about 381.5 MB for each RP-tree. Therefore, such a solution leads to a huge index of 47.7 GB given 128 RP-trees are included, or 95.4 GB given 256 RP-trees.

The large index size not only consumes a significant amount of storage but also prevents the system from scaling up as more and more documents are collected. One possible solution to reduce the index size is reusing the random vectors: we can generate a pool of random vectors once, and then randomly choose one from the pool each time one is needed. However, the immediate challenge emerges when we try to parallelise the tree building on multiple nodes, because we need to broadcast the pool of vectors onto every node, which causes significant network traffic.

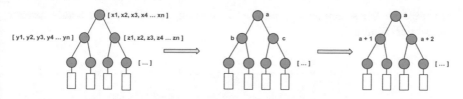

Fig. 16.6: Use a random seed to generate pseudo-random vectors on the fly

To address this challenge we use a pseudo random seed in building and storing the search index. Instead of maintaining a pool of random vectors, we just need a random seed for each RP-tree. As shown in Fig. 16.6, in a leaf cluster rather than store all the vectors, only the indices of vectors in the original data set are stored. The computation node can build all the random vectors on the fly from the given seed according to the random seed.

From the model building perspective, we can easily broadcast several random seeds with negligible traffic overhead compared to a large matrix. By doing so we improve computation efficiency and, from the storage perspective, we only need to store one 4-byte random seed for each RP-tree. In such a way, we are able to successfully reduce the storage overhead from 47.7 GB to 512 B for a search index consisting of 128 RP-trees (with cluster size 20), or from 95.4 GB to only 1 KB if 256 RP-trees are used.

16.6 Combining Multiple Trees

Let's consider further use of multiple RP-trees. In the design of PANNS, we have two design options to improve search accuracy. Given the size of the aggregated cluster taken as the union of the target clusters from every tree, we can either use fewer trees with larger leaf clusters, or use more trees with smaller leaf clusters. Increasing cluster size is intuitive: if we keep increasing it until it includes all the vectors, then it is totally accurate.

On the other hand, we expect that when using more trees the probability of a query point falling close to a splitting hyperplane should be reduced so it will be less likely for its nearest neighbours to lie in a different cluster. By reducing such misclassifications, the searching accuracy should improve. Although we are not aware of prior theoretical results in the field of nearest neighbour search algorithms that justify such a hypothesis, this concept can be thought of as a combination strategy similar to those from ensemble clustering, a very well established field of research. [17] [18] Similar to our case, ensemble clustering improve clustering solutions by fusing information from several data partitions.

Let's test the idea on Wikipedia data. In our tests, the data set contains 500,000 points and we always search for the 50 nearest neighbours of a query point. Then we measure the searching accuracy by calculating the amount of actual nearest neighbours found.

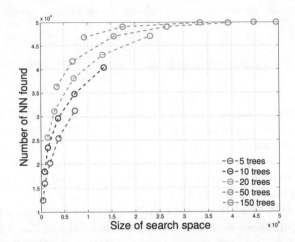

Fig. 16.7: The number of true nearest neighbours found for different number of trees

We query $1,000$ points in each run. The results presented in Fig. 16.7 correspond to the mean values of the aggregated nearest neighbours of the $1,000$ query points discovered by PANNS out of 100 runs. Note that x-axis represents the "size of search space" which is defined by the number of unique points within the union of all the leaf clusters that the query point falls in. Therefore, given the same search space size, using more trees indicates that the leaf clusters become smaller. As we can see in Fig. 16.7, for a given x value, the curves move upwards as we use more and more trees, indicating that the accuracy improves. As shown in the case of 50 trees, almost 80% of the actual nearest neighbours are found by performing a search over the 10% of the data set.

The empirical results clearly show *the benefits of using more trees instead of using larger clusters for improving search accuracy*. Moreover, regarding the searching performance, since searching can be easily parallelised, using more trees will not impact the searching time.

16.7 Critical Boundary Elimination

The classic RP trees generate a different random vector at each inner node of a tree. The computations in the child-branches cannot proceed without finishing the computation in the parent node, as show in the left figure of Fig. 16.8. The blue dotted lines represent the critical boundaries. Instead, we can use one identical random vector for all the sibling nodes of a tree. This small tweak does not affect

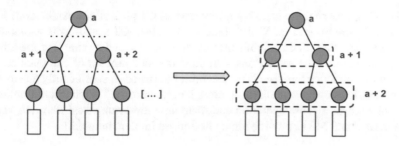

Fig. 16.8: The illustration of removing critical boundaries to parallelise the computation.

the accuracy at all, because a query point is routed down each of the trees only once. Hence, the query point is projected onto a random vector r_i sampled from the same distribution at each level of a tree. This means that we don't need all the inner non-leaf node to be independent random vectors. Instead, the query point is projected onto only l i.i.d. random vectors $r_1, \ldots, r_l$. An RP-tree has $2^l - 1$ inner nodes. Therefore, if each node of a tree had a different random vector as in classic RP-trees, $2^l - 1$ different random vectors would be required for one tree. However, when a single vector is used on each level, only l vectors are required. This reduces the amount of memory required by the random vectors from exponential to linear with respect to the depth of the trees.

A further benefit of using one random vector for one layer is that it speeds up the index construction significantly, since we can vectorise the computation. Let's first look at the projection of vector a on b. The projected length on b can be expressed as below.

$$\|a\| \cos \theta = a \cdot \frac{b}{\|b\|} \tag{16.1}$$

Here $\|a\|$ means the length of vector **a**. If we require that all the random vectors **b** has to be normalised, Eq. 16.1 becomes $a \cdot b$, the vector dot. Now we can perform the projection at this layer by computing Xb_l. Here X is the dataset, and each row is a document and each column is a feature; b_l is a random vector that we use for this layer. In this way, we don't have to wait for the left tree to finish to start cutting the right tree.

Now here is the tricky bit: we don't even have to wait for the upper layer to start cutting the lower layer! The reason is that, at each layer, we do random projection of *all the nodes* in the dataset on one single random vector b. We don't really care the random clustering result from the previous layer. Therefore, we can perform $Xb_1, Xb_2, \ldots, Xb_l$ at the same time. That means, the projected data set P can be computed directly from the dataset X and a random matrix B as $P = XB$ with only one pass of matrix multiplication. Here each column of B is just the random vector we use at a layer.

In this approach there is no boundary, and all the projections can be done in just one matrix multiplication. While some of the observed speed-up is explained by having to generate fewer random vectors, it is mostly due to enabling the efficient computation of all the projections. Although the total amount of computation stays the same, in practice this speeds up the index construction significantly due to cache effects and low-level parallelisation through vectorisation. The matrix multiplication is a basic linear algebra operation and many low level numerical libraries, such as OpenBLAS and MKL, provide highly optimised implementation.

16.8 Search Operations Parallelisation

By using RP-tree we already limit the search range from the whole text corpus to a cluster of a small number of documents (vectors), where we can do a linear search. We have also introduced several optimisations on the RP-tree itself, including using multiple trees, using a random seed to remove the storage of random vectors, improving computation efficiency etc. But we don't stop here: can we further improve the linear searching itself? It turns out, we can.

To select the best candidates from a cluster of points, we need to use the coordinates in the original space to calculate their relative distance to the query point. However, this first increases the storage overhead since we need to keep the original high dimensional data set which is usually huge; and second increases the query overhead since we need to access the data set. Performance is more severely degraded if the original dataset is too big to load into physical memory. Moreover, computing the distance between two points in the high dimensional space directly is very time-consuming.

Nonetheless, we will show that it is possible to completely remove the original dataset while keeping the accuracy at a satisfying level. The core idea of is simple. Let's look at the second subfigure in Fig. 16.3. Imagine that we add a new point to search for similar vectors. The normal approach is that we compute the distance between this node and A, B, C etc. But if you look at it close, all the existing nodes are already projected on the vector y, and we can also project the incoming query vector on y, and check to which of these points it is close to. Instead of computing the distances of two vectors, now we only compute the absolute value of subtraction of two numbers (since we can always project a vector onto another, obtaining a real number as result) as the distance. By replacing the original space with the projected one, we are able to achieve a significant reduction in storage and non-trivial gains in searching performance.

Of course, it is not always an accurate estimate. In the first subfigure of Fig. 16.3, a node can be physically close to A or B, but its projection could be closest to that of C. That again requires us to consider using multiple RP-trees. But instead of the actual vector content, in the leaf node of the trees we store only (index, projected value). Now for the input query vector, we run it in the N RP-trees and get N set of (index, value) pairs. Here each value is the absolute value of the difference of projected values between the vector in the tree and the query vector itself. Each vector of course is labelled by a unique index.

For each index, we use this metric: $\frac{\sum d_i}{\sum c_i}$ to measure how close it is to the query vector. Here d_i is the distance between node i and query node on projected space, and c_i is the count of total number of node i in all the candidate sets from all the RP-trees. Smaller values indicate lower distances. The intuition is that, if distance value of a node on the projected space is small, then it is possibly close to the query node; or, if a node appears many times from the candidate sets of different RP-trees, it is quite likely a possible close neighbour.

As a further improvement, we update this metric to $\frac{\sum d_i}{(\sum c_i)^3}$. By doing so, we give much more weight to the points which have multiple occurrences from different RP-trees by assuming that such points are more likely to be the true k-NN.

16.9 Code Implementation

Having introduced the theory, we will now show the implementation in OCaml. First, we show the simple random projection along a RP-tree.

```
let make_projection_matrix seed m n =
  Owl_stats_prng.init seed;
  Mat.gaussian m n |> Mat.to_arrays

let make_projected_matrix m n =
  Array.init m (fun _ -> Array.make n 0.);;
```

These two functions make the projection matrix and the matrix to save projected results, both returned as row vectors.

```
let project i j s projection projected =
  let r = ref 0. in
  Array.iter (fun (w, a) ->
    r := !r +. a *. projection.(w).(j);
  ) s;
  projected.(j).(i) <- !r;;
```

Based on these two matrices, the project function processes document i on the level j in the tree. The document vector is s. The projection is simply a dot multiplication between s and matrix projection.

```
let random seed cluster tfidf =
  let num_doc = Nlp.Tfidf.length tfidf in
  let vocab_len = Nlp.Tfidf.vocab_len tfidf in
  let level = Maths.log2 (float_of_int num_doc /. cluster)
    |> ceil |> int_of_float in
```

```
let projection = make_projection_matrix
  seed vocab_len level in
let projected = make_projected_matrix level num_doc in

Nlp.Tfidf.iteri (fun i s ->
  for j = 0 to level - 1 do
    project i j s projection projected;
  done;
) tfidf;

vocab_len, level, projected;;
```

The random function performs a random projection of a sparse data set based a TF-IDF model. Note that other vector models can be used similarly in the random projection process This function projects all the document vectors in the model to the projected matrix, level by level. Recall that the result only contains the projected value instead of the whole vector.

As we have explained in the 16.7 section, this process can be accelerated to use matrix multiplication. The code below shows this implementation for the random projection function. It also returns the shape of projection and the projected result.

```
let make_projection_matrix seed m n =
  Owl_stats_prng.init seed;
  Mat.gaussian m n

let random seed cluster data =
  let m = Mat.row_num data in
  let n = Mat.col_num data in
  let level = Maths.log2 (float_of_int m /. cluster)
    |> ceil |> int_of_float in

  let projection = make_projection_matrix seed n level in
  let projected = Mat.dot data projection |> Mat.transpose in

  n, level, projected, projection;;
```

After getting the projection result, we need to build a RP-tree accordingly. The following is about how to build the index in the form of a binary search tree. The tree type is defined as:

```
type t =
  | Node of float * t * t
  | Leaf of int array;;
```

An intermediate node includes three parts: split, left, right, and the leaves only contain document index.

```
let split_space_median space =
  let space_size = Array.length space in
  let size_of_l = space_size / 2 in
  let size_of_r = space_size - size_of_l in
  Array.sort (fun x y -> Pervasives.compare
    (snd x) (snd y)) space;
  let median =
    match size_of_l < size_of_r with
    | true  -> snd space.(size_of_l)
    | false -> (snd space.(size_of_l-1) +.
      snd space.(size_of_l)) /. 2.
  in
  let l_subspace = Array.sub space 0 size_of_l in
  let r_subspace = Array.sub space size_of_l size_of_r in
  median, l_subspace, r_subspace;;
```

The split_space_median function divides the projected space into subspaces to assign left and right subtrees. The passed in space is the projected values on a specific level. The criterion of division is the median value. The Array.sort function sorts the space into increasing order for median value.

```
let filter_projected_space level projected subspace =
  let plevel = projected.(level) in
  Array.map (fun (doc_id, _) ->
    doc_id, plevel.(doc_id)) subspace;;
```

Based on the document id of the points in subspace, filter_projected_space function filters the projected space. The purpose of this function is to update the projected value using a specified level so the recursion can continue. Both the space and the returned result are of the same format: (doc_id, projected value).

```
let rec make_subtree level projected subspace =
  let num_levels = Array.length projected in
  match level = num_levels with
  | true  -> (
      let leaf = Array.map fst subspace in
      Leaf leaf
    )
  | false -> (
      let median, l_space, r_space =
        split_space_median subspace in
      let l_space = match level < num_levels - 1 with
        | true  -> filter_projected_space
          (level+1) projected l_space
        | false -> l_space
      in
      let r_space = match level < num_levels - 1 with
```

```
        | true  -> filter_projected_space
          (level+1) projected r_space
        | false -> r_space
    in
    let l_subtree = make_subtree (level+1)
       projected l_space in
    let r_subtree = make_subtree (level+1)
       projected r_space in
    Node (median, l_subtree, r_subtree)
  );;
```

Based on these functions, `make_subtree` recursively grows the binary subtree to make a whole tree. The `projected` value is the projected points we get from the first step. It is of shape (level, document_number). The subspace is a vector of shape (1, document_number).

```
let grow projected =
  let subspace = Array.mapi
    (fun doc_id x -> (doc_id, x)) projected.(0) in
  let tree_root = make_subtree 0 projected subspace in
  tree_root;;
```

The `grow` function calls `make_subtree` to build the binary search tree. It initialises the first subspace at level 0, and then start recursively making the subtrees from level 0. Currently everything is done in memory for efficiency consideration.

```
let rec traverse node level x =
  match node with
  | Leaf n            -> n
  | Node (s, l, r) -> (
      match x.(level) < s with
      | true  -> traverse l (level+1) x
      | false -> traverse r (level+1) x
    );;
```

Now that the tree is built, we can perform search. The recursive `traverse` function traverses the whole tree to locate the cluster for a projected vector x starting from a given level.

```
let rec iter_leaves f node =
  match node with
  | Leaf n            -> f n
  | Node (s, l, r) -> iter_leaves f l; iter_leaves f r

let search_leaves node id =
  let leaf = ref [||] in
  (
```

```
try iter_leaves (fun l ->
  if Array.mem id l = true then (
    leaf := l;
    failwith "found";
  )
) node
with exn -> ()
);
Array.copy !leaf;;
```

Finally, `search_leaves` returns the leaves/clusters which have the given `id` inside it. It mainly depends on the `iter_leaves` function which iterates all the leaves in a tree and applies function, to perform this search.

All these steps are executed on one tree. When we collect the k-NN candidates from all the trees, we can aggregate and rank these candidates according to their similarity metrics before returning the results to a user.

16.10 Summary

In this chapter, we applied LSA to build a semantic search engine called Sofia. To scale up Sofia as a realistic Internet service, we addressed various technical challenges in the system implementation. Specifically, we utilised a parallel RP-tree algorithm and implemented stochastic SVD on Spark to tackle the performance challenges in index building and searching. We introduced the basic algorithm and how it can be optimised step by step, from storage to computation. These optimisations include aggregating results from multiple trees, replacing random variable with a single random seed, removing the projection computation boundary between different layers, using count to approximate vector distance, etc. For those who are interested in the NLP topic, we recommend continuing the studies on more advanced embedding techniques such as fastText [19], GloVe [20], BERT [21], etc.

References

1. Semantic Web Stack [Internet]. Wikipedia. Wikimedia Foundation; 2020 [cited 2021Dec22]. Available from: https://en.wikipedia.org/wiki/Semantic_Web_Stack
2. Goldberg D, Nichols D, Oki BM, Terry D. Using collaborative filtering to weave an information tapestry. Communications of the ACM. 1992 Dec 1; 35(12):61-70.
3. Su X, Khoshgoftaar TM. A survey of collaborative filtering techniques. Advances in artificial intelligence. 2009.
4. Koren Y, Bell R. Advances in collaborative filtering. Recommender systems handbook. 2015:77-118.
5. Fernández-Tobías I, Cantador I, Kaminskas M, Ricci F. Cross-domain recommender systems: A survey of the state of the art. Spanish conference on information retrieval 2012 Jun (pp. 1-12). sn.

6. Xing W, Ghorbani A. Weighted pagerank algorithm. Proceedings. Second Annual Conference on Communication Networks and Services Research, 2004. 2004 May 21 (pp. 305-314). IEEE.

7. Berry MW, Dumais ST, O'Brien GW. Using linear algebra for intelligent information retrieval. SIAM review. 1995 Dec;37(4):573-95.

8. Baeza-Yates R, Ribeiro-Neto B. Modern information retrieval. New York: ACM press; 1999 May 15.

9. Frakes WB, Baeza-Yates R, editors. Information retrieval: data structures and algorithms. Prentice-Hall, Inc.; 1992 Jun 1.

10. Van Rijsbergen C. Information retrieval: theory and practice. InProceedings of the Joint IBM/University of Newcastle upon Tyne Seminar on Data Base Systems 1979 Sep (pp. 1-14).

11. Wang L, Tasoulis S, Roos T, Kangasharju J. Kvasir: Scalable provision of semantically relevant web content on big data framework. IEEE Transactions on Big Data. 2016 Apr 28;2(3):219-33.

12. Hyvönen V, Pitkänen T, Tasoulis S, Jääsaari E, Tuomainen R, Wang L, Corander J, Roos T. Fast nearest neighbor search through sparse random projections and voting. IEEE International Conference on Big Data. 2016 Dec 5 (pp. 881-888).

13. Wang L, Tasoulis S, Roos T, Kangasharju J. Kvasir: Seamless integration of latent semantic analysis-based content provision into web browsing. Proceedings of the 24th International Conference on World Wide Web. 2015 May 18 (pp. 251-254).

14. Meng X, Bradley J, Yavuz B, Sparks E, Venkataraman S, Liu D, Freeman J, Tsai DB, Amde M, Owen S, Xin D. Mllib: Machine learning in apache spark. The Journal of Machine Learning Research. 2016 Jan 1;17(1):1235-41.

15. Halko N, Martinsson PG, Tropp JA. Finding structure with randomness: Probabilistic algorithms for constructing approximate matrix decompositions. SIAM review. 2011;53(2):217-88.

16. Dasgupta S, Sinha K. Randomized partition trees for exact nearest neighbor search. Conference on Learning Theory. 2013 Jun 13 (pp. 317-337).

17. Vega-Pons S, Ruiz-Shulcloper J. A survey of clustering ensemble algorithms. International Journal of Pattern Recognition and Artificial Intelligence. 2011 May; 25(03):337-72.

18. Fern XZ, Lin W. Cluster ensemble selection. Statistical Analysis and Data Mining: The ASA Data Science Journal. 2008 Nov 25;1(3):128-41.

19. Joulin A, Grave E, Bojanowski P, Mikolov T. Bag of tricks for efficient text classification. arXiv preprint arXiv:1607.01759. 2016 Jul 6.

20. Pennington J, Socher R, Manning CD. Glove: Global vectors for word representation. Proceedings of the 2014 conference on empirical methods in natural language processing (EMNLP) 2014 Oct (pp. 1532-1543).

21. Devlin J, Chang MW, Lee K, Toutanova K. Bert: Pre-training of deep bidirectional transformers for language understanding. arXiv preprint arXiv:1810.04805. 2018 Oct 11.

Appendix A
Conventions of Owl

All software systems have their own rules and conventions which with developers must comply, and Owl is no exception. In this appendix we cover function naming and various other conventions in the Owl library.

A.1 Pure vs Impure

The Ndarray module contains functions to manipulate and perform mathematical operations over multi-dimensional arrays. The **pure** (or immutable) functions are those that do not modify the provided variables but instead return a new one as a result. In contrast, **impure** (or mutable) functions are those which modify the passed-in variables in place.

Functional programming in general promotes the use of immutable data structures as impure functions make it more difficult to reason about the correctness of code. On the other hand, generating a fresh 1000×1000 matrix every time you modify a single element does not seem very practical.

The introduction of impure functions into Owl is thus only done under careful, practical consideration. In-place modification avoids expensive memory allocation and de-allocation which can significantly improve runtime performance of a numerical application, especially when large ndarrays and matrices are involved.

Thus, many pure functions in Ndarray module have corresponding impure versions indicated by an underscore "_" added at the end of the function name. For example, the following pure functions in the Arr module Arr.sin, Arr.add have corresponding impure functions, Arr.sin_, Arr.add_.

For unary operators such as Arr.sin_ x, the situation is rather straightforward when x will be modified in place. However, binary operators are more complex. For example, in Arr.add_scalar x a, x will be modified in place to store the final result – relatively straightforward as a is a scalar. In Arr.add_ x y, the question is where to store the final result when both inputs are ndarrays. Let's look at the type of Arr.add_ function:

L. Wang et al., *OCaml Scientific Computing*, Undergraduate Topics in Computer Science, https://doi.org/10.1007/978-3-030-97645-3

Table A.1: Alias of pure and impure binary math functions

Function Name	Pure	Impure
add	+	+=
sub	–	-=
mul	*	*=
div	/	/=
add_scalar	+$	+$=
sub_scalar	-$	-$=
mul_scalar	*$	*$=
div_scalar	/$	/$=

```
val Arr.add_ : ?out:Arr.arr -> Arr.arr -> Arr.arr -> unit
```

As we can see from the type signature, the output is specified in the optional out parameter. If out is not provided, the first operand (x) will be attempted to store the final result – as binary operators support broadcasting operations by default, when using impure functions every dimension of the first argument x must not be smaller than that of the second argument y. In other words, impure functions only allow broadcasting smaller y onto an x that is big enough to accommodate the result.

Most binary math functions have an associated a shorthand operator, such as +, -, *, and /. The impure versions also have associated operators, e.g., rather than Arr.(x + y) which returns the result in a new ndarray, you can write Arr.(x += y) which adds x and y and saves the result into x. These operators are listed in Tbl. A.1.

A.2 Ndarray vs Scalar

There are three general ndarray operators: *map*, *scan*, and *reduce*. Many Ndarray functions can be categorised as reduce operations, such as Arr.sum, Arr.prod, Arr.min, Arr.mean, and Arr.std. For example, we can sum all the elements in a ndarray using sum function.

```
let x = Arr.sequential [|3;3;3|];;
let a = Arr.sum x;;
```

The result is a one-element ndarray so to treat it as a scalar, you must retrieve the single value by calling get function.

```
let b = Arr.get a [|0|] in
b +. 10.;;
```

This becomes inconvenient in OCaml if we must always extract the scalar value from the return of reduce operations. In languages like Python and Julia, the return type is dynamically determined but OCaml's strong typing requires that we either use a unified type or implement another set of functions. In the end, Owl picked the latter in its design so every reduce operation has two versions:

- one allowing you to reduce along the specified axis or to reduce all the elements, but always returning an ndarray;
- one that reduces all the elements, always returning a scalar value.

The difference between the two is indicated by naming those returning a scalar with an extra "'" character in their names. For example, those functions returning an ndarray are named `Arr.sum`, `Arr.prod`, mean, etc, while those returning a scalar are named `Arr.sum'`, `Arr.prod'`, mean', etc.

A.3 Infix Operators

In each `Ndarray` and `Matrix` module, some frequently used functions have the corresponding infix operators as convenient aliases. Tbl. A.2 summarises the operators. In the table, both x and y represent either a matrix or an ndarray while a represents a scalar value.

We draw your attention to several things in this table:

- +\$ has its corresponding operator \$+ if we flip the order of parameters. However, it is important to be very careful about operator precedence in these cases: *OCaml determines precedence based on the first character of the infix operator.* +\$ preserves the precedence whereas \$+ does not. We thus recommend using \$+ with great care, and always using parentheses to explicitly specify precedence. The same applies to \$-, \$*, and \$/.
- For comparison operators, e.g. both = and =. compare all the elements in two variables x and y. The difference is that = returns a boolean value whereas =. returns a matrix or ndarray of the same shape and type as x and y. In the returned result, the value in a given position is 1 if the values of the corresponding position in x and y satisfy the predicate, otherwise it is 0.
- The comparison operators ending with \$, they are used to compare a matrix/ndarray to a scalar value.

A.4 Module Structures

Owl's `Dense` module contains modules supporting dense data structures such as `Dense.Matrix` supporting operations on dense matrices, and the `Sparse` module with modules supporting sparse data structures. We have four basic modules: dense ndarray, dense matrix, sparse ndarray, and sparse matrix. In this book we mostly use Dense modules and indeed, the matrix modules are a special case that's built upon the corresponding ndarray modules.

With Dense.Ndarray, we can create a dense *n*-dimensional array of no more than 16 dimensions. This constraint originates from the underlying Bigarray.Genarray module in OCaml. In practice, this constraint makes sense since the space requirement will explode as the dimension increases. If you need anything higher than 16 dimensions, you need to use Sparse.Ndarray to create a sparse data structure.

All four modules consists of five submodules to handle different types of numbers:

- S module supports single precision float numbers float32;
- D module supports double precision float numbers float64;
- C module supports single precision complex numbers complex32;
- Z module supports double precision complex numbers complex64;
- Generic module supports all aforementioned number types via a GADT.

Technically, S, D, C, and Z are wrappers of Generic module with explicit type information provided. Therefore you can save the type constructor which was passed into the Generic module if you use these submodules directly. In short, the Generic module can do everything that submodules can, but for some functions (e.g. creation) you must explicitly pass in the type information.

In practice, we often work with double precision numbers so Owl provides shortcuts to the data structures of double precision floating point numbers. Arr is equivalent to double precision real Dense.Ndarray.D, and similarly, Mat is equivalent to double precision real Dense.Matrix.D. These two modules are frequently used in this book.

You can cast one value from one type to another one by using the cast_* functions in Generic module. For example, Generic.cast_s2d casts from float32 tofloat64, and Generic.cast_c2z casts from complex32 to complex64.

Many functions in the Generic module can handle aforementioned four different number types. This polymorphism is achieved by pattern matching and general abstract data type in OCaml. In the code below, we use the sum function in Dense.Matrix.Generic module as an example.

```
open Owl;;

let x = Dense.Matrix.S.eye 5 in
Dense.Matrix.Generic.sum x;;

let x = Dense.Matrix.D.eye 5 in
Dense.Matrix.Generic.sum x;;

let x = Dense.Matrix.C.eye 5 in
Dense.Matrix.Generic.sum x;;

let x = Dense.Matrix.Z.eye 5 in
Dense.Matrix.Generic.sum x;;
```

As we can see, no matter what kind of numbers are held in an identity matrix, we can always pass it to Dense.Matrix.Generic.sum function. Similarly, we can do the same thing for other modules (`Dense.Ndarray`, `Sparse.Matrix`, etc.) and other functions (`add`, `mul`, `neg`, etc.).

A.5 Operator Extension

The operators above do not allow interoperation on different number types, e.g., you cannot add a `float32` matrix to `float64` matrix. However, some people just like the Pythonic way of working so `Owl.Ext` module is specifically designed to make prototyping faster and easier. Once you open the module, `Ext` immediately provides a set of operators to allow you to inter-operate on different number types, as in Tbl. A.3. It automatically casts types for you if necessary.

You may have noticed that the operators ending with `$` (e.g., `+$`, `-$` ...) disappeared from the table, which is simply because we can add/sub/mul/div a scalar with a matrix directly and we do not need these operators any more. Similar for comparison operators, because we can use the same `>` operator to compare a matrix to another matrix, or compare a matrix to a scalar, we do not need `>$` any longer. Allowing inter-operation makes the operator table much shorter.

Currently, the operators in `Ext` only support inter-operation on dense structures. Besides binary operators, `Ext` also implements most of the common math functions which can be applied to float numbers, complex numbers, matrices, and ndarray. These functions are:

`im`; `re`; `conj`, `abs`, `abs2`, `neg`, `reci`, `signum`, `sqr`, `sqrt`, `cbrt`, `exp`, `exp2`, `expm1`, `log`, `log10`, `log2`, `log1p`, `sin`, `cos`, `tan`, `asin`, `acos`, `atan`, `sinh`, `cosh`, `tanh`, `asinh`, `acosh`, `atanh`, `floor`, `ceil`, `round`, `trunc`, `erf`, `erfc`, `logistic`, `relu`, `softplus`, `softsign`, `softmax`, `sigmoid`, `log_sum_exp`, `l1norm`, `l2norm`, `l2norm_sqr`, `inv`, `trace`, `sum`, `prod`, `min`, `max`, `minmax`, `min_i`, `max_i`, `minmax_i`.

Note that `Ext` contains its own `Ext.Dense` module which further contains the submodules such as the dense single precision float ndarray `Ext.Dense.Ndarray.S` or the dense single precision complex matrix `Ext.Dense.Matrix.C`. These modules are simply the wrappers of the original modules in `Owl.Dense` module so they provide most of the APIs already implemented. The extra thing these wrapper modules do is automatically pack and unpack the raw number types for you. You can certainly use the raw data types and then use the constructors defined in `Owl_ext_types` to wrap them up by yourself. Let's see some examples to understand how convenient it is to use `Ext` module.

```
open Owl.Ext;;

let x = Dense.Matrix.S.uniform 5 5;;
let y = Dense.Matrix.C.uniform 5 5;;
let z = Dense.Matrix.D.uniform 5 5;;

x + F 5.;;
x * C Complex.({re = 2.; im = 3.});;
```

```
x - y;;
x / y;;
x *@ y;;

x > z;;
x >. z;;
(x >. z) * x;;
(x >. F 0.5) * x;;
(F 10. * x) + y *@ z;;

round (F 10. * (x *@ z));;
sin (F 5.) * cos (x + z);;
```

As we draw this appendix to a close, we must make one caveat. Ext tries to mimic dynamic languages like Python but with unified types. This prevents the OCaml compiler from doing type checking in compilation phase and introduces extra overhead in calling functions. Therefore, besides fast experimenting in toplevel, we do not recommend use of the Ext module in production code.

Table A.2: Infix operators in ndarray and matrix modules

Operator	Example	Operation	Dense/Sparse	Ndarray/Matrix
+	x + y	element-wise add	both	both
–	x - y	element-wise sub	both	both
*	x * y	element-wise mul	both	both
/	x / y	element-wise div	both	both
+$	x +$ a	add scalar	both	both
-$	x -$ a	sub scalar	both	both
*$	x *$ a	mul scalar	both	both
/$	x /$ a	div scalar	both	both
$+	a $+ x	scalar add	both	both
$-	a $- x	scalar sub	both	both
$*	a $* x	scalar mul	both	both
$/	a $/ x	scalar div	both	both
=	x = y	comparison	both	both
!=	x != y	comparison	both	both
<>	x <> y	same as !=	both	both
>	x > y	comparison	both	both
<	x < y	comparison	both	both
>=	x >= y	comparison	both	both
<=	x <= y	comparison	both	both
=.	x =. y	element-wise cmp	Dense	both
!=.	x !=. y	element-wise cmp	Dense	both
<>.	x <>. y	same as !=.	Dense	both
>.	x >. y	element-wise cmp	Dense	both
<.	x <. y	element-wise cmp	Dense	both
>=.	x >=. y	element-wise cmp	Dense	both
<=.	x <=. y	element-wise cmp	Dense	both
=$	x =$ y	comp to scalar	Dense	both
!=$	x !=$ y	comp to scalar	Dense	both
<>$	x <>$ y	same as !=	Dense	both
>$	x >$ y	compare to scalar	Dense	both
<$	x <$ y	compare to scalar	Dense	both
>=$	x >=$ y	compare to scalar	Dense	both
<=$	x <=$ y	compare to scalar	Dense	both
=.$	x =.$ y	element-wise cmp	Dense	both
!=.$	x !=.$ y	element-wise cmp	Dense	both
<>.$	x <>.$ y	same as !=.$	Dense	both
>.$	x >.$ y	element-wise cmp	Dense	both
<.$	x <.$ y	element-wise cmp	Dense	both
>=.$	x >=.$ y	element-wise cmp	Dense	both
<=.$	x <=.$ y	element-wise cmp	Dense	both
=~'	x =~ y	approx =	Dense	both
=~$	x =~$ y	approx =$	Dense	both
=~.	x =~. y	approx =.	Dense	both
=~.$	x =~.$ y	approx =.$	Dense	both
%	x % y	mod divide	Dense	both
%$	x %$ a	mod divide scalar	Dense	both
**	x ** y	power function	Dense	both
*@	x *@ y	matrix multiply	both	Matrix
/@	x /@ y	solve linear system	both	Matrix
**@	x **@ a	matrix power	both	Matrix
min2	min2 x y	element-wise min	both	both
max2	max2 x y	element-wise max	both	both
@=	x @= y	concatenate vertically	Dense	both
@|	x @| y |	concatenate horizontally	Dense	both

Table A.3: Operator extensions used in Ext module.

Operator	Example	Operation
+	x + y	add
-	x - y	sub
*	x * y	mul
/	x / y	div
=	x = y	comparison, return bool
!=	x != y	comparison, return bool
<>	x <> y	same as !=
>	x > y	comparison, return bool
<	x < y	comparison, return bool
>=	x >= y	comparison, return bool
<=	x <= y	comparison, return bool
=.	x =. y	element wise comparison
!=.	x !=. y	element wise comparison
<>.	x <>. y	same as !=.
>.	x >. y	element wise comparison
<.	x <. y	element wise comparison
>=.	x >=. y	element wise comparison
<=.	x <=. y	element wise comparison
%	x % y	element wise mod divide
**	x ** y	power function
*@	x *@ y	matrix multiply
min2	min2 x y	element-wise min
max2	max2 x y	element-wise max

Appendix B
Visualisation

We have used several different types of plot in this book, created by using Owl's cross-platform `Plot` module. It is designed to make reasonably complex plots with minimal coding effort, and is built on Plplot[1] but hides much of its complexity. The module is since `Plplot` calls the underlying graphics device driver to plot. Code for several of the figures in this book was included already; in this appendix we will use many other examples to demonstrate how to draw various typical plots in OCaml.

B.1 Plotting in Owl

B.1.1 Create Plots

Let's start with the standard way to create a plot using `Plot.create` . Here is its type definition in `Owl_plot.mli`:

```
val create : ?m:int -> ?n:int -> string -> handle
```

For example,

```
# let f x = Maths.sin x /. x in
  let h = Plot.create "plot_001.png" in

  Plot.set_title h "Function: f(x) = sine x / x";
  Plot.set_xlabel h "x-axis";
  Plot.set_ylabel h "y-axis";
  Plot.set_font_size h 8.;
  Plot.set_pen_size h 3.;
  Plot.plot_fun ~h f 1. 15.;;
```

---

[1] Plplot library: http://plplot.sourceforge.net/index.php

© The Author(s), under exclusive license to Springer Nature Switzerland AG 2022
L. Wang et al., *OCaml Scientific Computing*, Undergraduate Topics in Computer Science,
https://doi.org/10.1007/978-3-030-97645-3

For all plots, we obtain a handle h from the `create` function. The underlying image type is inferred automatically from the file name provided. In this case, we want to create a PNG image that plots the sine function so we will use the `plot_fun` function. We also set the x- and y-axis labels and figure title, together with the font and line size. You will find that these commands are similar to those in other plotting tools such as Matplotlib. Once you call `Plot.output`, the plot will be "sealed" and written into the final file. The generated figure is shown below.

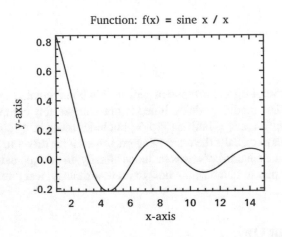

Fig. B.1: Basic function plot

B.1.2 Specification

For most high-level plotting functions in Owl, there is an optional parameter, `spec`. This takes a list of specifications to exercise fine control over the plot's appearance. Every function has a set of slightly different parameters. If you pass in some parameters that a function cannot use, they will simply be ignored. If you pass in the same parameter multiple times, only the last instance will take effects.

In the following, we will provide some examples using the `spec` parameter to fine tune Owl's plots. The first example in Fig. B.2 shows how to configure the `mesh` plot using `ZLine`, `Contour`, and other `spec` parameters.

```
# let x, y = Mat.meshgrid (-2.5) 2.5 (-2.5) 2.5 50 50 in
  let z = Mat.(sin ((x * x) + (y * y))) in
  let h = Plot.create ~m:2 ~n:3 "plot_020.png" in

  Plot.subplot h 0 0;
```

```
Plot.(mesh ~h ~spec:[ ZLine XY ] x y z);

Plot.subplot h 0 1;
Plot.(mesh ~h ~spec:[ ZLine X ] x y z);

Plot.subplot h 0 2;
Plot.(mesh ~h ~spec:[ ZLine Y ] x y z);

Plot.subplot h 1 0;
Plot.(mesh ~h ~spec:[ ZLine Y; NoMagColor ] x y z);

Plot.subplot h 1 1;
Plot.(mesh ~h ~spec:[ ZLine Y; Contour ] x y z);

Plot.subplot h 1 2;
Plot.(mesh ~h ~spec:[ ZLine XY; Curtain ] x y z);

Plot.output h;;
```

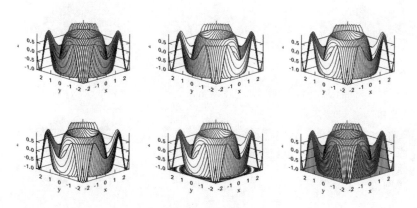

Fig. B.2: Plot specification

The second example in Fig. B.3 shows how to tune the surf plotting function when drawing a 3D surface.

```
# let x, y = Mat.meshgrid (-1.) 1. (-1.) 1. 50 50 in
  let z = Mat.(tanh ((x * x) + (y * y))) in
  let h = Plot.create ~m:2 ~n:3 "plot_021.png" in

  Plot.subplot h 0 0;
  Plot.(surf ~h ~spec:[ ] x y z);
```

```
Plot.subplot h 0 1;
Plot.(surf ~h ~spec:[ Faceted ] x y z);

Plot.subplot h 0 2;
Plot.(surf ~h ~spec:[ NoMagColor ] x y z);

Plot.subplot h 1 0;
Plot.(surf ~h ~spec:[ Contour ] x y z);

Plot.subplot h 1 1;
Plot.(surf ~h ~spec:[ Curtain ] x y z);

Plot.subplot h 1 2;
Plot.(surf ~h ~spec:[ Altitude 10.; Azimuth 125. ] x y z);

Plot.output h;;
```

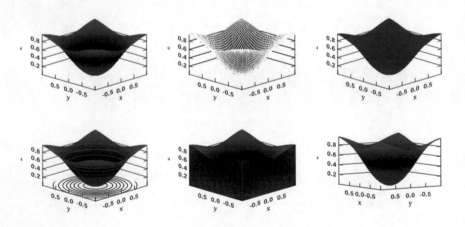

Fig. B.3: Surf plot

B.1.3 Subplots

You might have already spotted another feature in the previous example: subplots.
You can change the number of rows and columns of the subgraph layout by varying
the values of the m and n parameters in the create function.

```
# let f p i = match i with
    | 0 -> Stats.gaussian_rvs ~mu:0. ~sigma:0.5 +. p.(1)
    | _ -> Stats.gaussian_rvs ~mu:0. ~sigma:0.1 *. p.(0)
  in
  let y = Stats.gibbs_sampling f [|0.1;0.1|] 5_000
    |> Mat.of_arrays in
  let h = Plot.create ~m:2 ~n:2 "plot_002.png" in
  Plot.set_background_color h 255 255 255;

  (* subplot at 0,0 *)
  Plot.subplot h 0 0;
  Plot.set_title h "Bivariate model";
  Plot.scatter ~h (Mat.col y 0) (Mat.col y 1);

  (* subplot at 0,1 *)
  Plot.subplot h 0 1;
  Plot.set_title h "Distribution of y";
  Plot.set_xlabel h "y";
  Plot.set_ylabel h "Frequency";
  Plot.histogram ~h ~bin:50 (Mat.col y 1);

  (* subplot at 1,0 *)
  Plot.subplot h 1 0;
  Plot.set_title h "Distribution of x";
  Plot.set_ylabel h "Frequency";
  Plot.histogram ~h ~bin:50 (Mat.col y 0);

  (* subplot at 1,1 *)
  Plot.subplot h 1 1;
  Plot.set_foreground_color h 255 0 0;
  Plot.set_title h "Sine function";
  Plot.(plot_fun ~h ~spec:[ LineStyle 2 ] Maths.sin 0. 28.);
  Plot.autocorr ~h (Mat.sequential 1 28);

  Plot.output h;;
```

B.1.4 Multiple Lines

You can certainly plot multiple functions in the same plot. Fig. B.5a is an example with both sine and cosine lines in one plot.

```
# let h = Plot.create "plot_024.png" in
  Plot.(plot_fun ~h ~spec:[ RGB (0,0,255); Marker "#[0x2299]";
```

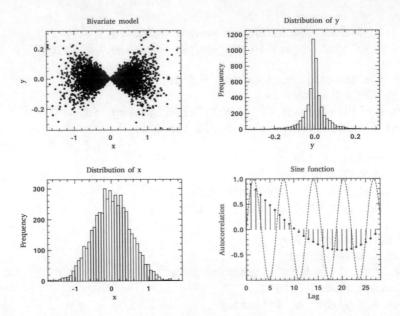

Fig. B.4: Subplots

```
   MarkerSize 8. ] Maths.sin 0. 9.);
Plot.(plot_fun ~h ~spec:[ RGB (255,0,0); Marker "#[0x0394]";
   MarkerSize 8. ] Maths.cos 0. 9.);
Plot.legend_on h [|"Sine function"; "Cosine function"|];
Plot.output h;;
```

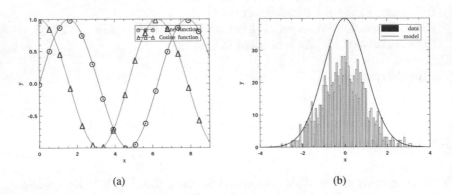

Fig. B.5: (a) Plot multiple lines. (b) Mix line plot and histogram.

Fig. B.5b is another example showing both the histogram and line plots in one figure.

```
# (* generate data *)
  let g x = (Stats.gaussian_pdf x ~mu:0. ~sigma:1.) *. 100. in
  let y = Mat.gaussian ~mu:0. ~sigma:1. 1 1000 in

  (* plot multiple data sets *)
  let h = Plot.create "plot_025.png" in
  Plot.set_background_color h 255 255 255;
  Plot.(histogram ~h ~spec:[ RGB (255,0,50) ] ~bin:100 y);
  Plot.(plot_fun ~h ~spec:[ RGB (0,0,255); LineWidth 2. ]
    g (-4.) 4.);
  Plot.legend_on h [|"data"; "model"|];

  Plot.output h;;
```

So as long as you "hold" the plot without calling Plot.output, you can repeatedly add datasets to the same figure.

B.1.5 Legend

The plot legend can be turned on and off by calling Plot.legend_on and Plot.legend_off respectively. When you call Plot.legend_on, you must also provide an array of legend names as well as the legend position. There are eight default positions in Plot:

```
type legend_position =
    North | South | West | East |
    NorthWest | NorthEast | SouthWest | SouthEast;;
```

The following example shows how to control the legend, and the resulting figure is shown in Fig. B.6a.

```
# (* generate data *)
  let x = Mat.(uniform 1 20 *$ 10.) in
  let y = Mat.(uniform 1 20) in
  let z = Mat.gaussian 1 20 in

  (* plot multiple data sets *)
  let h = Plot.create "plot_026.png" in
  Plot.(plot_fun ~h ~spec:[ RGB (0,0,255); LineStyle 1;
    Marker "*" ] Maths.sin 1. 8.);
  Plot.(plot_fun ~h ~spec:[ RGB (0,255,0); LineStyle 2;
    Marker "+" ] Maths.cos 1. 8.);
```

```
Plot.scatter ~h x y;
Plot.stem ~h x z;

let u = Mat.(abs(gaussian 1 10 *$ 0.3)) in
Plot.(bar ~h ~spec:[ RGB (255,255,0); FillPattern 3 ] u);

let v = Mat.(neg u *$ 0.3) in
let u = Mat.sequential 1 10 in
Plot.(area ~h ~spec:[ RGB (0,255,0); FillPattern 4 ] u v);

(* set up legend *)
Plot.(legend_on h ~position:NorthEast
  [|"test 1"; "test 2"; "scatter"; "stem"; "bar"; "area"|]);
Plot.output h;;
```

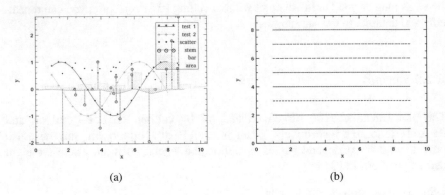

(a)   (b)

Fig. B.6: (a) Plot with legends. (b) Different line patterns.

B.1.6 Drawing Patterns

The plotting module supports multiple line patterns. Fig. B.6b demonstrates them.

```
# let h = Plot.create "plot_004.png" in
  Plot.set_background_color h 255 255 255;
  Plot.set_pen_size h 2.;
  Plot.(draw_line ~h ~spec:[ LineStyle 1 ] 1. 1. 9. 1.);
  Plot.(draw_line ~h ~spec:[ LineStyle 2 ] 1. 2. 9. 2.);
  Plot.(draw_line ~h ~spec:[ LineStyle 3 ] 1. 3. 9. 3.);
  Plot.(draw_line ~h ~spec:[ LineStyle 4 ] 1. 4. 9. 4.);
```

```
Plot.(draw_line ~h ~spec:[ LineStyle 5 ] 1. 5. 9. 5.);
Plot.(draw_line ~h ~spec:[ LineStyle 6 ] 1. 6. 9. 6.);
Plot.(draw_line ~h ~spec:[ LineStyle 7 ] 1. 7. 9. 7.);
Plot.(draw_line ~h ~spec:[ LineStyle 8 ] 1. 8. 9. 8.);
Plot.set_xrange h 0. 10.;
Plot.set_yrange h 0. 9.;
Plot.output h;;
```

Similarly, we can also fill rectangles with different patterns, as shown in Fig. B.7a.

```
# let h = Plot.create "plot_005.png" in
  Array.init 9 (fun i ->
    let x0, y0 = 0.5, float_of_int i +. 1.0 in
    let x1, y1 = 4.5, float_of_int i +. 0.5 in
    Plot.(draw_rect ~h ~spec:[ FillPattern i ] x0 y0 x1 y1);
    Plot.(text ~h ~spec:[ RGB (0,255,0) ] 2.3 (y0-.0.2)
      ("pattern: " ^ (string_of_int i)))
  ) |> ignore;

  Plot.output h;;
```

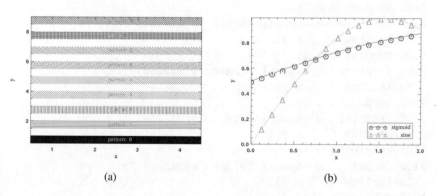

(a)   (b)

Fig. B.7: (a) Fill patterns. (b) Line plot with customised marker.

B.2 Plot Types

Having introduced these plotting elements, we will now demonstrate some of the different supported plot types.

Line plot (Fig. B.7b) is the most basic type. You can specify the colour, marker, and line style in the function.

```
# let x = Mat.linspace 0. 2. 100 in
  let y0 = Mat.sigmoid x in
  let y1 = Mat.map Maths.sin x in
  let h = Plot.create "plot_022.png" in
  Plot.(plot ~h ~spec:[ RGB (255,0,0); LineStyle 1;
    Marker "#[0x2299]"; MarkerSize 8. ] x y0);
  Plot.(plot ~h ~spec:[ RGB (0,255,0); LineStyle 2;
    Marker "#[0x0394]"; MarkerSize 8. ] x y1);
  Plot.(legend_on h ~position:SouthEast [|"sigmoid"; "sine"|]);
  Plot.output h;;
```

Next is the **Scatter Plot**. As with the line plot you can specify the marker type and marker size. The example below in Fig. B.8 shows the various patterns of markers, referenced by different ids.

```
# let x = Mat.uniform 1 30 in
  let y = Mat.uniform 1 30 in
  let h = Plot.create ~m:3 ~n:3 "plot_006.png" in
  Plot.set_background_color h 255 255 255;
  Plot.subplot h 0 0;
  Plot.(scatter ~h ~spec:[ Marker "#[0x2295]";
    MarkerSize 5. ] x y);
  Plot.subplot h 0 1;
  Plot.(scatter ~h ~spec:[ Marker "#[0x229a]";
    MarkerSize 5. ] x y);
  Plot.subplot h 0 2;
  Plot.(scatter ~h ~spec:[ Marker "#[0x2206]";
    MarkerSize 5. ] x y);
  Plot.subplot h 1 0;
  Plot.(scatter ~h ~spec:[ Marker "#[0x229e]";
    MarkerSize 5. ] x y);
  Plot.subplot h 1 1;
  Plot.(scatter ~h ~spec:[ Marker "#[0x2217]";
    MarkerSize 5. ] x y);
  Plot.subplot h 1 2;
  Plot.(scatter ~h ~spec:[ Marker "#[0x2296]";
    MarkerSize 5. ] x y);
  Plot.subplot h 2 0;
  Plot.(scatter ~h ~spec:[ Marker "#[0x2666]";
    MarkerSize 5. ] x y);
  Plot.subplot h 2 1;
  Plot.(scatter ~h ~spec:[ Marker "#[0x22a1]";
    MarkerSize 5. ] x y);
  Plot.subplot h 2 2;
  Plot.(scatter ~h ~spec:[ Marker "#[0x22b9]";
    MarkerSize 5. ] x y);
```

```
Plot.output h;;
```

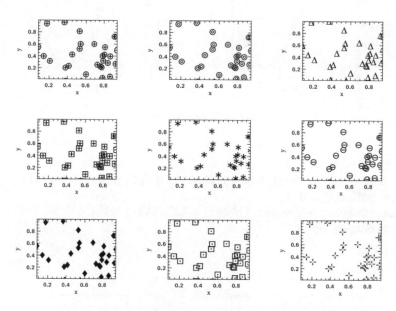

Fig. B.8: Scatter plot

The **Stairs plot**, or stairstep plot, draws the elements in a given ndarray as a staircase-like curve (Fig. B.8).

```
# let x = Mat.linspace 0. 6.5 20 in
  let y = Mat.map Maths.sin x in
  let h = Plot.create ~m:1 ~n:2 "plot_007.png" in
  Plot.set_background_color h 255 255 255;
  Plot.subplot h 0 0;
  Plot.plot_fun ~h Maths.sin 0. 6.5;
  Plot.(stairs ~h ~spec:[ RGB (0,128,255) ] x y);
  Plot.subplot h 0 1;
  Plot.(plot ~h ~spec:[ RGB (0,0,0) ] x y);
  Plot.(stairs ~h ~spec:[ RGB (0,128,255) ] x y);
  Plot.output h;;
```

The **box plot** as in Fig. B.10 shows datasets in terms of their quartiles, and is often used in descriptive statistics.

```
# let y1 = Mat.uniform 1 10 in
  let y2 = Mat.uniform 10 100 in
  let h = Plot.create ~m:1 ~n:2 "plot_008.png" in
```

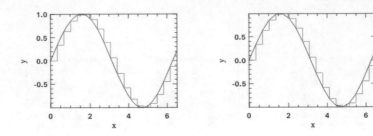

Fig. B.9: Stairs plot

```
Plot.subplot h 0 0;
Plot.(bar ~h ~spec:[ RGB (0,153,51); FillPattern 3 ] y1);
Plot.subplot h 0 1;
Plot.(boxplot ~h ~spec:[ RGB (0,153,51) ] y2);
Plot.output h;;
```

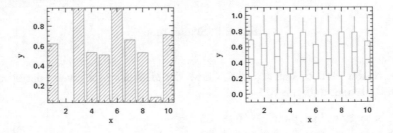

Fig. B.10: Box plot

Fig.B.11 shows a stem plot.

```
# let x = Mat.linspace 0.5 2.5 25 in
  let y = Mat.map (Stats.exponential_pdf ~lambda:0.1) x in
  let h = Plot.create ~m:1 ~n:2 "plot_009.png" in
  Plot.set_background_color h 255 255 255;
  Plot.subplot h 0 0;
  Plot.set_foreground_color h 0 0 0;
  Plot.stem ~h x y;
  Plot.subplot h 0 1;
  Plot.(stem ~h ~spec:[ Marker "#[0x2295]";
    MarkerSize 5.; LineStyle 1 ] x y);
  Plot.output h;;
```

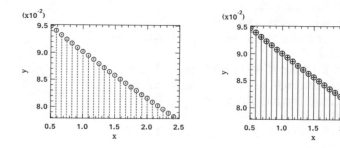

Fig. B.11: Stem plot

As it is often used to show the autocorrelation of a variable, the `Plot` module includes the `autocorr` function for your convenience. An example is shown in Fig. B.12.

```
# let x = Mat.linspace 0. 8. 30 in
  let y0 = Mat.map Maths.sin x in
  let y1 = Mat.uniform 1 30 in
  let h = Plot.create ~m:1 ~n:2 "plot_010.png" in
  Plot.subplot h 0 0;
  Plot.set_title h "Sine";
  Plot.autocorr ~h y0;
  Plot.subplot h 0 1;
  Plot.set_title h "Gaussian";
  Plot.autocorr ~h y1;
  Plot.output h;;
```

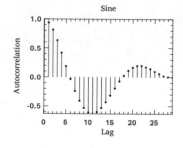
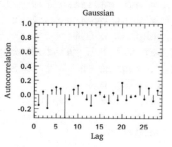

Fig. B.12: Stem plot with autocorrelation

The **Area Plot** is similar to line plot except that it fills the space between the line and the x-axis (Fig. B.13).

```
# let x = Mat.linspace 0. 8. 100 in
  let y = Mat.map Maths.atan x in
  let h = Plot.create ~m:1 ~n:2 "plot_011.png" in
  Plot.subplot h 0 0;
  Plot.(area ~h ~spec:[ FillPattern 1 ] x y);
  let x = Mat.linspace 0. (2. *. 3.1416) 100 in
  let y = Mat.map Maths.sin x in
  Plot.subplot h 0 1;
  Plot.(area ~h ~spec:[ FillPattern 2 ] x y);
  Plot.output h;;
```

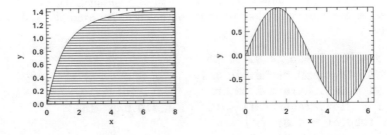

Fig. B.13: Area plot

The **Histogram Plot** and **CDF Plot** are very commonly used when visualising a series of measurements. You can easily plot the histogram and empirical cumulative distribution of the data using the histogram and ecdf functions. Fig. B.14 shows two examples.

```
# let x = Mat.gaussian 200 1 in
  let h = Plot.create ~m:1 ~n:2 "plot_012.png" in
  Plot.subplot h 0 0;
  Plot.set_title h "histogram";
  Plot.histogram ~h ~bin:25 x;
  Plot.subplot h 0 1;
  Plot.set_title h "empirical cdf";
  Plot.ecdf ~h x;
  Plot.output h;;
```

In Owl, it is straightforward to choose to use log-scale on either or both x- and y-axes, as the code below and Fig. B.15 show.

```
# let x = Mat.logspace (-1.5) 2. 50 in
  let y = Mat.map Maths.exp x in
  let h = Plot.create ~m:2 ~n:2 "plot_013.png" in
```

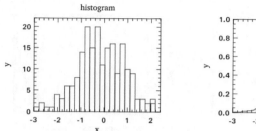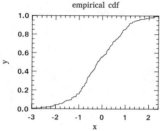

Fig. B.14: Histogram plot and CDF

```
Plot.subplot h 0 0;
Plot.set_xlabel h "Input Data X";
Plot.set_ylabel h "Input Data Y";
Plot.(loglog ~h ~spec:[ RGB (0,255,0); LineStyle 2;
  Marker "+" ] ~x:x y);

Plot.subplot h 0 1;
Plot.set_xlabel h "Index of Input Data Y";
Plot.set_ylabel h "Input Data Y";
Plot.(loglog ~h ~spec:[ RGB (0,0,255); LineStyle 1;
  Marker "*" ] y);

Plot.subplot h 1 0;
Plot.set_xlabel h "Input Data X";
Plot.set_ylabel h "Input Data Y";
Plot.semilogx ~h ~x:x y;

Plot.subplot h 1 1;
Plot.set_xlabel h "Index of Input Data Y";
Plot.set_ylabel h "Input Data Y";
Plot.semilogy ~h y;

Plot.output h;;
```

We have already shown some examples of **3D plots** earlier. There are four functions in the Plot module related to 3D plot: surf, mesh, heatmap, and contour. We look first at the mesh and surf functions (Fig. B.16).

```
# let x, y = Mat.meshgrid (-2.5) 2.5 (-2.5) 2.5 100 100 in
  let z0 = Mat.(sin ((x **$ 2.) + (y **$ 2.))) in
  let z1 = Mat.(cos ((x **$ 2.) + (y **$ 2.))) in
  let h = Plot.create ~m:2 ~n:2 "plot_014.png" in
```

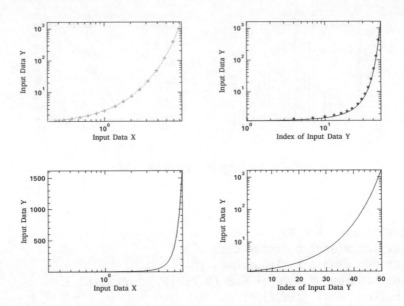

Fig. B.15: Change plot scale on x- and y-axis to log

```
Plot.subplot h 0 0;
Plot.surf ~h x y z0;
Plot.subplot h 0 1;
Plot.mesh ~h x y z0;
Plot.subplot h 1 0;
Plot.surf ~h x y z1;
Plot.subplot h 1 1;
Plot.mesh ~h x y z1;
Plot.output h;;
```

You can control the viewpoint with the altitude and azimuth parameters. The code below shows an example (Fig. B.17).

```
# let x, y = Mat.meshgrid (-2.5) 2.5 (-2.5) 2.5 100 100 in
  let z = Mat.(sin ((x * x) + (y * y))) in
  let h = Plot.create ~m:1 ~n:3 "plot_015.png" in
  Plot.subplot h 0 0;
  Plot.(mesh ~h ~spec:[ Altitude 50.; Azimuth 120. ] x y z);
  Plot.subplot h 0 1;
  Plot.(mesh ~h ~spec:[ Altitude 65.; Azimuth 120. ] x y z);
  Plot.subplot h 0 2;
  Plot.(mesh ~h ~spec:[ Altitude 80.; Azimuth 120. ] x y z);
  Plot.output h;;
```

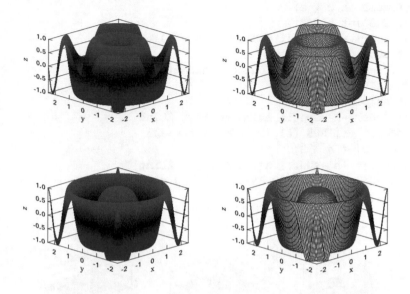

Fig. B.16: 3D plot

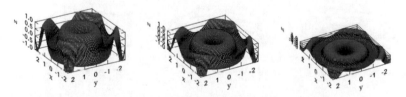

Fig. B.17: Customised 3D Plot, example 1

Fig. B.10 shhows a second example using different functions.

```
# let x, y = Mat.meshgrid (-3.) 3. (-3.) 3. 50 50 in
  let z = Mat.(
    3. $* ((1. $- x) **$ 2.) * exp (neg (x **$ 2.) -
    ((y +$ 1.) **$ 2.)) - (10. $* (x /$ 5. -
    (x **$ 3.) - (y **$ 5.)) * (exp (neg (x **$ 2.) -
    (y **$ 2.)))) - ((1./.3.) $* exp (neg ((x +$ 1.) **$ 2.) -
    (y **$ 2.))))
  in
  let h = Plot.create ~m:2 ~n:3 "plot_016.png" in
  Plot.subplot h 0 0;
  Plot.surf ~h x y z;
  Plot.subplot h 0 1;
```

```
Plot.mesh ~h x y z;
Plot.subplot h 0 2;
Plot.(surf ~h ~spec:[ Contour ] x y z);
Plot.subplot h 1 0;
Plot.(mesh ~h ~spec:[ Contour; Azimuth 115.;
  NoMagColor ] x y z);
Plot.subplot h 1 1;
Plot.(mesh ~h ~spec:[ Azimuth 115.; ZLine X;
  NoMagColor; RGB (61,129,255) ] x y z);
Plot.subplot h 1 2;
Plot.(mesh ~h ~spec:[ Azimuth 115.; ZLine Y;
  NoMagColor; RGB (130,255,40) ] x y z);
Plot.output h;;
```

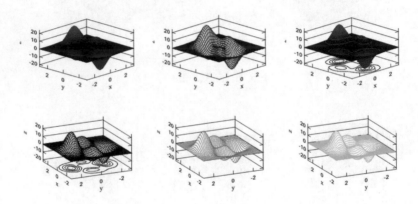

Fig. B.18: Customised 3D Plot, example 2

Finally, let's look at the heatmap and contour plot using the same functions as before. The generated figures are shown in Fig. B.10.

```
# let x, y = Mat.meshgrid (-3.) 3. (-3.) 3. 100 100 in
  let z = Mat.(
    3. $* ((1. $- x) **$ 2.) * exp (neg (x **$ 2.) -
    ((y +$ 1.) **$ 2.)) - (10. $* (x /$ 5. - (x **$ 3.) -
    (y **$ 5.)) * (exp (neg (x **$ 2.) - (y **$ 2.)))) -
    ((1./.3.) $* exp (neg ((x +$ 1.) **$ 2.) - (y **$ 2.))))
  in
  let h = Plot.create ~m:2 ~n:2 "plot_017.png" in
  Plot.subplot h 0 0;
  Plot.(mesh ~h ~spec:[ Contour ] x y z);
  Plot.subplot h 0 1;
```

```
Plot.heatmap ~h x y z;
Plot.subplot h 1 0;
Plot.mesh ~h x y z;
Plot.subplot h 1 1;
Plot.contour ~h x y z;
Plot.output h;;
```

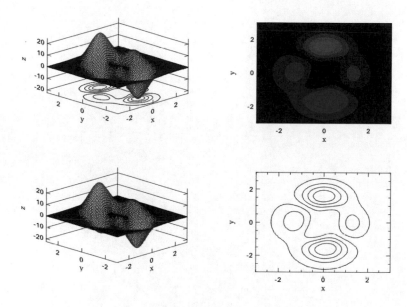

Fig. B.19: Heatmap and contour plot

Index

Printed in the United States
by Baker & Taylor Publisher Services